Advances in Rock Dynamics and Applications

Advances in Rock Dynamics and Applications

Contributors

Tsuyoshi Ishikawa, Masaharu Tanimizu et al.

AURIS
Reference

www.aurisreference.com

Advances in Rock Dynamics and Applications

Contributors: Tsuyoshi Ishikawa, Masaharu Tanimizu et al.

Published by Auris Reference Limited

www.aurisreference.com

United Kingdom

Advances in Rock Dynamics and Applications

ISBN: 978-1-78154-838-7

British Library Cataloguing in Publication Data
A CIP record for this book is available from the British Library

Printed in the United Kingdom

Exclusively distributed by CBS Publishers & Distributors Pvt. Ltd.

Sales & Distribution Rights only for India, Pakistan, Bangladesh, Sri Lanka, Nepal and Bhutan. This book is not to be sold outside these territories.

Contents

List of Abbreviations .... *vii*

List of Contributors.. ....*ix*

Preface.. ....*xv*

Chapter 1   **Coseismic Fluid–Rock Interactions at High Temperatures in the Chelungpu Fault**... 1

Chapter 2   **Fractures and Fracturing: Hydraulic Fracturing in Jointed Rock**........ 13

Chapter 3   **Hydraulic and Sleeve Fracturing Laboratory Experiments on 6 Rock Types** 51

Chapter 4   **Three-Dimensional Numerical Model of Hydraulic Fracturing in Fractured Rock Masses**.. 63

Chapter 5   **Model Test of Anchoring Effect on Zonal Disintegration in Deep Surrounding Rock Masses** 77

Chapter 6   **Petrological and Geochemical Characteristics of Mafic Granulites Associated with Alkaline Rocks in the Pan-African Dahomeyide Suture Zone, Southeastern Ghana** 109

Chapter 7   **Roughness Research of Center Profile Curve on Rock Fracture Surface Based on Statistical Method** 131

Chapter 8   **Failure Probability Model Considering the Effect of Intermediate Principal Stress on Rock Strength**.................................... 147

Chapter 9   **Rock Mass Blastability Classification Using Fuzzy Pattern Recognition and the Combination Weight Method** 161

Chapter 10   **Effects of Rock Mass Conditions and Blasting Standard on Fragmentation Size at Limestone Quarries**..................................... 187

Chapter 11   **Study on Hysteretic Fracture of Naturally Cracked Surrounding Rock**... 199

Chapter 12   **Deformation and Failure Characteristics of the Rock Masses Around Deep Underground Caverns**.. 223

Chapter 13   Application of Base Force Element Method on Complementary
               Energy Principle to Rock Mechanics Problems.................................. 249

Chapter 14   Geochemistry of Hydrothermal Alteration in Volcanic Rocks 285

               Citations 317

               Index.. 321

List of Abbreviations

AE	Acoustic Emission
AHP	Analysis hierarchy process
BISA	Barites-iron-sand cementation analogical
BPM	Bonded particle model
CWM	Combination weight method
CNC	Computerized numerical control
CI	Consistency index
CR	Consistency ratio
DAS	Data acquisition system
DTA	Differential thermal analysis
DFN	Discrete fracture network
EDS	Energy dispersive spectrometry
EGS	Enhanced Geothermal Systems
EM	Entropy method
FBG	Fiber bragg grating
GT	Game theory
HF	Hydraulic Fracturing
IAT	Island arc theoleiitic
JRC	Joint Roughness Coefficients
KC	Kpong complex
LEFM	Linearly Elastic Fracture Mechanics
MTS	Mechanics Test Systems
NFRs	Naturally fractured reservoirs
PFC	Particle flow code
PFC code	Particle Flow Code
REE	Rare earth elements
STS	Signal translating system
SJM	Smooth joint model
SRM	Synthetic rock mass
TG	Thermogravimetric
TSB	Trans-Saharan belt
UCS	Unconstrained compressive strength
WAC	West African craton
XRD	X-ray diffraction
GP	Geoenvironment Protection
GP	Geohazard Prevention
RBCTS	Rock Biaxial Compressive Test System

List of Contributors

Tsuyoshi Ishikawa
Kochi Institute for Core Sample Research, Japan Agency for Marine-Earth Science and Technology (JAMSTEC), 200 Monobe-otsu, Nankoku, Kochi 783-8502, Japan

Masaharu Tanimizu
Kochi Institute for Core Sample Research, Japan Agency for Marine-Earth Science and Technology (JAMSTEC), 200 Monobe-otsu, Nankoku, Kochi 783-8502, Japan

Kazuya Nagaishi
Marine Works Japan Ltd, Nankoku, Kochi 783-8502, Japan

Jun Matsuoka
Marine Works Japan Ltd, Nankoku, Kochi 783-8502, Japan

Osamu Tadai
Marine Works Japan Ltd, Nankoku, Kochi 783-8502, Japan

Masumi Sakaguchi
Marine Works Japan Ltd, Nankoku, Kochi 783-8502, Japan

Tetsuro Hirono
Department of Earth and Space Science, Graduate School of Science, Osaka University, Toyonaka, Osaka 560-0043, Japan

Toshiaki Mishima
Research Center for Inland Seas, Kobe University, Kobe, Hyogo 657-8501, Japan

Wataru Tanikawa
Kochi Institute for Core Sample Research, Japan Agency for Marine-Earth Science and Technology (JAMSTEC), 200 Monobe-otsu, Nankoku, Kochi 783-8502, Japan

Weiren Lin
Kochi Institute for Core Sample Research, Japan Agency for Marine-Earth Science and Technology (JAMSTEC), 200 Monobe-otsu, Nankoku, Kochi 783-8502, Japan

Hiroyuki Kikuta
Kochi Institute for Core Sample Research, Japan Agency for Marine-Earth Science and Technology (JAMSTEC), 200 Monobe-otsu, Nankoku, Kochi 783-8502, Japan

Wonn Soh
Kochi Institute for Core Sample Research, Japan Agency for Marine-Earth Science and Technology (JAMSTEC), 200 Monobe-otsu, Nankoku, Kochi 783-8502, Japan

Sheng-Rong Song
Department of Geosciences, National Taiwan University, Taipei 10617, Taiwan

Charles Fairhurst
Senior Consultant, Itasca Consulting Group, Inc, Minneapolis Minnesota, USA
Professor Emeritus, University of Minnesota, Minneapolis, Minnesota, USA

Sebastian Brenne
Ruhr-University Bochum, Germany

Michael Molenda1
Ruhr-University Bochum, Germany

Ferdinand Stöckhert1
Ruhr-University Bochum, Germany

Michael Alber1
Ruhr-University Bochum, Germany

B. Damjanac
Itasca Consulting Group, Inc., Minneapolis, Minnesota, USA

C. Detournay
Itasca Consulting Group, Inc., Minneapolis, Minnesota, USA

P.A. Cundall
Itasca Consulting Group, Inc., Minneapolis, Minnesota, USA

Varun
Itasca Consulting Group, Inc., Minneapolis, Minnesota, USA

Xu-Guang Chen
Institute of Tunnel and Urban Railway Engineering, Hohai University, Nanjing 210098, Key Laboratory of Ministry of Education for Geomechanics and Embankment Engineering, Hohai Univ., Nanjing 210098, China
State Key Laboratory for GeoMechanics and Deep Underground Engineering, China University of Mining & Technology, Xuzhou 221000, China
Research Center of Geotechnical and Structural Engineering, Shandong University, Jinan 250061, China

Qiang-Yong Zhang
Research Center of Geotechnical and Structural Engineering, Shandong University,

Jinan 250061, China

Yuan Wang
Institute of Tunnel and Urban Railway Engineering, Hohai University, Nanjing 210098, Key Laboratory of Ministry of Education for Geomechanics and Embankment Engineering, Hohai Univ., Nanjing 210098, China

De-Jun Liu
Research Center of Geotechnical and Structural Engineering, Shandong University, Jinan 250061, China

Ning Zhang
Research Center of Geotechnical and Structural Engineering, Shandong University, Jinan 250061, China

Prosper M. Nude
Department of Earth Science, University of Ghana, Legon-Accra, Ghana

Kodjopa Attoh
Department of Earth &Atmospheric Sciences, Cornell University, Ithaca, NY 14853, USA

John W. Shervais
Department of Geology, Utah State University, Logan UT 84322, USA

Gordon Foli
Department of Earth and Environmental sciences, University for Development studies, Navrongo Campus Ghana

Xuezai Pan
School of Mathematics, Nanjing Normal University, Taizhou College, Taizhou, China
Faculty of Science, Jiangsu University, Zhenjiang, China

Zhigang Feng
State Key Laboratory of Coal Resources and Safe Mining, China University of Mining and Technology, Beijing, China
Faculty of Science, Jiangsu University, Zhenjiang, China

Guoxing Dai
Faculty of Science, Jiangsu University, Zhenjiang, China

Hongguang Liu
Faculty of Civil Engineering and Mechanics, Jiangsu University, Zhenjiang, China

Yonglai Zheng
College of Civil Engineering, Tongji University, Shanghai 200092, China

Shuxin Deng
College of Civil Engineering, Tongji University, Shanghai 200092, China

Shuangshuang Xiao
State Key Laboratory of Coal Resources and Safe Mining, School of Mines, China University of Mining

Kemin Li
State Key Laboratory of Coal Resources and Safe Mining, School of Mines, China University of Mining

Xiaohua Ding
State Key Laboratory of Coal Resources and Safe Mining, School of Mines, China University of Mining

Tong Liu
State Key Laboratory of Coal Resources and Safe Mining, School of Mines, China University of Mining

Takashi Sasaoka
Department of Earth Resources Engineering, Faculty of Engineering, Kyushu University, Fukuoka, Japan

Yoshiaki Takahashi
Department of Earth Resources Engineering, Faculty of Engineering, Kyushu University, Fukuoka, Japan

Wahyudi Sugeng
Department of Earth Resources Engineering, Faculty of Engineering, Kyushu University, Fukuoka, Japan

Akihiro Hamanaka
Center of Environmental Science and Disaster Mitigation for Advanced Research, Muroran Institute Technology, Muroran, Japan

Hideki Shimada
Department of Earth Resources Engineering, Faculty of Engineering, Kyushu University, Fukuoka, Japan

Kikuo Matsui
Department of Earth Resources Engineering, Faculty of Engineering, Kyushu University, Fukuoka, Japan

Shiro Kubota
National Institute of Advanced Industrial Science and Technology, Ibaraki, Japan

Zhibin Zhong
School of Civil Engineering, Southwest Jiaotong University, Chengdu, Sichuan 610031, China

Ronggui Deng
School of Civil Engineering, Southwest Jiaotong University, Chengdu, Sichuan 610031, China

Fang Lin
School of Civil Engineering, Southwest Jiaotong University, Chengdu, Sichuan 610031, China

Ying Zhang
School of Civil Engineering, Southwest Jiaotong University, Chengdu, Sichuan 610031, China

Lei Lv
Chengdu Southwest Jiaoda Yaosen Engineering Technology Co., Ltd., Chengdu, Sichuan 610031, China

Jing Yin
School of Civil Engineering, Southwest Jiaotong University, Chengdu, Sichuan 610031, China

Xiaomin Fu
State Key Laboratory of Geohazard Prevention and Geoenvironment Protection, Chengdu University of Technology, Chengdu, Sichuan 610059, China

Chong Zhang
College of Electronic and Information Engineering, Henan University of Science and Technology, 263 Kaiyuandadao, Luolong, Luoyang, Henan 471023, China

Zhechao Wang
Geotechnical and Structural Engineering Research Center, Shandong University, 17923 Jingshi Road, Jinan, Shandong 250061, China

Qi Wang
Geotechnical and Structural Engineering Research Center, Shandong University, 17923 Jingshi Road, Jinan, Shandong 250061, China

Yijiang Peng

The Key Laboratory of Urban Security and Disaster Engineering, Ministry of Education, Beijing University of Technology, Beijing 100124, China

Qing Guo,
The Key Laboratory of Urban Security and Disaster Engineering, Ministry of Education, Beijing University of Technology, Beijing 100124, China

Zhaofeng
The Key Laboratory of Urban Security and Disaster Engineering, Ministry of Education, Beijing University of Technology, Beijing 100124, China

Zhang
The Key Laboratory of Urban Security and Disaster Engineering, Ministry of Education, Beijing University of Technology, Beijing 100124, China

Yanyan Shan
The Key Laboratory of Urban Security and Disaster Engineering, Ministry of Education, Beijing University of Technology, Beijing 100124, China

Silvina Marfil
Department of Geosciences, National Taiwan University, Taipei 10617, Taiwan

Pedro Maiza
Universidad Nacional del Sur – INGEOSUR- CIC de la Provincia de Buenos Aires – CONICET Argentina

Preface

The study of rock dynamics is important because many rock mechanics and rock engineering problems involve dynamic loading ranging from earthquakes to vibrations and explosions. The text Advances in Rock Dynamics and Applications deals with the distribution and propagation of loads, dynamic responses, and processes of rocks and rate-dependent properties, coupled with the physical environment. First chapter discusses coseismic fluid–rock interactions at high temperatures in the chelungpu fault. Second chapter emphasizes the central role of fractures in rock, primarily natural fractures developed on a wide spectrum of scales over many tectonic epochs and many millions of years. Hydraulic and sleeve fracturing laboratory experiments on 6 rock types have been focused in third chapter. Three-dimensional numerical model of hydraulic fracturing in fractured rock masses has been presented in fourth chapter. In fifth chapter, a 3D geomechanics model test has been conducted to research the anchoring effect of zonal disintegration. Sixth chapter presents petrological and geochemical data on the nepheline-bearing mafic rocks previously referred to as mafic nepheline gneiss at the contact zone between the high-pressure (HP) mafic granulites and the Kpong complex (KC) rocks. Seventh chapter focuses on roughness research of center profile curve on rock fracture surface based on statistical method. Failure probability model considering the effect of intermediate principal stress on rock strength has been developed in eighth chapter. In ninth chapter, a model for rock mass blastability classification has been developed in combination with a fuzzy pattern recognition method. Tenth chapter summarizes the results of blasting tests and describes the impacts of rock mass conditions and blasting standard on the size of fragmented rocks. In eleventh chapter, experimental and numerical simulation are conducted to evaluate the deformation and fracture modes of rock with an inverted U-shaped opening in natural original cracked and noncracked rhyolite, respectively. Twelfth chapter identifies the deformation and failure characteristics of a deep cavern under different ground stress conditions using model test and theoretical analysis methods. In thirteenth chapter, the base force element method (BFEM) on complementary energy principle has been used to analyze the engineering problems of rock mechanics. The purpose of last chapter is to evaluate the relation between the chemical composition of major, minor and trace elements, the stable isotopes in kaolin and the mineralogical alteration zonation to confirm the hydrothermal genesis of the deposits, working with volcanic rocks from Patagonia Argentina.

Chapter 1

COSEISMIC FLUID–ROCK INTERACTIONS AT HIGH TEMPERATURES IN THE CHELUNGPU FAULT

Tsuyoshi Ishikawa[1], Masaharu Tanimizu[1], Kazuya Nagaishi[2], Jun Matsuoka[2], Osamu Tadai[2], Masumi Sakaguchi[2], Tetsuro Hirono[3], Toshiaki Mishima[4], Wataru Tanikawa[1], Weiren Lin[1], Hiroyuki Kikuta[1], Wonn Soh1 & Sheng-Rong Song[5]

[1]Kochi Institute for Core Sample Research, Japan Agency for Marine-Earth Science and Technology (JAMSTEC), 200 Monobe-otsu, Nankoku, Kochi 783-8502, Japan

[2]Marine Works Japan Ltd, Nankoku, Kochi 783-8502, Japan

[3]Department of Earth and Space Science, Graduate School of Science, Osaka University, Toyonaka, Osaka 560-0043, Japan

[4]Research Center for Inland Seas, Kobe University, Kobe, Hyogo 657-8501, Japan

[5]Department of Geosciences, National Taiwan University, Taipei 10617, Taiwan

ABSTRACT

Aqueous fluids are thought to have an essential role in faulting and the dynamic propagation of earthquake rupture. Fluid overpressure can affect earthquake nucleation[1, 2] and in a process termed thermal pressurization, pore fluid pressure produced by frictional heating can reduce the effective normal stress acting on the fault surface[3, 4, 5]. This may lead to a marked reduction in fault strength during slip. However, the coseismic presence of fluids within slip zones and the role of fluids in dynamic fault weakening is still a matter of debate. Here we present compositions of major and trace elements as well as isotope ratios of core samples representing relatively undamaged as well as very fine-grained deformed material from three active zones of the Chelungpu fault, Taiwan. Depth profiles across the most intensely sheared bands that range in thickness from 2–15 cm exhibit sharp compositional peaks of fluid-mobile elements and of strontium isotopes. We suggest that high-temperature fluids (>350 °C) derived from heating of sediment pore fluids during the earthquake interacted with material within the fault zone and mobilized the elements. The coseismic presence of high-temperature fluids under conditions of low hydraulic diffusivity[6] within the fault zone is favourable for thermal pressurization. This effect may have caused a dynamic decrease of friction along the Chelungpu fault during the 1999 magnitude 7.6 Chi-Chi earthquake.

INTRODUCTION

The Chi-Chi earthquake, the largest inland earthquake in Taiwan in the twentieth century, produced an 85-km-long surface rupture along the north–south trending Chelungpu thrust fault. During the earthquake, the rupture propagated from south to north, and seismic observations showed that ground velocities and displacements in the northern part of the fault, up to 3 m s^{-1} and 8 m, respectively, were larger than those in the southern part, whereas ground accelerations were higher in the south[7, 8]. It has been proposed that these differences, together with the marked absence of high-frequency radiation in the north, may be attributable to a dynamic decrease of friction in the northern part of the fault caused by thermal pressurization[9] or elasto-hydrodynamic lubrication[10]. The Taiwan Chelungpu-fault Drilling Project (TCDP) was undertaken to investigate the faulting mechanism of the Chi-Chi earthquake in the northern part of the fault system (Supplementary Information, Fig. S1). The core samples analysed in this study were collected from TCDP Hole B, in which three active fault zones were recognized at depths of 1,134–1,137, 1,194–1,197 and 1,242–1,244 m (hereafter, FZB1136, FZB1194 and FZB1243, respectively)[11, 12]. These fault zones consist of a black gouge zone (BGZ), a grey gouge zone (GGZ), a breccia zone (BrZ) and a fracture-damaged zone (FDZ) (Fig. 1). The BGZs are considered to have resulted from intense shearing with frictional heating[11, 12, 13, 14, 15, 16]. The shallowest fault zone (FZB1136) is most likely to have been the site of the main fault rupture during the Chi-Chi earthquake[13, 17]. The major element composition and mineralogy of the analysed samples are typical of hemipelagic-neritic shale and sandy shale (see the Methods section and Supplementary Information, Table S1).

FZB1136 (upper), FZB1194 (middle) and FZB1243 (lower). The magnetic susceptibility data are from ref. 11. BM=a 2–3-cm-thick disc-shaped, relatively stiff black material observed in the BGZs of FZB1194 and FZB1243[11, 14]. Because the vertical thickness of each sample was 3–5 cm, the trace-element and isotope data are plotted with a +2 cm offset relative to the depth of the upper surface of the sample shown in Supplementary Information, Table S1.

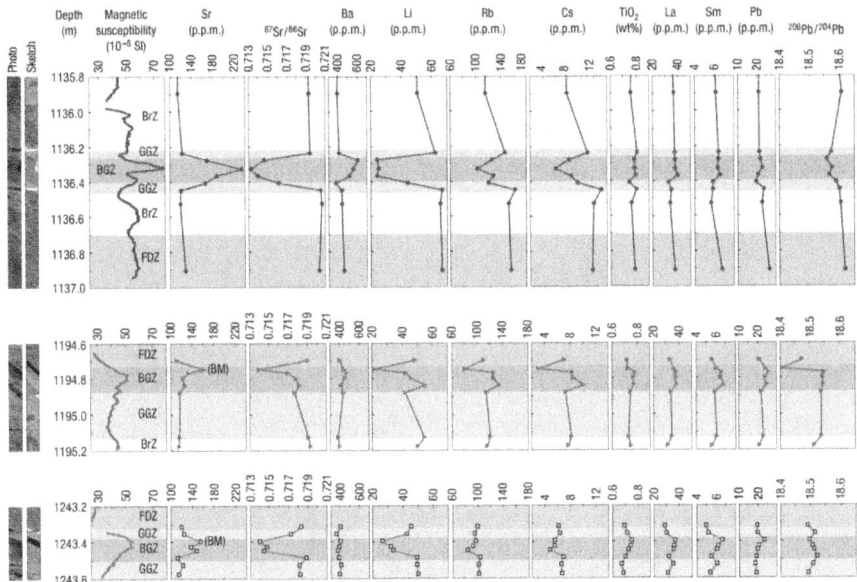

Figure 1: Depth profiles of magnetic susceptibility, trace-element concentrations and Sr and Pb isotope ratios across the three fault zones.

Depth profiles of trace-element concentrations and isotope ratios across the three fault zones (Fig. 1) exhibit marked increases of Sr and decreases of Li, Rb, Cs and $^{87}Sr/^{86}Sr$ in all three BGZs. In contrast, concentrations of Ti, La, Sm and Pb show no systematic variation across the fault zones, whereas the Ba concentration and Pb isotope ratios increase and decrease, respectively, only in FZB1136 and FZB1194. Values normalized with respect to those in the adjacent GGZ, BrZ and FDZ show 20–90% enrichment of Sr, 20–60% depletion of Cs, Rb and Li, and a 5–9‰ decrease of the $^{87}Sr/^{86}Sr$ ratio in the BGZs (Fig. 2). Notably, the BGZs in FZB1136 and FZB1194 exhibit similar degrees of depletion in Cs, Rb and Li. In FZB1243, the deviations of the Cs, Rb, Sr and Li values from background values are about half those observed in FZB1194. These systematic variations in trace-element and isotope compositions strongly suggest that a similar process controlled the chemical characteristics of all three BGZs. Moreover, the disappearance of carbonate minerals from the BGZs (ref. 16) indicates that the higher Sr contents, and the lower $^{87}Sr/^{86}Sr$ ratios observed in the BGZs, did not result from a concentration of Sr-rich carbonate minerals with $^{87}Sr/^{86}Sr<0.709$ (ref. 18). Similarly, the depletions of Cs, Rb and Li cannot be explained by simple losses of alkali-element-hosting aluminosilicate minerals such as illite and K-feldspar: this is proved by the lack of correlation of Al_2O_3 content with Cs, Rb and Li contents and the much larger degree of depletion than would be expected from the K_2O values. Relatively homogeneous Al_2O_3 and TiO_2 contents within each fault zone (<7% r.s.d.) suggest that any gain or loss of volume associated with gouge formation was unimportant for producing the observed trace-element variations. **a,b**, The calculated compositions (solid and dashed lines) were obtained using the D values at 350 °C from the data of You et al.[19] (**a**) and James et al.[20] (**b**).

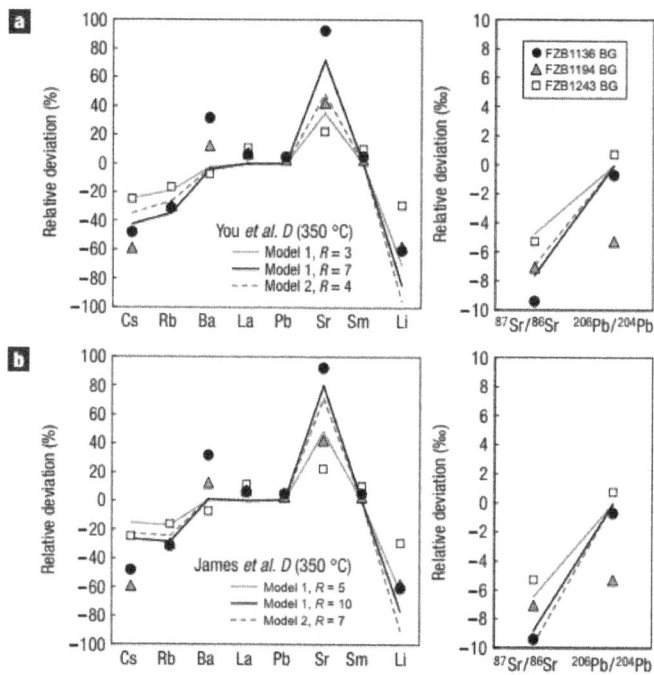

Figure 2: Trace-element and isotope compositions of the black gouges at 1,136.31 m, 1,194.73 m and 1,243.43 m and calculated gouge compositions.

The gouge compositions are expressed as relative deviations from the reference values, which are the averages of those from sample pairs of 1,136.22 m and 1,136.51 m (from FZB1136), 1,194.68 m and 1,195.11 m (FZB1194) and 1,243.30 m and 1,243.51 m (FZB1243). Note that the observed enrichments and depletions of elements do not correlate with the mobility of the elements in melting processes (the elements listed on the left are generally more mobile).

Sr, Cs, Rb and Li are generally known to be fluid-mobile elements, and high-temperature solid–fluid interactions can affect their bulk compositions. Hydrothermal experiments at 25–350 °C using hemipelagic mudstones[19, 20] have shown that (1) Cs, Rb and Li in sediment are significantly mobilized in fluids at >300 °C, whereas mobilization of La, Sm and Ti is limited or negligible, (2) Sr in fluids decreases at ~300 °C as a result of uptake into albite plagioclase formed at that temperature. These experimental results at 300–350 °C are consistent with the chemical characteristics of the BGZ observed in this study. The fixation of fluid-derived Sr in albite is supported by the correlated increases of Na and Sr in the BGZs, similar distributions of Na and Al in the black gouges indicated by electron-microprobe chemical mapping[21] and the presence of sediment pore fluids with low $^{87}Sr/^{86}Sr$ ratios shown by $^{87}Sr/^{86}Sr=0.7100–0.7126$ observed in the H_2O-soluble components extracted from the sediments.

To quantify the effect of high-temperature solid–fluid interaction, we made simple model calculations (see Supplementary Information, Methods). Assuming chemical equilibrium, the concentration of a trace element in the gouge after the interaction with fluid can be calculated as

$$C_s = D \frac{C_{s0} + RC_{f0}}{R+D},$$

where C_{s0}, C_s and C_{f0} are the element concentrations in the initial solid (gouge), solid equilibrated with fluid and initial fluid, and R and D are the fluid/gouge mass ratio and the solid/fluid bulk distribution coefficient of the element, respectively. Values of D were estimated on the basis of hydrothermal experimental results of You et al.[19] and Jameset al.[20], and the values of C_{f0} and C_{s0} were based on sediment pore-water data[19, 22] and the reference values for FZB1136 (Fig. 2), respectively. In our calculations, we assumed interaction of solid with a single batch of fluid (model 1), or with multiple batches of fluids (model 2). The results of these calculations show that at 350 °C, model 1 with R=7–10 (and to lesser extent, model 2 with R=4–7) successfully reproduced the trace-element and isotope spectra of the BGZs of FZB1136 and FZB1194 (Figs 2 and 3), and that the data of FZB1243 can be explained at the same temperature with lower R. In particular, the D values from the data of You et al.[19], in which hemipelagic sediments of comparable mineralogy to that of the TCDP sediments were used, provided better results. At 300 °C or lower temperatures, however, it becomes difficult to reconcile the calculated values with the observed values because the rapid increases of D_{Cs} and D_{Rb} values with decreasing temperature suppressed the extraction of these elements from the gouges, even at large R values (Fig. 3). This result strongly suggests that the chemical characteristics of the BGZ were produced by the interaction with aqueous fluids derived from sediment pore water at temperatures as high as 350 °C. Possibly this temperature represents the lower limit because the values of D_{Cs} and D_{Rb} can become even lower at >350 °C. The present-day ambient temperature in the fault zones is 46–49 °C (ref. 17), suggesting that a high temperature of >350 °C is unlikely to be attained without earthquake-associated frictional heating of the slip zone or influx of a hot transient pulse of fluid, which could result from post-seismic discharge from deep rupture zones. In situhydraulic diffusivity along FZB1136 (ref. 6) is estimated to be 7×10^{-5} m² s⁻¹, which is too low to allow the latter process to occur, although the change in bulk chemical compositions of BGZs requires some local flow or circulation of fluids. On the other hand, elevated magnetic susceptibility observed in the BGZs (Fig. 1) is attributable to frictional heating and the resultant formation of ferrimagnetic mineral grains at >400 °C (refs 15,23), which is consistent with our results. Thus, we conclude that the occurrence of high-temperature solid–fluid interaction was coseismic. a–d, The calculations are based on model 1 at 350 °C (a), model 2 at 350 °C (b), model 1 at 300 °C (c) and model 1 at 250 °C (d).

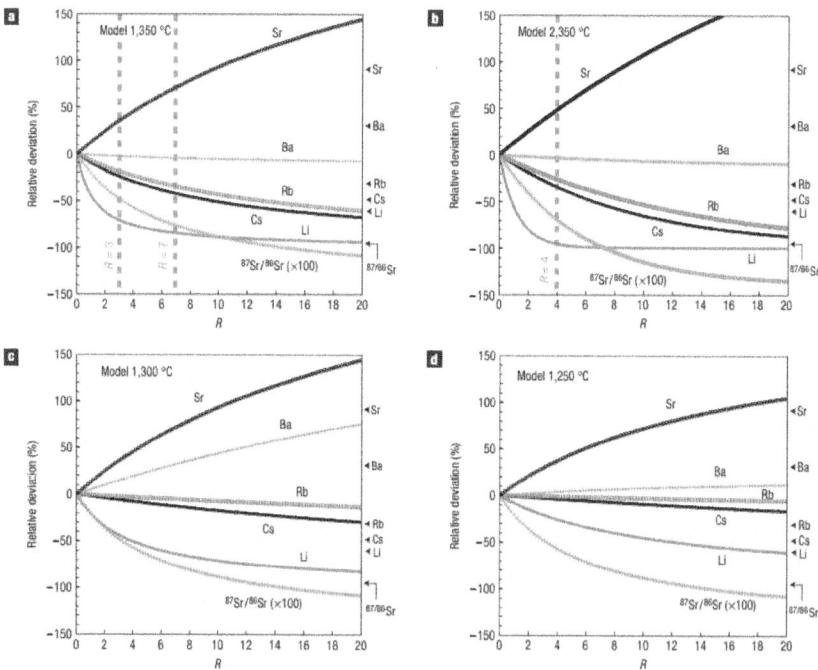

Figure 3: Calculated compositions of the black gouge in FZB1136 as a function of the fluid/gouge mass ratio.

The gouge compositions are expressed as relative deviations from the reference values of FZB1136. D values from the data of You et al.[19]were used. The values indicated by the arrowheads at the right side correspond to the peak values of the signals observed in the depth profiles (Fig. 1). The dotted vertical lines in **a** and **b** indicate the R values, for which calculated trace-element and isotope spectra are shown in Fig. 2.

Here, we focus on the averaged chemico-physical effects during the period from the beginning of frictional heating to the cooling of the slip zone to below ~200 °C, which may have lasted for a few tens of seconds[17]. In this case, kinetic effects must be taken into consideration for a rigorous evaluation of the conditions during gouge–fluid interaction. The degree to which solid–fluid chemical (isotopic) equilibrium is achieved (F) is given by[24]:

$$F = 1 - e^{-\frac{C_s}{C_f}\frac{A}{M}kt},$$

where k is the reaction rate, t is time, C_s/C_f is the ratio of trace-element concentration in the solid to that in the fluid and A/M is the ratio of solid surface area to fluid mass. An electron microscope observation for the slip zone found in Hole C (a side-track from Hole B)[13] gave A=3.23×10^7 m^2 m^{-3}. We estimated A/M ratios of 2.2×10^5, 2.5×10^5 and 4.5×10^5 m^2 kg^{-1} for 300, 350 and 400 °C, respectively, assuming P=30 MPa and 20% porosity[12]. Only limited data are available for k values associated with the trace-element equilibration. The reported k values for recrystallization of calcite and Sr isotope exchange at 300, 350 and 400 °C are 10$^{-8.33}$, 10$^{-7.55}$ and 10$^{-6.88}$ mol m^{-2} s^{-1}, respectively[24]. These k values, with a constant C_s/C_f ratio of 60 (ref. 24) and the aboveA/M ratios, give 99% equilibrium (F=0.99) after 75 s at 300 °C, after 10 s at 350 °C and after 1.2 s at 400 °C. Thus, near-equilibrium can be achieved during a coseismic event if the net k values are the same order of magnitude as those in calcite recrystallization, and the temperature is high enough (in this case, T>350 °C). Although element partitioning within the fault zones is controlled by interaction of fluids with multiple mineral phases, a gouge composition consistent with solid–fluid near-equilibrium can result from the very high A/M ratio associated with production of ultrafine mineral grains in the slip zone, and from enhanced chemical reaction rates because of mechano-chemical effects induced by mechanical friction[23]. It should be noted that incomplete equilibration results in a smaller change of fluid-mobile element composition in BGZ, which gives underestimated temperature and/or R value in our approach.

The data presented here show that frictional heating of the slip zone took place under fluid-oversaturated conditions in FZB1136 (and in FZB1194 and

FZB1243). According to previous numerical modelling[5, 25], hydraulic diffusivity estimated for FZB1136 ($7{\times}10^{-5}$ m^2 s^{-1}) (ref. 6) is low enough to allow thermal pressurization to occur. Synchronous peaks observed in the trace-element and magnetic susceptibility profiles across the fault zones (Fig. 1) also suggest that high-temperature fluid was effectively sealed within or near the slip zone, because the samples with elevated magnetic susceptibility represent the site of frictional heating, possibly at temperatures >400 °C (refs 15,23). These observations strongly suggest that thermal pressurization and the resultant dynamic decrease of friction occurred during the 1999 Chi-Chi earthquake and during previous earthquakes associated with the northern Chelungpu fault. Although melt-like structures of submicrometre scale are rarely observed in 2–3-cm-thick, relatively stiff black material found in BGZs of FZB1194 and FZB1243 ('BM' in Fig. 1)[14], such a microstructure seems to have resulted from a rather minor and localized process and is unimportant for the control of bulk chemical and physical properties of the slip zone, because such black materials show strong fluid-induced geochemical signals. A detailed structural analysis showed that the BGZ consists of multiple slip zones, each 2–3 cm thick[13]. Thus, the shapes and intensities of the compositional peaks across each fault zone represent the integrated signal recorded by repeated earthquakes during the history of the Chelungpu fault. The Chi-Chi earthquake can be regarded as a subduction zone earthquake in a tectonic sense[26], and the chemico-physical process in the slip zone investigated here will help us to understand the faulting process occurring in shallow subduction zones, including those related to the tsunami earthquakes.

METHODS

The samples analysed in this study were recovered from a borehole five years after the 1999 Chi-Chi earthquake. The SiO_2, Al_2O_3 and MgO contents of the analysed samples (see Supplementary Information, Table S1) are typical of hemipelagic-neritic shale and sandy shale, and the contents of organic carbon (plus graphite) are <0.5% (ref. 16). X-ray diffraction data and microscopic observations show that the samples consist predominantly of quartz, feldspars (plagioclase±K-feldspar), clay minerals (illite>chlorite>kaolinite, smectite), calcite and pyrite, although the BGZs lack kaolinite, calcite and pyrite[16, 21, 27]. Before the chemical analyses of this study were carried out, hydrogen and oxygen isotope ratios and chlorine and sulphate concentrations in pore fluids extracted from 1,135.97–1,136.22 m depth in FZB1136 and 1,194.88–1,195.08 m depth in FZB1194 were analysed[28]. The pore fluids were found to contain high SO_4^{2-} concentrations of about 53 mmol kg^{-1}, presumably derived from dissolution of pyrite in the fault zones[28]. These results confirm that there was minimum contamination of the core samples from the drilling water, which was river and tap water containing 0.4–0.6 mmol kg^{-1} SO$_4{}^{2-}$. All chemical analyses in this study were carried out at Kochi Core Center, Japan. Major-element concentrations were determined by X-ray fluorescence spectrometry using the conventional glass-bead method. For trace-element and isotope analyses, pieces from the interior of the core sample were collected with a ceramic knife with care taken to avoid contamination, and 30 mg of each sample was analysed after digestion with HF. Trace-element concentrations were measured by inductively coupled plasma mass spectrometry, and Sr and Pb isotope ratios were determined by thermal ionization mass spectrometry[29] and multiple-collector inductively coupled plasma mass spectrometry[30], respectively. Replicate analyses of NIST SRM 987 Sr and NIST SRM 981 Pb isotope standards gave mean values ± 2SD as follows: ^{87}Sr/ ^{86}Sr=0.710252±0.000011, ^{206}Pb/^{204}Pb=16.9308 ±0.0010, ^{207}Pb/^{204}Pb=15.4839±0.0011 and^{208}Pb/^{204}Pb=36.6743±0.0030. A rock reference sample, GSJ JB-2, that was analysed along with the TCDP samples gave concentrations of Li=7.71 p.p.m., Rb=6.04 p.p.m., Sr=178 p.p.m., Cs=0.71 p.p.m., Ba=229 p.p.m., La=2.24 p.p.m., Sm=2.28 p.p.m., Pb=5.08 p.p.m., ^{87}Sr/^{86}Sr=0.703682, ^{206}Pb/^{204}Pb=18.3320, ^{207}Pb/^{204}Pb=15.5464 and^{208}Pb/^{204}Pb=38.2261. Apart from the analyses described above, Sr and Pb isotope ratios of the H$_2$O-soluble components in the same samples were determined. About 2 g of each sample was agitated in 10 ml of 18 MΩ H$_2$O. After recovery of the >2 μm and <2 μm fractions of solids for X-ray diffraction analyses, the remaining solution was acidified with HNO$_3$. 0.5–1 ml of the resultant 0.15 M–HNO$_3$ solution was then used for the isotope analyses.

AUTHOR CONTRIBUTIONS

T.I., paper writing, project planning and isotope analysis; M.T., project planning, sample collection and isotope analysis; K.N., J.M., O.T., M.S., major- and trace-element and isotope analyses; T. H., T.M., W.T., W.L., H.K., project planning and sample collection; W.S., S.-R.S., project planning.

In the abstract of the version of this Letter originally published online, the word 'undamages' should have read 'undamaged'. This has now been corrected for all versions of the Letter.

ACKNOWLEDGEMENTS

We thank the TCDP Hole B research group, including K. Aoike, K. Fujimoto, Y. Hashimoto, M. Murayama, T. Fukuchi, M. Ikehara, H. Ito, M. Kinoshita, K. Masuda, T. Matsubara, O. Matsubayashi, K. Mizoguchi, N. Nakamura, K. Otsuki, T. Shimamoto, H. Sone and M. Takahashi.

Received 4 July 2008; Accepted 14 August 2008; Published online 14 September 2008; Corrected 15 September 2008.

REFERENCES

1.   Sibson, R. H. Implication of fault-valve behaviour for rupture nucleation and recurrence. Tectonophysics 211, 283–293 (1992).

2.   Collettini, C., Chiaraluce, L., Pucci, F., Barchi, M. R. & Cocco, M. Looking at fault reactivation matching structural geology and seismological data. J. Struct. Geol.27, 937–942 (2005).

3.   Sibson, R. H. Interaction between temperature and pore-fluid pressure during earthquake faulting—a mechanism for partial or total stress relief. Nature Phys. Sci. 243, 66–68 (1973).

4.   Andrews, D. J. A fault constitutive relation accounting for thermal pressurization of pore fluid. J. Geophys. Res. 107, 2363 (2002).

5.   Rice, J. R. Heating and weakening of faults during earthquake slip. J. Geophys. Res. 111, B05311 (2006).

6.   Doan, M. L., Brodsky, E. E., Kano, Y. & Ma, K. F. In situ measurement of the hydraulic diffusivity of the active Chelungpu Fault, Taiwan. Geophys. Res. Lett. 33, L16317 (2006).

7.   Shin, T. C., Kuo, K. W., Lee, W. H. K., Teng, T. L. & Tsai, Y. B. A preliminary report on the 1999 Chi-Chi (Taiwan) earthquake. Seismol. Res. Lett. 71, 24–30 (2000).

8.   Chung, J. K. & Shin, T. C. Implications of the rupture process from the displacement distribution of strong ground motions recorded during the 21 September 1999 Chi-Chi, Taiwan earthquake. Terr. Atmos. Ocean. Sci. 10, 777–786 (1999).

9.   Andrews, D. J. Thermal pressurization explains enhanced long-period motion in the Chi-Chi earthquake. Eos Trans. AGU 86, S34A-04 (2005) Fall Meet. Suppl., Abstract.

10.   Ma, K. F. et al. Evidence for fault lubrication during the 1999 Chi-Chi, Taiwan, earthquake (Mw7.6). Geophys. Res. Lett. 30, 1244 (2003).

11.   Hirono, T. et al. High magnetic susceptibility of fault gouge within Taiwan Chelungpu-fault: Nondestructive continuous measurements of physical and chemical properties in fault rocks recovered from Hole B, TCDP. Geophys. Res. Lett. 33, L15303 (2006). ChemPort |

12.   Hirono, T. et al. Nondestructive continuous physical property measurements of core samples recovered from Hole B, Taiwan Chelungpu-fault Drilling Project. J. Geophys. Res. 112, B07404 (2007).

13. Ma, K.-F. et al. Slip zone and energetics of a large earthquake from the Taiwan Chelungpu-fault Drilling Project. Nature 444, 473–476 (2006).

14. Hirono, T. et al. Evidence of frictional melting from disk-shaped black material, discovered within the Taiwan Chelungpu fault system. Geophys. Res. Lett. 33, L19311 (2006).

15. Mishima, T., Hirono, T., Soh, W. & Song, S.-R. Thermal history estimation of the Taiwan Chelungpu fault using rock-magnetic methods. Geophys. Res. Lett. 33, L23311 (2006).

16. Ikehara, M. et al. Low total and inorganic carbon contents within the Chelungpu fault System. Geochem. J. 41, 391–396 (2007).

17. Kano, Y. et al. Heat signature on the Chelungpu fault associated with the 1999 Chi-Chi, Taiwan earthquake. Geophys. Res. Lett. 33, L14306 (2006).

18. Burke, W. H. et al. Variation of seawater $^{87}Sr/^{86}Sr$ throughout Phanerozoic time.Geology 10, 516–519 (1982). ISI

19. You, C.-F., Castillo, P. R., Gieskes, J. M., Chan, L. H. & Spivack, A. J. Trace element behavior in hydrothermal experiments: Implication for fluid processes at shallow depths in subduction zones. Earth Planet. Sci. Lett. 140, 41–52 (1996). ChemPort |

20. James, R.H., Allen, D. E. & Seyfried, W. E.Jr. An experimental study of alteration of oceanic crust and terrigenous sediments at moderate temperatures (51 to 350 °C): Insights as to chemical processes in near-shore ridge-flank hydrothermal systems. Geochim. Cosmochim. Acta 67, 681–691 (2003). ChemPort |

21. Hirono, T. et al. Characterization of slip zone associated with the 1999 Taiwan Chi-Chi earthquake: X-ray CT image analyses and microstructural observations of the Taiwan Chelungpu fault. Tectonophysics 449, 63–84 (2008).

22. Kharaka, Y. K. & Hanor, J. S. in Surface and Ground Water, Weathering, and Soils(ed. Drever, J. I.) 499–540 (Treatise on Geochemistry, Holland and Turekian, Elsevier, Oxford, 2003).

23. Tanikawa, W. et al. High magnetic susceptibility produced in high-velocity frictional tests on core samples from the Chelungpu fault in Taiwan. Geophys. Res. Lett. 34, L15304 (2007).

24. Beck, J. R., Berndt, M. E. & Seyfried, W. E.Jr. Application of isotopic doping techniques to evaluation of reaction kinetics and fluid/mineral distribution coefficients: An experimental study of calcite at elevated temperatures and pressures. Chem. Geol. 97, 125–144 (1992). ChemPort |

25. Bizzarri, A. & Cocco, M. A thermal pressurization model for the spontaneous dynamic rupture propagation on a three-dimensional fault: 1. Methodological approach. J. Geophys. Res. 111, B05303 (2006).

26. Seno, T. The 21 September, 1999 Chi-Chi earthquake in Taiwan: Implications for tsunami earthquakes. Terr. Atmos. Ocean. Sci. 11, 701–708 (2000).

27. Hirono, T. et al. Clay mineral reactions caused by frictional heating during an earthquake: An example from the Taiwan Chelungpu fault. Geophys. Res. Lett. 35, doi: doi:10.1029/2008GL034476 (2008).

28. Hirono, T. et al. Chemical and isotopic characteristics of interstitial fluids within the Taiwan Chelungpu fault. Geochem. J. 41, 97–102 (2007).

29. Yoshikawa, M. & Nakamura, E. Precise isotope determination of trace amounts of Sr in magnesium-rich samples. J. Min. Petr. Econ. Geol. 88, 548–561 (1993).

30. Tanimizu, M. & Ishikawa, T. Determination of rapid and precise Pb isotope analytical techniques using MC-ICP-MS and new results for GSJ rock reference samples. Geochem. J. 40, 121–133 (2006). ChemPort |

Chapter 2

FRACTURES AND FRACTURING: HYDRAULIC FRACTURING IN JOINTED ROCK

Charles Fairhurst[1, 2]

[1] Senior Consultant, Itasca Consulting Group, Inc, Minneapolis Minnesota, USA

[2] Professor Emeritus, University of Minnesota, Minneapolis, Minnesota, USA

ABSTRACT

Rock in situ is arguably the most complex material encountered in any engineering discipline. Deformed and fractured over many millions of years and different tectonic stress regimes, it contains fractures on a wide variety of length scales from microscopic to tectonic plate boundaries.

Hydraulic fractures, sometimes on the scale of hundreds of meters, may encounter such discontinuities on several scales. Developed initially as a technology to enhance recovery from petroleum reservoirs, hydraulic fracturing is now applied in a variety of subsurface engineering applications. Often carried out at depths of kilometers, the fracturing process cannot be observed directly.

Early analyses of the hydraulic fracturing process assumed that a single fracture developed symmetrically from the packed off-pressurized interval of a borehole in a stressed elastic continuum. It is now recognized that this is often not the case. Pre-existing fractures can and do have a significant influence on fracture development, and on the associated distributions of increased fluid pressure and stresses in the rock.

Given the usual lack of information and/or uncertainties concerning important variables such as the disposition and mechanical properties of pre-existing fracture systems and properties, rock mass permeabilities, in-situ stress state at the depths of interest, fundamental questions as to how a propagating fracture is affected by encounters with pre-existing faults, etc., it is clear that design of hydraulic fracturing treatments is not an exact science.

Fractures in fabricated materials tend to occur on a length of scale that is small; of the order of the 'grain size' of the material. Increase in the size of the structure does not introduce new fracture sets.

Numerical modeling of fracture systems has made significant advances and is being applied to attempt to assess the extent of these uncertainties and how they may affect the outcome of practical fracturing programs. Geophysical observations including both micro-seismic activity and P- and S-wave velocity changes during and after stimulation are valuable tools to assist in verifying model predictions and development of a better overall understanding of the process of hydraulic fracturing on the field scale. Fundamental studies supported by laboratory investigations can also contribute significantly to improved understanding.

Given the widening application of hydraulic fracturing to situations where there is little prior experience (e.g., Enhanced Geothermal Systems (EGS), gas extraction from 'tight shales' by fracturing in essentially horizontal wellbores, etc.) development of a greater understanding of the mechanics of hydraulic fracturing in naturally fractured rock masses should be an industry-wide imperative. HF 2013 International Conference for Effective and Sustainable Hydraulic Fracturing is very timely!

This lecture will describe examples of some current attempts to address these uncertainties and gaps in understanding. And, it is hoped, it will stimulate discussion of how to achieve more effective practical design of hydraulic fracturing treatments.

INTRODUCTION

The term 'rock' covers a wide variety of materials and widely different rheological properties often proximate to each other in the subsurface. Tectonic and gravitational forces, sustained over millions of years, have deformed and fractured the rock on many scales. These forces are transmitted in part through the solid skeleton of the rock, and in part through the fluids under pressure in the pore spaces. Long-term circulation through rock at high temperatures at depth involves dissolution and precipitation along the fluid pathways, producing changes in the chemical composition of the fluids and modifying the overall fluid circulation.

Rock in situ is 'pre-loaded' and in a state of changing equilibrium. Any engineering activity changes this equilibrium (see Appendix 1). Often the changes can be accommodated in stable fashion, but serious instabilities can develop.

The rock mass is opaque. Although geophysics is making impressive advances in defining large structures such as faults and bedding planes, most of the features that influence the rock response to engineering activities remain hidden. Mining and civil engineering activities allow three-dimensional access to the underground and direct observation of smaller features such as fracture networks, but most of the newer engineering applications involve essentially one-dimensional access by borehole. Rock engineering problems fall into the 'data –limited' category, as defined by Starfield and Cundall (1988), and strategies to address them must follow a different strategy than engineering problems where detailed and precise design information is available.

Faced with such complexity and lack of structural details, traditional subsurface engineering design has been guided by empirical procedures developed and refined through long experience.

Projects are now venturing well beyond current experience, and for many, 'novel' applications now considered (e.g., Enhanced Geothermal Systems, Carbon Sequestration, see Appendix 1). There is little experience, few guiding rules and very little data to guide the engineering approach.

Such obstacles notwithstanding, subsurface processes, both long–term geological and short term responses, to engineering activities do obey the laws of Newtonian Mechanics.

Classical continuum mechanics has long been used to guide some aspects of design, but considerable care is required in practical application, due to the need to simplify the representation of the real conditions in order to obtain analytical solutions.

The remarkable developments in high-speed computation and associated modeling techniques over the past one to two decades provide an important new tool, which complemented by the appropriate field instrumentation, can augment the classical continuum analyses and help overcome the lack of prior experience. Some empiricism and general practical guidelines may still be useful for the design engineer, but these can and should be mechanics-informed.

This lecture attempts to illustrate the 'mechanics-informed' approach with respect to the practical application of hydraulic fracturing and related engineering procedures to rock engineering.

HYDRAULIC FRACTURING

Hydraulic fracturing first was used successfully in the late 1940's to increase production from petroleum reservoirs (Howard and Fast, 1970). The technology has evolved since and is now a major, essential technique in oil and gas production. This and other impressive oil industry developments, such as directional drilling, have attracted interest in application of these technologies to a variety of other subsurface engineering operations. Enhanced Geothermal Energy (EGS) is a notable example. Geothermal Energy is a huge resource. Commenting on the EGS resource in the USA, Tester et al. (2005), state:

"....we have estimated the total EGS resource base to be more than 13 million exajoules (EJ)[1] - . Using reasonable assumptions regarding how heat would be mined from stimulated EGS reservoirs, we also estimated the extractable portion to exceed 200,000 EJ or about 2,000 times the annual consumption of primary energy in the United States in 2005. With technology improvements, the economically extractable amount of useful energy could increase by a factor of 10 or more, thus making EGS sustainable for centuries." [2] -

"At this point, the main constraint is creating sufficient connectivity within the injection and production well system in the stimulated region of the EGS reservoir to allow for high per-well production rates without reducing reservoir life by rapid cooling." [3] -

Field experiments to extract geothermal energy from rock at depth by hydraulic fracturing were started in 1970 by scientists of the Los Alamos National Laboratory, USA. Two boreholes were drilled into crystalline rock (one 2.8 km deep, rock temperature 195°C; the other 3.5 km rock, 235°C) at Fenton Hill, New Mexico. Hydraulic fracturing was used to develop fractures from the boreholes in order to create a fractured region through which water could be circulated to extract heat from the rock. The experiment was terminated in 1992. Commenting on what was learned from the Fenton Hill study, Duchane and Brown (2002) note:

"The idea that hydraulic pressure causes competent rock to rupture and create a disc-shaped fracture was refuted by the seismic evidence. Instead, it came to be understood that hydraulic stimulation leads to the opening of existing natural joints that have been sealed by secondary mineralization. Over the years additional evidence has been generated to show that the joints oriented roughly orthogonal to the direction of the least principal stress open first, but that as the hydraulic pressure is increased, additional joints open."

This is an early indication that pre-existing fractures mass significantly affect how hydraulic fractures propagate in a rock mass.

INFLUENCE OF FRACTURES AND DISCONTINUITIES ON THE STRENGTH OF BRITTLE MATERIALS

Hydraulic fracturing can be considered as a technique to overcome the strength of a rock mass in situ, initiation and propagation of a crack through a system of pre-existing fractures, essentially planar discontinuities (e.g., bedding planes), and intact rock.

In examining the fracture propagation process, the pioneering work of Griffith (1921, 1924) is a logical point of departure. Griffith had identified planar discontinuities, or flaws, in fabricated materials as the reason why the observed technical strength of brittle materials was about three orders of magnitude lower than the theoretical inter-atomic cohesive (tensile) strength. [4] - Using an analytical solution by Inglis (1913) for the elastic stresses generated around an elliptical crack in a plate, Griffith observed that the maximum tensile stress at the tip of the crack $\sigma_t = \sigma_0 (1 + 2a/b)$, where a and b are the major and minor semi-axes of the ellipse, and as the ellipse degenerated to a sharp crack or flaw (i.e., as the ratio a/b became very high)[5] - , the stress σ_t could rise to a value high enough to reach the inter-atomic cohesive strength sufficient to cause the original crack to start to extend.

But would the crack continue to extend and lead to macroscopic failure? To address this question, Griffith invoked the *Theorem of Minimum Potential Energy,* which may be stated as "The stable equilibrium state of a system is that for which the potential energy of the system is a minimum." For the particular application of this theorem to brittle rupture, Griffith added the statement, "The equilibrium position, if equilibrium is possible, must be one in which rupture of the solid has occurred, if the system can pass from the unbroken to the broken condition by a process involving a continuous decrease of potential energy."[6] -

Griffith's classical work has provided the foundation for the field of "Fracture Mechanics" [Knott (1973); Anderson (2005)] responsible for major continuing advances in the development of high-performance fabricated materials.

Since we will make reference later to this specific definition by Griffith, it is useful to re-state it here.

THEOREM OF MINIMUM POTENTIAL ENERGY

"The stable equilibrium state of a system is that for which the potential energy of the system is a minimum. The equilibrium position, if equilibrium is possible, must be one in which rupture of the solid has occurred, if the system can pass from the unbroken to the broken condition by a process involving a continuous decrease of potential energy."

Although much of classical Fracture Mechanics has emphasized applications to problems of Linearly Elastic Fracture Mechanics (LEFM) it is important to recognize that the theorem of minimum potential applies equally to inelastic problems.

MECHANICS OF HYDRAULIC FRACTURING

As used classically in petroleum engineering, hydraulic fracturing involves sealing off an interval of a borehole at depth in an oil or gas bearing horizon, subjecting the interval to increasing fluid pressure until a fracture is generated, injecting some form of granular proppant into the fracture as it extends a considerable distance from the borehole into the petroleum bearing formation, and then releasing the pressure. This causes the sides of the fracture to compress onto the proppant, creating a high-permeability pathway to allow oil and/or natural gas to flow back to the well and to the surface.

Figure 1 shows a simple two-dimensional cross-section through an idealized hydraulic fracture. The borehole injection point is at the center of the fracture, which is assumed to be a narrow ellipse that has extended in a plane normal to the direction of the maximum[7] - (least compressive) in-situ stress.

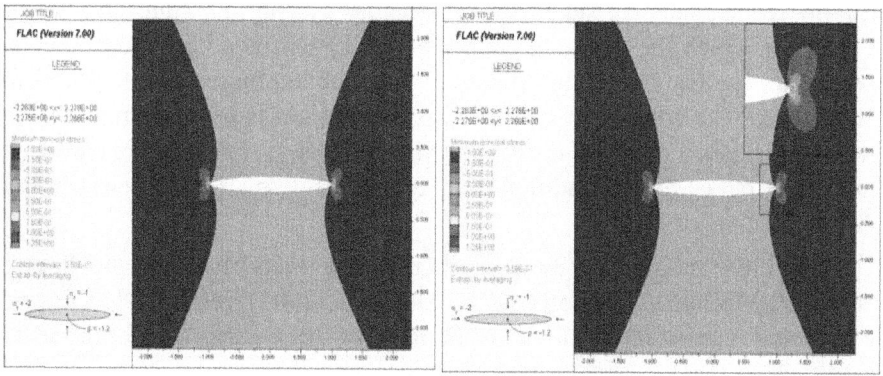

Figure 1: Left) Major and (right) minor principal stresses in the vicinity of an internally pressurized elliptical crack in an impermeable rock.

In the case shown, the crack major/minor axis ratio a/b is 10:1. The internal fluid pressure p = 1.2, while the least compressive principal stress $\sigma_x = 1.0$. This results in a tensile stress concentration at the crack tip. The magnitude of the elastic stress concentration at the crack tip increases directly with 2a/b, (Inglis, 1913). Hence for the case of a>>b, i.e., a 'sharp' crack[8] - , the concentration is very high, and the crack will extend essentially as soon as the fluid pressure exceeds the magnitude of the least compressive principal stress (σ_x in Figure 3) it begins to extend, and there will be a pressure gradient from the injection point towards the crack tip as the fluid flows towards the tips. This gradient will depend on the fluid viscosity. Also, since the rock will exhibit some level of permeability, fluid will also flow (or 'leak–off') into the formation as it flows under pressure along the fracture; the rock has a finite strength, or 'toughness' so that energy will be required to extend the crack.

An analytical solution for the stresses in the elastic medium and the crack-opening displacement along the crack was first published by Inglis (1913) and served as the basis for early applications to hydraulic fracturing and fracture treatment design. The Perkins, Kern (1961) and Nordgren (1972) (PKN) andGeertsma and de Klerk (1969) (GDK) models are still used, although numerical models and combinations are now popular. Details of the PKN and GDK models can be found on the SPE website: http://petrowiki.spe.org/ Fracture_propagation_models. Several differences between the stationary crack assumed by Inglis (1913) and a hydraulic fracture introduce significant difficulties in developing an accurate model of the fracturing process. Thus, the fracture is generated by application of an increasing fluid pressure until the fracture is initiated and extends away from the injection point. Flow of fluid in the fracture is governed by classical fluid flow equations of Poiseuille and Reynolds (lubrication); the pressure drop along the fracture depends on the viscosity of the fluid, and the permeability of the rock (leading to fluid 'leak-off'); the fracture aperture depends on the stiffness of the rock mass and the fluid pressure distribution along the crack; and fracture extension depends on the mechanical energy supplied to the region around the crack tip. The tip may propagate ahead of the fluid, leading to a 'lag,'a dry region between the crack tip and fluid front.

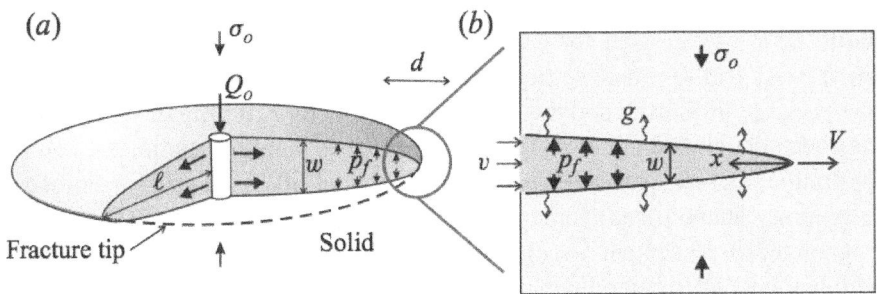

Figure 2: Radial Model of Axi-symmetric Flow and Deformation associated with Hydraulic Fracturing.

Figure 2 illustrates these features for the classical Radial Model in which it is assumed that the fracture propagates symmetrically away from the borehole in a plane normal to the minimum (least compressive) principal in-situ stress, σ0.

Development of efficient and robust Hydraulic Fracturing (HF) simulators is central to successful practical HF treatment of petroleum reservoirs. As noted earlier, competing physical processes are operative during the fracturing operation. This has led to a sustained effort over many years to understand and map the multi-scale nature of the tip asymptotics that arise as a result of these competing physical processes in fluid-driven fracture. These asymptotics solutions are critical to the construction of efficient and robust HF simulators. For example, in an impermeable medium, the viscous energy dissipation associated with driving fluid through the fracture competes with the energy required to break the solid material. Breaking of the bonds corresponds to the familiar asymptotic form of linear elastic fracture mechanics (LEFM), i.e., the opening in the tip region is of the form, e.g., (Rice, 1968), with denoting the distance from the tip. However, under conditions where viscous dissipation dominates, the coupling between the fluid flow and solid deformation leads to (Spence and Sharp, 1985; Lister, 1990; Desroches et al., 1994), on a scale that is considerably larger than the size of the LEFM-dominated region, but still small relative to the overall fracture size. In other words, in the viscosity-dominated regime, the zone governed by the LEFM asymptote is negligibly small compared to the crack length. Thus, in the viscosity-dominated regime, the HF simulator should embed a 2/3 power law asymptote rather than the classic 1/2 asymptote of LEFM. Garagash et al.(2011) discuss the generalized asymptotics near the tip an advancing hydraulic fracture, an extension of two particular asymptotics obtained at Schlumberger Cambridge Research Laboratory in the early 1990's (Desroches et al., 1994; Lenoach, 1995).

Three classes of numerical algorithms for HF simulators have now been built: (i) a moving grid for KGD, radial, PKN and P3D fracture simulators; (ii) a fixed grid for plane strain and axisymmetric HF with allowance for a lag between the fluid front and the crack tip, and fracture curving (a versatile code has been developed at CSIRO[9] - Melbourne to simulate the interaction of a hydraulic fracture with other discontinuities); and (iii) fixed grid for simulating a arbitrary shape planar fracture in a homogenous elastic rock. These codes rely on the displacement discontinuity method (Crouch and Starfield, 1983) for solving the elastic component of the problem, i.e., the relationship between the fracture aperture and the fluid pressure.

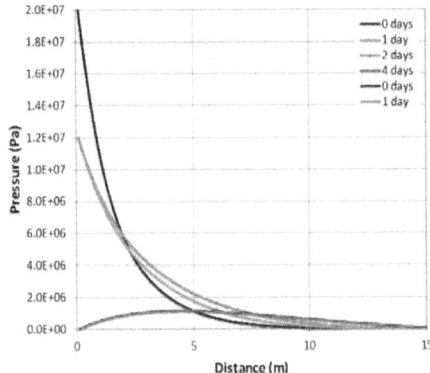 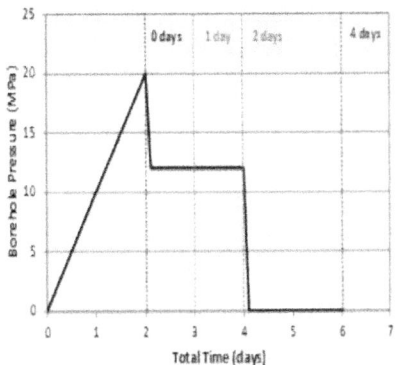

Figure 3: Fluid Pressure Distribution along the Central Axis (Ox) of Figure 1 for a permeable rock due to pressurization and de-pressurization of the borehole.

Figure 3 is presented to illustrate that the fluid pressure in a permeable rock can continue to flow away from the point of injection even after the borehole pressure is reduced to zero. The example shows the distribution of fluid pressure in the rock mass (permeability 5 mD) after (i) 2 days of pressurization up to the peak pressure of 20 MPa in the fracture; (ii) stop pumping and reduce fluid pressure quickly to 12MPa at the point of injection; (iii) hold the pressure constant for 2 days; and (iv) drop the pressure to zero.

It is seen that the pressure in the rock (red curve) has a maximum at some distance from the borehole such that fluid continues to flow into the rock for some time after the pressure in the borehole is reduced to zero. Different combinations of rock permeability, pumping rates and durations can lead to higher peak pressure values in the rock, and longer periods during which fluid can continue to flow away from the well. Such flow may contribute to slip on pre-existing fractures after the pressure in the borehole is reduced to zero.

HYDROSHEAR

Hydraulic fracturing is considered to be initiated from a packed–off interval borehole when the net state of stress around the well bore reaches the tensile strength of the rock. It is important to recognize that fluid pressurization of a well in permeable rock will result in flow of the fluid into the rock as soon as the fluid pressure stimulation process is started. This changes the effective stress state in the rock mass and can lead to slip on pre-existing fractures at fluid pressures below the pressure required to crate and extend a hydraulic fracture. This process of inducing slip on pre-existing fractures is termed 'Hydro-shear'. Flow of pressurized fluid into the rock reduces the effective normal stress ($\sigma n - p$) everywhere in the rock { σn = normal stress at any point; p = fluid pressure.] If c and μ respectively represent the cohesion and coefficient of friction acting across the surfaces of a fracture in the rock, then the effective resistance of the fracture to (shear) sliding, τr, will be:

$$\tau r = c + \mu\ (\sigma n - p) \tag{1}$$

Thus, if the pressure p is raised progressively then τr will be reduced correspondingly until it reaches the limit at which sliding will occur. The situation is illustrated graphically in Figure 3. The rock is subjected to a three-dimensional state of stress represented by the principal stresses $\sigma 1$, $\sigma 2$, $\sigma 3$ and the fluid pressure p. The series of points 'X' indicate the effective state of stress on an array of pre-existing fractures in the rock. As illustrated in Figure 5, the effect of increasing the fluid pressure in the medium is to move the stress state on these cracks close to the limiting shear resistance, i.e., to the limiting value represented by the Mohr-Coulomb limit.

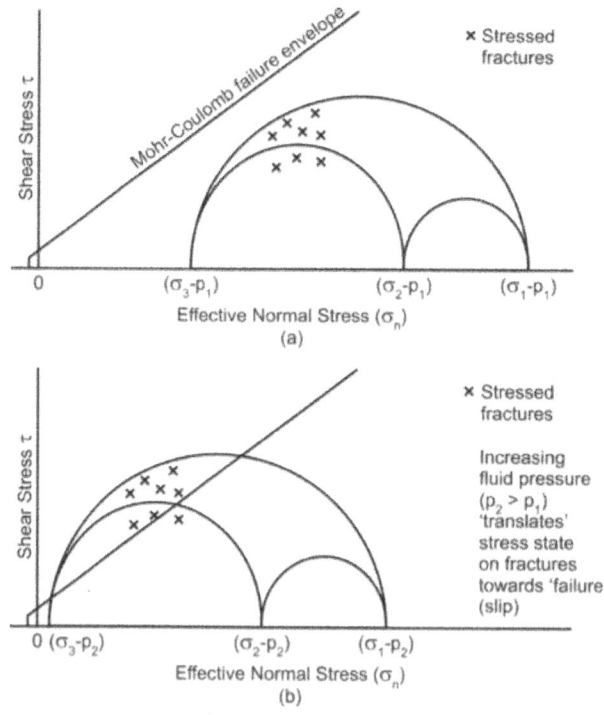

Figure 4: Hydro-shearing — a procedure to generate slip on pre-existing fractures by increasing the fluid pressure to a level below that required to generate a hydraulic fracture.

As the stress state reaches this limit, the cracks will slip. In order to initiate a hydraulic fracture, the fluid pressure would need to be increased further, until the limiting Mohr circle reaches the tensile strength limit of the failure envelope. Since crack surfaces are often not smooth, shear slip will tend to result in crack dilation, and an associated increase in fluid conductivity. It is suggested that hydro-shearing could be more effective than hydraulic fracturing as a stimulation technique in certain applications, e.g., in stimulation of high-temperature geothermal reservoirs. Cladouhos et al. (2011) discuss the application of hydro-shearing as a geothermal stimulation technique. The possibility that silica proppant may dissolve in the aggressive high-temperature fluid environment of some geothermal reservoirs whereas slip on rough fractures develops aperture increase without the need for proppant is also presented as an argument in favor of hydroshearing.

DEFORMATION AND FAILURE OF ROCK IN SITU

As with fabricated materials, the deformation and failure of brittle rock is also dependent strongly on fractures and discontinuities. In a rock mass, however, the fractures occur over a very wide range of scales from sub-microscopic to the size of tectonic plates. A large specimen of rock will probably include some large fractures, and as the scale of the rock mass increases, fractures from different tectonic epochs.

Study of fracture systems underground in mines and in civil engineering projects allow systems of fractures to be identified and classified statistically into discrete fracture networks (DFN's). The network will include intersecting sets of planar fractures, but individual fractures will tend to be of different lengths, and though organized in two or three spatial orientations, of variable, finite length and not collinear.

Figure 7 presents a two-dimensional illustration of the application of DFN's to the numerical modeling of a fractured rock mass. The in-situ rock mass is considered as a large specimen of intact rock that has been transected by the DFN determined from field observations and fracture mapping underground or at surface outcrops. The properties of the intact rock are built into a Bonded Particle Model of the rock (using the Particle Flow Code (*PFC*) code) based on results of laboratory tests of the intact rock deformability and strength. The intact rock representation is shown on the left of Figure 6. The DFN (shown on the upper right in Figure 6) then is superimposed onto the intact rock.

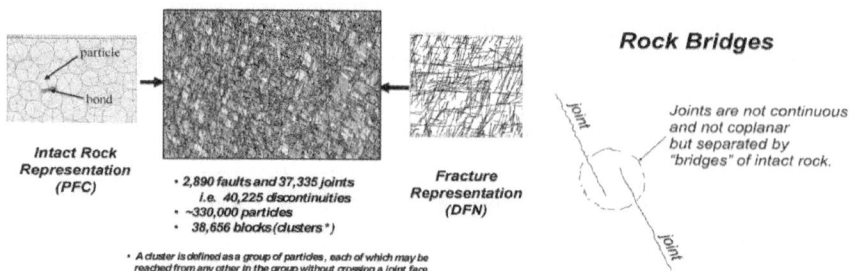

Figure 5: The Synthetic Rock Mass (SRM) representation of a fractured rock mass (in two dimensions).Damjanac et al. (2013) present a discussion of the 'construction' of an SRM in three dimensions.Pierce (2011) presents a comprehensive discussion of practical guidelines and factors involved in the construction of DFN's.

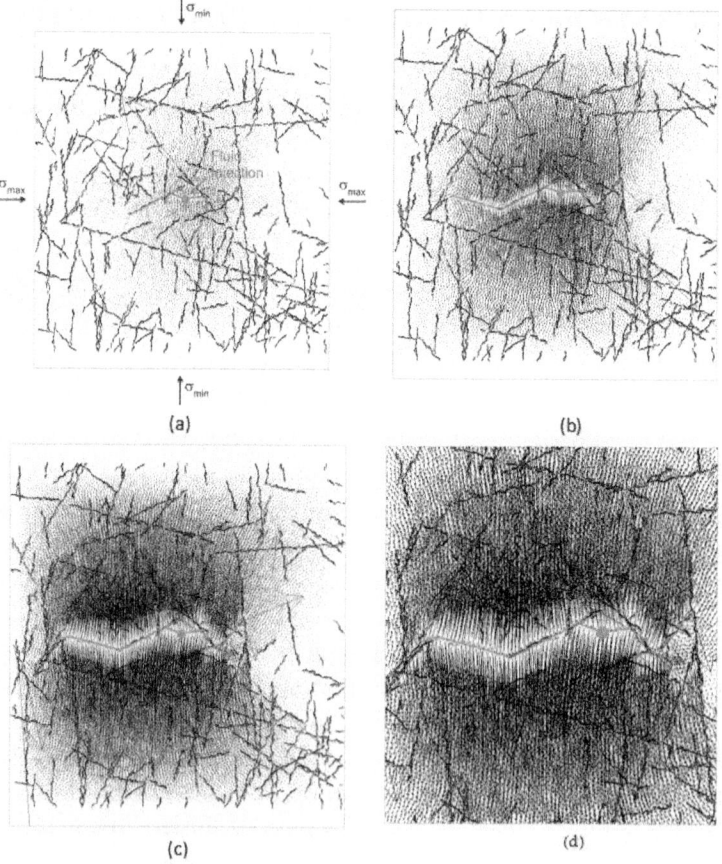

Figure 6: Extracts from simulation of the propagation of a hydraulic fracture in a two-dimensional impermeable SRM (Synthetic Rock Mass). (The horizontal stress σ_{max} is

29 MPa and the vertical stress σ_{min} is 12 MPa – Figure 5(a)). Note that the intact rock between the fractures has a finite strength and can break by rupture of the cemented bonded particles shown in Figure 5. The pressure required to propagate the fracture after breakdown was approximately 10 MPa above the minimum (i.e., least compressive) principal.

Cohesion and friction values are assigned to the joint planes.[10] - The 'unconfined' strength of a typical large SRM is of the order of a few percent of an intact rock specimen of the same rock (Cundall, 2008). Much of the in-situ strength is derived, of course, from the in-situ stresses imposed on the SRM in situ. One of the consequences of the finite length and lack of collinearity of joint sets in DFN's is the formation of bridges of intact rock Figure 4 within the SRM. These bridges provide regions of intact rock, and of stress concentration, in the SRM and account for a significant part of the overall strength of the rock mass. Earlier models of a rock mass, considered to consist of several sets of through-going fractures, exhibited much lower rock mass strength (Hoek and Brown, 1980).

Figure 5 presents selected extracts from a two–dimensional *PFC* simulation of the development of a hydraulic fracture in a jointed Synthetic Rock Mass. The SRM model was developed following the procedure outlined in Figure 5. The joint distribution was based on a DFN obtained at the Northparkes Mine in Australia.[11] - Figure 5(a) shows the location of a vertical borehole that was pressurized by fluid until a hydraulic fracture was initiated. The rock mass is assumed to be impermeable. (The path of the fracture has been traced in blue for clarity.) Displacements in the rock mass produced by the hydraulic fracture are shown as vectors on each side of the fracture. It is seen that the fracture started more or less symmetrically on each side of the borehole, but propagation of the right wing was arrested when the hydraulic fracture encountered an adversely oriented pre-existing joint (Figure 5(b)). With increasing pressure, in the borehole, the hydraulic fracture continued to extend asymmetrically towards the left (Figures 5(c) and 5(d) Figure 5(d) is simply an enlarged view of Figure 5(c)). It is seen that the propagating fracture extended partially by opening existing fractures and partially by developing new fractures through intact rock. Although local deviations occur, the overall path of fracture growth is approximately perpendicular to the direction of the minimum compression stress. The existing fractures introduce an asymmetry to the rock mass. In terms of the idealized symmetric crack of Figure 2, the system in Figure 3 can be considered as two cracks, one extending to the right and one to the left of the borehole with a higher 'fracture toughness' on the right compared to the left, etc.

Jeffrey et al. (2009) conducted an underground test in the Northparkes Mine, Australia to observe the propagation of a hydraulic fracture in naturally fractured tock. Figure 7 shows part of the path of the fracture, as seen in a tunnel excavated into the fractured rock. The fracture path shows similar characteristics to those shown in the *PFC* simulation in Figure 6.

Figure 7: Hydraulic fracture (green plastic) crossing a shear zone on the face of a tunnel excavated through the fracture. "The arrows indicate the trace of the fracture with green plastic contained in it. There is no clear fracture between points 1 and 2 but the fracture may have crossed this zone either deeper into the rock or in the rock that has been excavated. Approximately 2 m of fracture extent is visible" (Jeffrey et al., 2009).

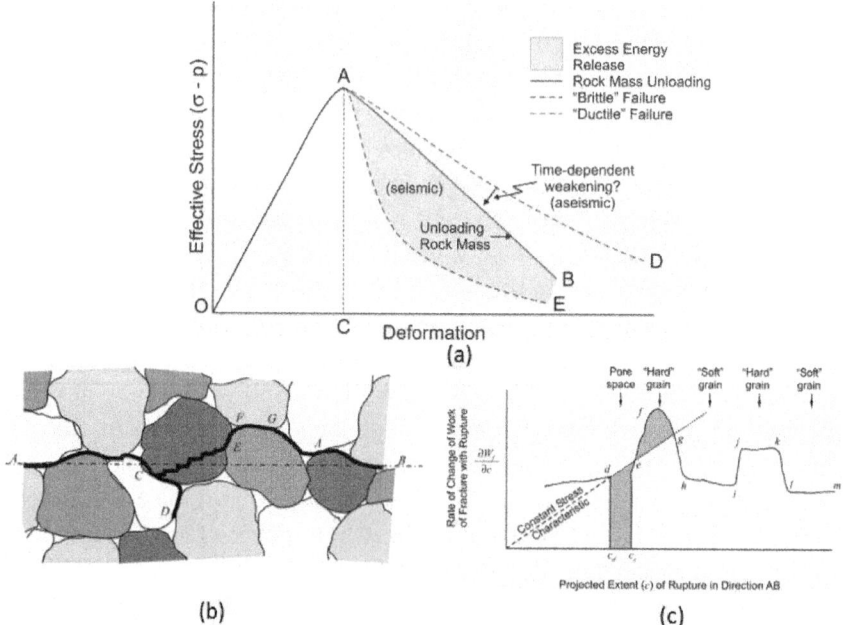

Figure 8: Energy changes during propagation of a fracture through heterogeneous rock.

The energy required to initiate crack propagation is represented by the area OAC in Figure 7(a). Whether or not the crack will extend depends on the energy that becomes available from the intact rock around the crack. If the energy released from the rock mass, represented by the area under the red curve AB, is greater than the energy required to extend the crack, represented by the area under curve AE, then the crack will extend; the excess energy represented by the shaded area serves to accelerate the crack and release seismic energy. If the energy required to extend the crack is represented by the area under the green curve AD, it is greater than the energy that would be released from the rock mass, and hence the crack would not extend. It is possible that the crack could exhibit some form of time-dependent weakening (e.g., due to fluid flow to the crack, viscous behavior, etc.) such that the energy required to extend the crack would be reduced. This could lead to crack extension, i.e., as the slope AD increased to overlap AB, but with no excess energy to produce seismicity. Figures 7(b) and 7(c)[12] - illustrate another feature of crack extension on the granular scale. The energy required to extend a crack through or around a grain will be variable; the fracture may encounter pore spaces where no crack energy is required. Application of a constant load to such a heterogeneous system will result in local acceleration and deceleration of the crack-producing bursts of microseismicity. Similar effects can arise in rock fracture propagation at all scales.

It is worth noting that all of these processes of fracture propagation, albeit complex, develop in accordance with the principle of seeking the minimum potential energy of the system.

Much of the preceding discussion has focused on two-dimensional analysis or models. In reality, we are dealing with three- dimensional space (as noted in Figure 6), plus the influence of time (e.g., with respect to fluid flow, or time-dependent rock properties). Figure 8 provides an example from an actual record of hydraulic fracture propagation.

Figure 8 shows the sequence of microseismic events observed during hydraulic fracture stimulation ('treatment' in Figure 8(a)) of a borehole. Early time events are shown as green dots; later events are in red. The microseismic pattern indicates that fracturing started on both sides of the borehole at the injection horizon, but then moved up some 100 m to a higher horizon. As pumping continued, fracturing continued (red locations) on both horizons. It was concluded that the initial fracture in the lower horizon had intercepted a high-angle fault, allowing injection fluid to move to the higher level where it opened up and extended another fracture. Continued pumping led to fracture extension on both horizons. Numerical analysis Figure 8(b) indicated that initial fracture propagation at the lower level resulted in induced tension on the fault above the horizon, but compression on the fault below the lower injection horizon. This explains why injection fluid did not penetrate along the fault below the horizon, and provides a good illustration of the benefit of combining numerical analysis with field observation in understanding fracturing processes.

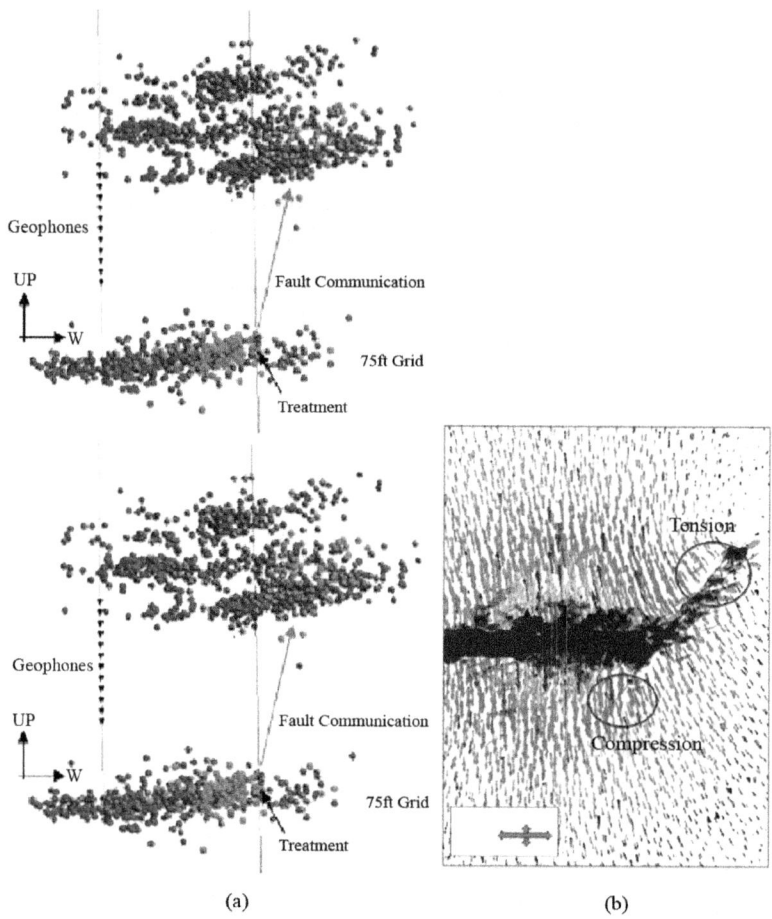

(a)   (b)

Figure 9: a) Microseismicity observed during hydraulic fracturing in a deep borehole; (b) numerical 'explanation' of the behavior observed in (a).

MICROSEISMICITY AS AN INDICATOR OF SLIP ON FRACTURES

Microseismicity stimulated during hydraulic fracturing and associated stimulation techniques (e.g., hydroshear) is often used to indicate slip and deformation on fractures in the rock. In some cases, it is tacitly assumed that absence of microseismicity indicates absence of slip or deformation. In fact, there is growing evidence that microseismicity does not present a complete picture of deformations induced by stimulation or other effects leading to stress change. Figure 9, reproduced from Cornet (2012) (with permission from the author), shows P-wave velocity changes observed by 4D (time-dependent) tomography during the stimulation of the borehole GPK2 in the year 2000. A detailed discussion of the procedure used to observe and determine the P-wave changes is presented by Calo et al. (2012). It is seen that the region of detected microseismicity (the cloud of black dots is small compared to the region where the P-wave velocity is reduced by as much as 20% in some regions). Some of the changes in velocity were temporary, suggesting that they may be related to temporal changes in fluid pressure; other changes appeared to be more permanent deformation that occurred aseismically. These observations indicate that microseismicity, although a valuable indicator of the response of a rock mass to stimulation by fluid injection, does not identify the complete region influenced by a stimulation.

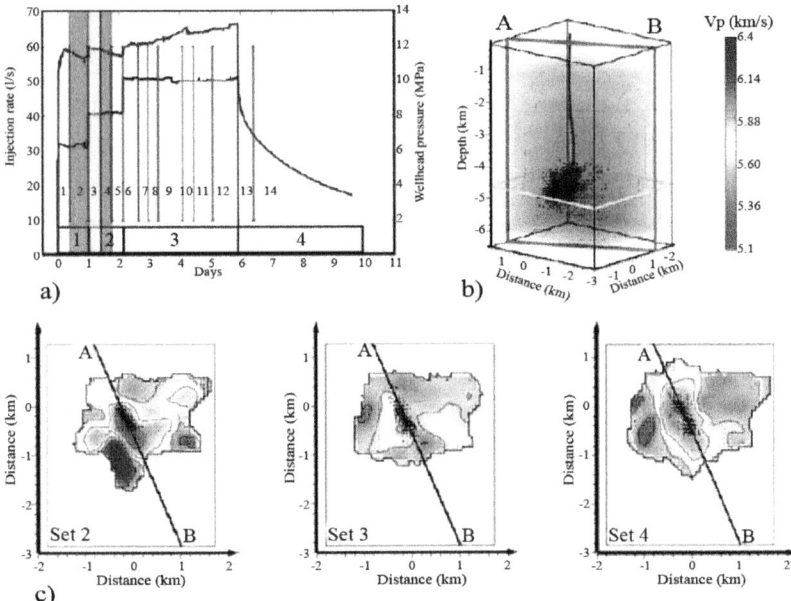

Figure 10: Aseismic slip induced by forced fluid flow as detected by P-wave tomography. (Soultz- sous- Fôrets, France. (a) The injection program (black curve is flow rate, blue curve is well head pressure, horizontal axis is time in days); (b) 3D view of the seismic cloud with respect to the GPK2 borehole. Vertical axis is depth and horizontal axes are distances respectively toward the north and toward the east; and (c) horizontal projections corresponding to the yellow horizontal plane. The vertical green plane is

shown as line AB in the plots of part c. P-wave velocity tomography for sets 2, 3 and 4 are indicated respectively by orange, yellow and green colors in the injection program. The vertical axis corresponds to North.

IN-SITU STRESS

As already noted, hydraulic fractures tend to develop in a more or less planar fashion, extending normal to the minimum regional principal stress. Determining the direction, and perhaps the magnitude, of the regional minimum stress is an important element of hydraulic fracturing strategy, especially with the development of directional drilling, which allows borehole to be drilled in the direction considered most favorable for fracturing with respect to stress direction. (see e.g., Figure 15and related discussion).

Determination of the in-situ stress state also can be a significant challenge.

Stress in rock is distributed throughout the mass, and is influenced by the complicated structure of the mass[13] - . Most techniques of stress determination rely on what are essentially 'point' determinations. One difficulty of determining the regional stress is illustrated by the simple, albeit somewhat artificial, example of Figure 11. This shows a two-dimensional numerical model of the stress distribution in an elastic plate containing several finite frictional fractures.

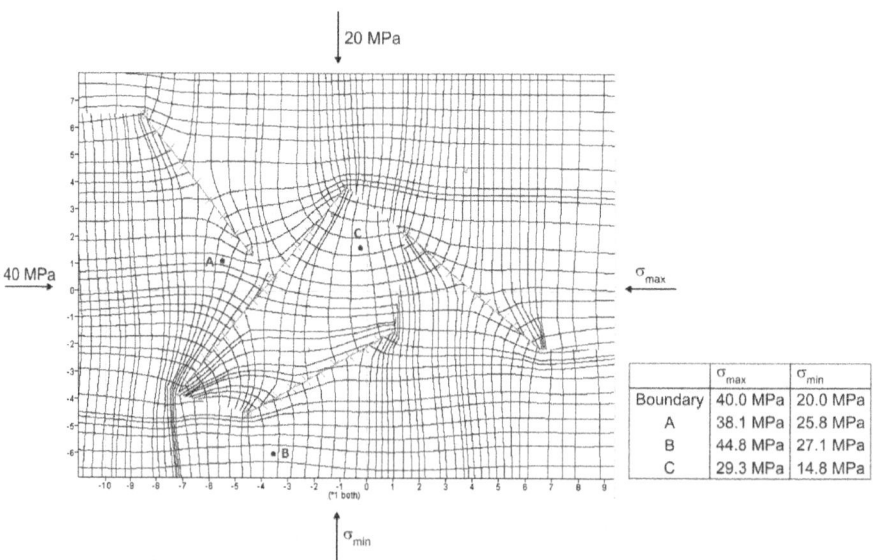

	σ_{max}	σ_{min}
Boundary	40.0 MPa	20.0 MPa
A	38.1 MPa	25.8 MPa
B	44.8 MPa	27.1 MPa
C	29.3 MPa	14.8 MPa

Figure 11: Influence of frictional cracks on the distribution and orientation of principal stresses, illustrative example.

The exercise serves to illustrate the difficulty of making stress determinations from local point measurements, be they in a borehole or on the surface. Stresses can change in orientation and magnitude locally due to geological inhomogeneities, fractures, faults, etc., many of which may be hidden or cannot be observed from the measurement location. Although determinations made at points A and B are reasonably close to the boundary values, point C is considerably different, and the directions of principal stress, as indicated by the principal stress trajectories, can be very different from the (regional) orientations, i.e., at the model boundary.

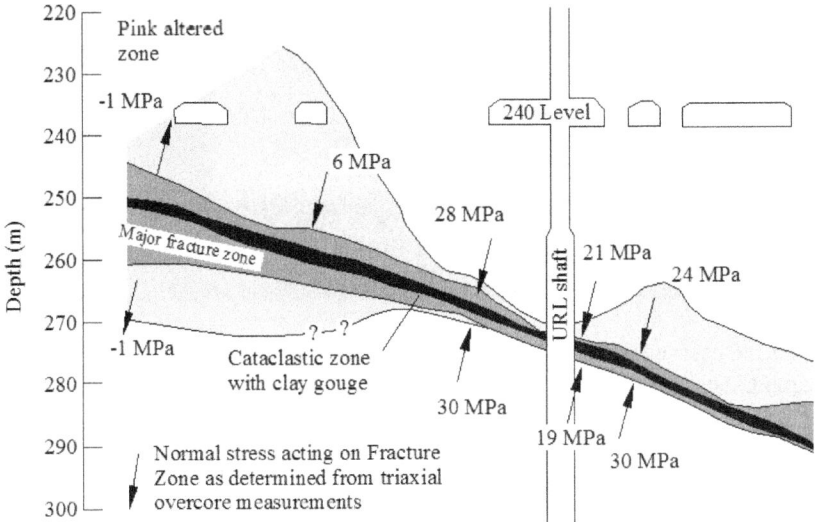

Observed Variability of Normal Stress Across a Thrust Fault Underground Research Laboratory Pinawa, Canada.

Figure 12: Normal stress variation across a thrust fault, Underground Research Laboratory, Canada.

Figure 12 provides an actual example of the variability of stress over relatively short distances. (The vertical and horizontal scales are equal in Figure 12). In this case, the main interest was to assess how normal stresses were affected by the thickness of gouge in the plane of the thrust fault.

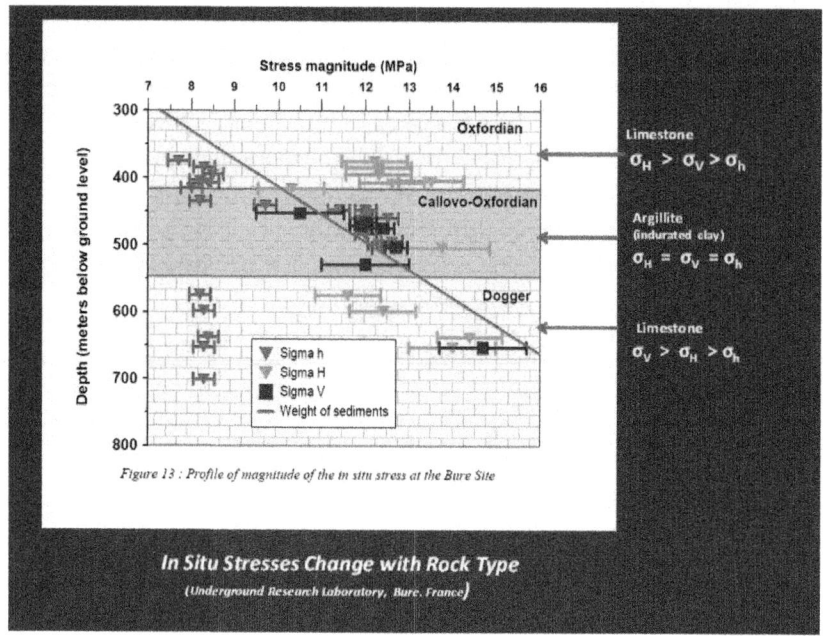

Figure 13 : Profile of magnitude of the in situ stress at the Bure Site

In Situ Stresses Change with Rock Type
(Underground Research Laboratory, Bure, France)

Figure 13: Observed stress distributions in argillite and limestones at the Underground Research Laboratory, Bure, France.

Figure 13 illustrates another important geological influence on stress distribution, changing lithology. This example is from the French Underground Research Laboratory (URL) [14] - at Bure in NE France. Laboratory tests on specimens of the Callovo-Oxfordien Argillite indicate a long-term viscosity of this rock suggesting that any imposed deviatoric stresses would tend towards an isotropic stress state over the order of 10 million years.

Test specimens from the limestones above and below the argillite do not appear to exhibit such viscosity. The stress distributions determined from field measurements support such differences in rheological characteristics of the rock formations.

Commenting on the in-situ stresses observations at Bure (i.e., as shown in Figure 13) Cornet (2012) notes as follows:

"Further, the complete absence of microseismicity in the Paris Basin (Grünthal and Wahlström, 2003, Fig. 4) and the absence of large scale horizontal motion as detected by GPS monitoring (Nocquet and Calais, 2004) indicate that no significant horizontal large-scale active deformation process exists today in this area.

"The important conclusion here is that the natural stress field measured on a 100 km² area at depth ranging between 300 m and 700 m does not vary linearly with depth and is not controlled by friction on preexisting well- oriented faults. Rather, the stress magnitudes seem to be controlled by the creeping characteristics of the various layers rather than by their elastic characteristics, with a loading mechanism that remains to be identified but which is neither related directly to gravity nor apparently to present tectonics.

"It is concluded here that the smoothing out of stress variations with depth into linear trends may be convenient for gross extrapolation to greater depth. But it should not be taken as a demonstration that vertical stress profiles in sedimentary rocks are governed by friction along optimally oriented faults, given the absence of both microseismicity and actively creeping fault. It should not be used for integrating together stress tensor components obtained within layers with different rheological characteristics."

Other examples could be cited, but the message is clear. Determination of in-situ stress in rock is an extremely challenging task, with results subject to considerable variability and uncertainty.

Stress orientations can be estimated from consideration of regional tectonics, faulting and interpretation of evidence from local structural geology supported in some cases by evidence based on borehole logs (e.g., tensile fractures induced along the well bore). Stress magnitudes are, in general, more difficult to determine and usually less significant, except as indicators of how stresses may be distributed across a site where the geology and engineering design are complex. In such cases, interpretation of stress distribution is best done in conjunction with a numerical model of the site, preferably one that includes the influence of important uncertainties and discussion with structural geologists familiar with the area under study.

'CRITICAL STRESS STATE' IN THE EARTH'S CRUST

It is sometimes asserted that the Earth's crust is everywhere close to a 'critical state of stress,' i.e., that a small change in the devatoric stress in the rock is likely to produce slip on one or more faults with associated seismic activity. The current global interest in development of major resources of natural gas, the central role of hydraulic fracturing in this development, and the public apprehension that hydraulic fracturing will 'trigger earthquakes' has led to strong opposition to fracturing, and even legislation to ban the use of hydraulic fracturing in some countries and some States in the USA.

As illustrated by Figure 14, the seismic hazard, (i.e., probability of a damaging earthquake) varies very considerably from place to place. Thus, an earthquake of a given magnitude is 1000 times more likely to occur in Southern California than it is in the Eastern United States. The hazard is even lower in regions such as Texas, North Dakota and in the stable Canadian Shield region of the North American tectonic plate. While many earthquakes are initiated at depths considerably greater than depths where hydraulic fracturing is applied, it seems plausible to suggest that there may be less potential for fracturing to induce seismic activity in regions that have low seismic hazard. Also, as indicated by the comments of Cornet in the previous section of this paper, there is evidence that the critical stress hypothesis warrants detailed scrutiny, at least. This could have major implications for development of the world's major natural gas and EGS (enhanced geothermal systems) resources. Two recent studies,National Research Council (2012) and Royal Society – Royal Academy of Engineering (2012), have each concluded that the risk that hydraulic fracturing as used in development of energy resources would trigger significant seismic activity is small, but it would be valuable to examine the critical stress hypothesis more rigorously than has been done to date.

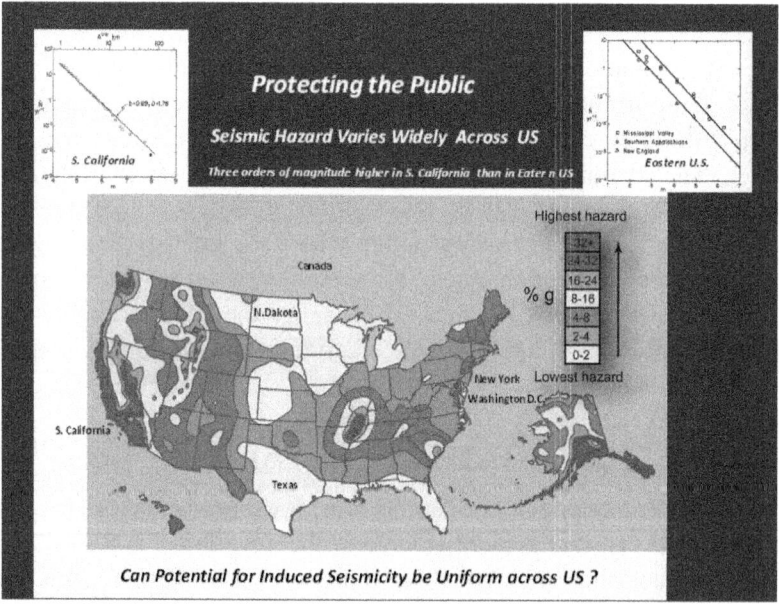

Figure 14: Seismic hazard map of the United States — US Geological Survey.

HYDRAULIC FRACTURING IN TIGHT SHALES

The development of inclined and horizontal drilling (see Appendix 1 - Figure A1-2) has helped stimulate intense activity to develop natural gas production from so-called tight shale, i.e., rock in which natural gas is held tightly within the very fine pore structure of the rock. Figure 15 illustrates the procedure used to stimulate these shales. The well is drilled horizontally in the gas-bearing formation, more or less in the direction of the minimum principal stress. Hydraulic fractures are generated (and propped) at intervals along the well to generate a network of connected flow paths that will allow the gas to flow to the well. Depth (i.e., extent) and spacing of the fractures should be optimized to produce the formations effectively. Bunger et al. (2012) discuss the factors in the design of an effective fracture strategy.

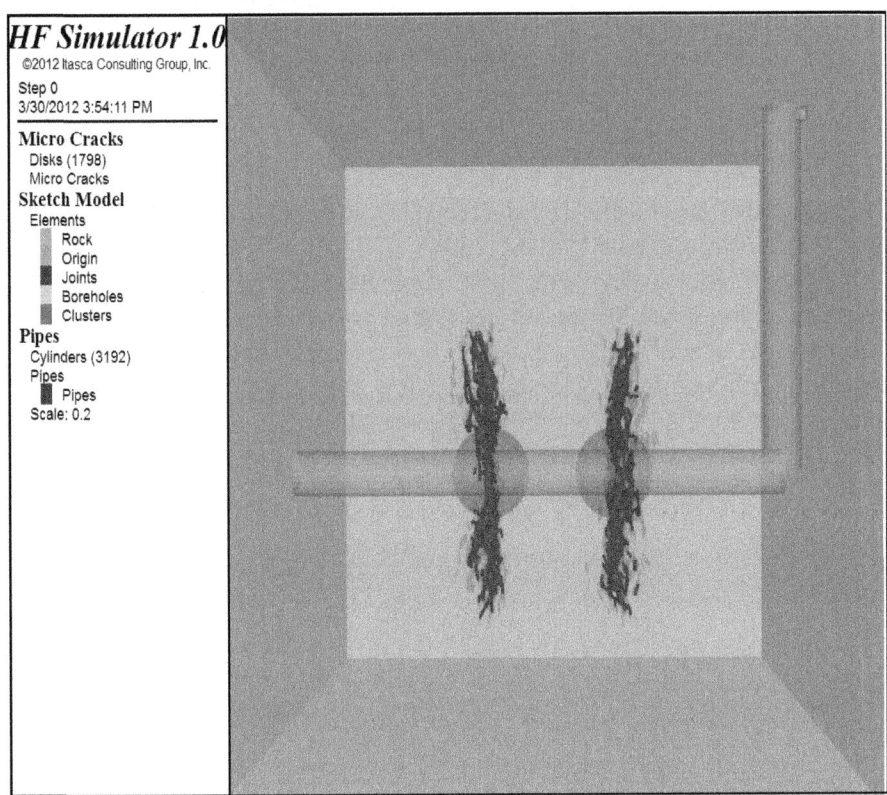

Figure 15: Staged hydraulic fracturing in a horizontal well. There may be many such wells along the horizontal well.

Why Doesn't Microseismicity Correlate With Production?

The Total Rock Volume Affected by Microseismicity Accounts for Less Than 1% of Gas Production in First 6 Months

Figure 16: The volume of rock defined by microseismicity is a very small fraction of the volume producing gas.

Figure 16 shows a slide from a recent presentation by Prof. Mark Zoback, who kindly agreed to allow the author to include it here. Although on a somewhat smaller scale, the fact that considerable deformation and fracturing must be taking place that is not associated with detected microseismicity is similar to the phenomena discussed in connection with Figure 10. Prof. Zoback refers to such aseismic deformation as slow slip, and is conducting research to understand the underlying mechanisms, including the possible influence of the clay content of the shale. As can be seen in Figure 17 (courtesy of Prof. Zoback), the clay content can be large.

 Average Shale Properties

	BARNETT	MARCELLUS	EAGLE FORD	FLOYD
Depth (ft)	3 - 9,000	2 - 9,500	4 - 13,500	6 - 13,000
TOC (%)	1 - 10	1 - 15	2 - 7	1 - 7
RO (%)	0.7 - 2.3	0.5 - 4+	0.5 - 1.7	0.7 - 2+
Porosity (%)	2 - 14	2 - 15	6 - 14	1 - 12
Qtz + Calcite (%)	40 - 50	40 - 60	50 - 80	20 - 30
Clay (%)	20 - 40	30 - 50	15 - 35	45 - 65
Areal Extent (mi²)	22,000	60,000	15,000	6,000
Resource Size (Tcf)	25 - 250	50 - 500	10 - 100	<< 1

How many Floyd Shales are There?

Figure 17: Clay content of some typical 'tight' gas shales.

Figure 18 illustrates the very fine, micron scale, pore structure of a typical tight shale. Although the mechanism(s) by which flow pathways are established in such a fine structure is not clear, the level of microseismic energy release associated with brittle breakage of one or a few bonds will be very small and of high frequency (such that the radiated energy would be rapidly attenuated), and hence, not detectable by any geophone. Thus, absence of microseismicity may not indicate an absence of breakage of brittle bonds. Some mechanism must be operative that generates flow pathways. Intuitively, it might be expected that the clay content of the shale might lead to ductile and viscous deformation that could tend to close the pathways.

(a)

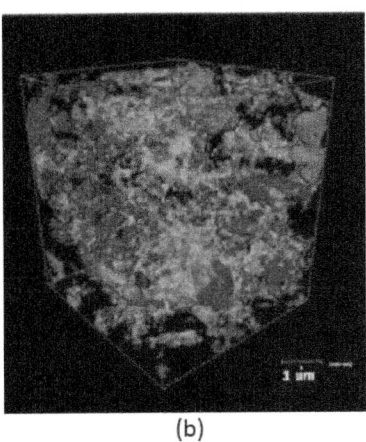

(b)

Figure 18: a) Outer surface of a FIB-SEM (Focused Ion Beam- Scanning Electron Microscope) volume of Eagle Ford Shale; (b) Transparency view of the distribution of connected pores (blue), isolated pores (red) and organic matter (green). (Courtesy of Prof. Amos Nur and J. Wallis (see Wallis et al., (2012) for details of technology.)

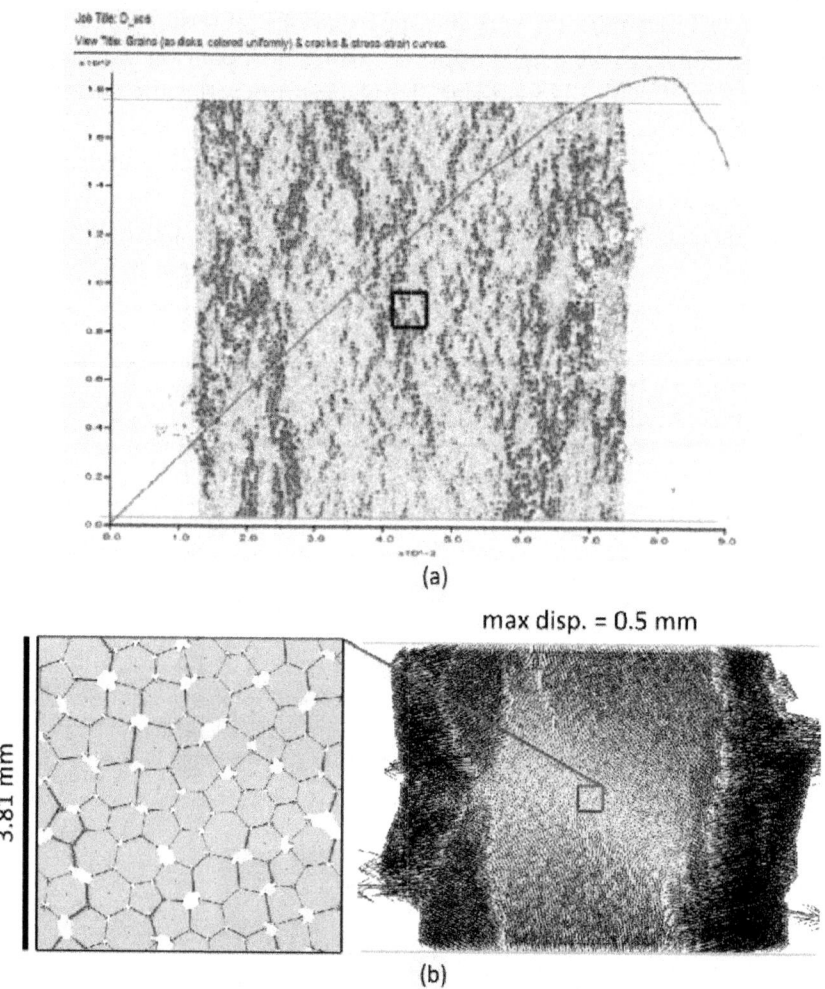

(a)

max disp. = 0.5 mm

3.81 mm

(b)

Figure 19: Micro-rupture of bonds within a *PFC* model of a rock loaded to failure, and beyond, in uniaxial compression. The darker red regions in (a) indicate coalescence of smaller groups of bonds that have ruptured. Eventually these larger regions develop to provide a mechanism that leads to collapse of the specimen. It is seen that bond breakage occurs throughout the specimen as the load is increased. The larger dark red regions will release larger amplitude, lower frequency waves that can be detected, whereas the smaller 'pathways' cannot be detected seismically. The load-deformation curve is shown as an 'overlay' on the specimen.

FRACTURE NETWORK ENGINEERING

This paper has emphasized the central role of fractures in rock, primarily natural fractures developed on a wide spectrum of scales over many tectonic epochs and many millions of years. These fractures and fracture systems are of special significance with respect to hydraulic fracturing and related techniques of fluid injection into rock since the fluid will tend to seek out those fractures that can be more readily opened against the local in-situ stress field as the fluid is injected. Given the complexity and lack of information on the fracture system, stress environment, etc., how can the engineering of hydraulic fracturing and related fluid injection programs advance most effectively?

Confronted with the same complexity of rock in situ, civil engineers and mining engineers have tended to adopt the 'Observational Approach' (Peck, 1969). In essence, this approach involves developing an initial engineering design for the problem, based on a first assessment/estimate of the rock (or soil) properties. Observe the actual performance and modify the initial design as needed to arrive at the desired performance. An example of the Observational Approach (as used in the New Austrian Tunnelling Method) is discussed in Fairhurst and Carranza-Torres (2002), see pp. 24-30.

Application of the Observational Approach to Hydraulic Fracturing and related fluid injection techniques faces some disadvantages and some advantages. We do not have 3D access to the engineering site. We do have powerful numerical modeling tools to help make a more informed initial estimate of how the system will perform; and we have sensing systems, both downhole and remote.Figure 20 illustrates a procedure that tries to apply the Observational Approach to hydraulic fracturing and related systems. The illustration describes an application to the extraction of Geothermal Energy.

In this application, an initial design approach is developed based on a numerical modeling study incorporating any available data, insight, etc., on the site. This model provides an initial prediction of the performance. Instrumentation, both downhole and on-surface observes the initial response of the system and compares it with the prediction. This triggers a feedback signal to modify the design input to move the performance closer to the one desired. This iteration continues, changing progressively towards the performance desired.

Stones have begun to speak, because an ear is there to hear them. .....

Cloos, Conversations with the Earth (1954), 4

Microseismicity —predicted and observed.

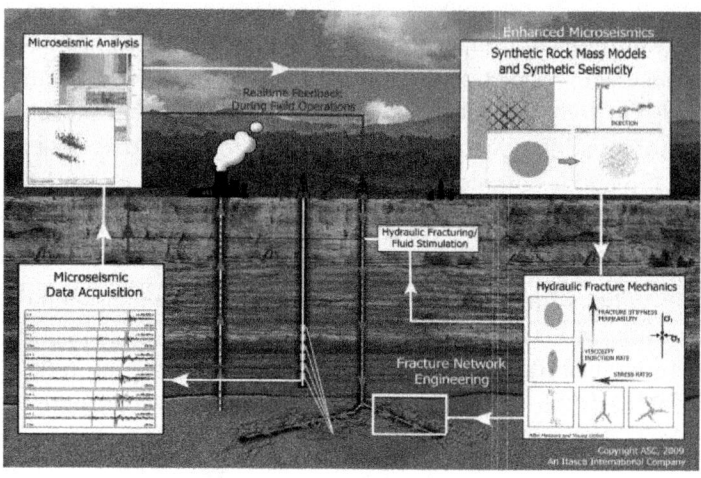

Fracture Network Engineering. *Synthetic Rock Mass and Synthetic Seismicity Models are compared with observed microseismic signals for **real time control of fracture network development**. (Enhanced Geothermal Systems.)*

Figure 20: Fracture network engineering system.

Although the writer knows of no such Fracture Network Engineering system currently in operation, many of the components are available and it is time to start.

Conclusions

Expectations for higher living standards of a rising world population, and the associated demand for Earth's resources of energy, minerals and water, lead inevitably to greater focus on resources of the subsurface.

This focus includes the need to develop improved technology to develop these resources, and a better understanding of the nature of the subsurface environment as an engineering material.

Earthquakes and dynamic releases of energy are a daily reminder that on the global scale, Earth is critically stressed, and constantly trying to adjust seeking to achieve a condition of minimum potential energy for the entire system. On going for many, many millions of years, such adjustments have resulted in the heterogeneous assembly of blocks of rock bounded by essentially planar surfaces; fault, fractures and similar 'discontinuities' varying in scale from tectonic plates and continents down to micron and even nanometers.

Some of these volumes are critically stressed; others are far from a critical condition. National maps of seismic hazards provide evidence of this heterogeneity on a larger scale.

Although Earth Resource Engineering activities may be kilometers in extent, they are small-scale within the larger Earth context. Subsurface engineering in a critically stressed region can be a much different challenge than in a stable region. It is important to assess the initial conditions carefully for each case, and especially where fluid injection is a main component of a project.

The sub-surface is opaque in several ways. Details of the key features that can control the response to an engineering activity in the sub-surface are often unknown. Problems are data-limited. This is particularly the case when the engineering is based on deep borehole systems, as in hydraulic fracturing and related fluid injection technologies.

Although operating in ways that may appear complex, the response of the subsurface to stimulation does obey the laws of Newtonian mechanics, and it is clear that pre-existing natural discontinuities have a major influence on how the subsurface responds to engineered changes.

The advent of powerful computers and developments in numerical modeling provide a potentially major tool to help develop better-informed strategies of subsurface engineering. Used interactively in close conjunction with instrumentation, both downhole and surface based, it should be possible to progressively develop a mechanics-informed understanding and path forward for more effective subsurface engineering.

Much as the field of Fracture Mechanics has led, and continues to lead, to major technological improvements for fabricated materials, so can development of the field of Rock Fracture Mechanics be of transformative value to subsurface engineering, and to society in general.

Hydraulic fracturing and related injection-stimulation systems will certainly be a central element in the future of Earth Resource Engineering. The organizers of HF 2013 are to be commended for focusing attention on this critically important topic.

Appendix 1

EARTH RESOURCES ENGINEERING

In 2006, the US Academy of Engineering introduced the term 'Earth Resources Engineering' to replace 'Petroleum, Mining and Geological Engineering' in recognition of the broader range of engineering activities and concerns associated with use of the subsurface. The new title, it is hoped, will also stimulate important synergies between the various disciplines involved. Mining and civil engineers, for example, have direct three-dimensional access to the subsurface not available to colleagues in other subsurface activities. This access provides a major opportunity to conduct research and gain understanding of the mechanics of subsurface processes under actual in-situ conditions, as exemplified by Jeffrey et al. (2009), see Figure A1-1.

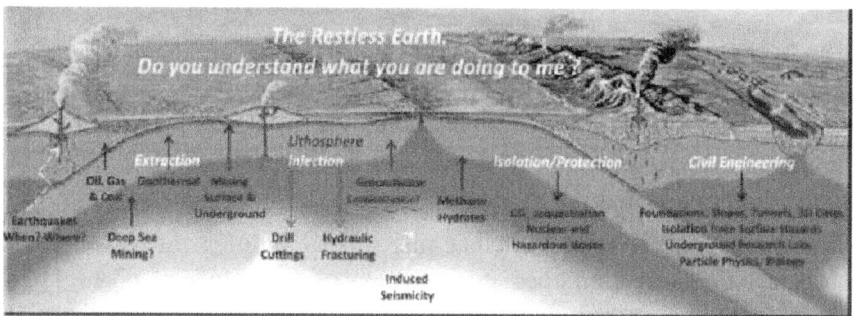

Figure A1-1: The restless Earth. Earth Resource Engineering activities are all confined to a very shallow part of the 40 km -700 km thick Earth's solid crust (lithosphere). Deepest borehole ~ 12 km; mine ~ 4km. Rock stress increases vertically σv ~ 27MPa/km; laterally σh~ (0.5- 3.0).σv: Pore water pressure p = 10 MPa /km; temperature increase ~25°C /km depth.

Study of slip on active faults is a good example.

"The physics of earthquake processes has remained enigmatic due partly to a lack of direct and near-field observations that are essential for the validation of models and concepts. DAFSAM[15] -proposes to reduce significantly this limitation by conducting research in deep mines that are unique laboratories for full-scale analysis of seismogenic processes. The mines provide a 'missing link› that bridges between the failure of simple and small samples in laboratory experiments, and earthquakes along complex and large faults in the crust. There is no practical way to conduct such analyses in other environment. To unravel the complexity of earthquake processes, this project is designed as integrated multidisciplinary studies of specialists from seismology, structural geology, mining and rock engineering, geophysics, rock mechanics, geochemistry and geobiology. The scientific objectives of the project are the characterization of near-field behavior of active faults before, during and after earthquakes".[16] - See also http://www.iris.edu/hq/instrumentation_meeting/files/pdfs/IRIS_ Johnston.pdf

Petroleum engineers can now reach depths in excess of 6 km and have developed advanced drilling control technologies that allow precise access to locations extending horizontally to more than 10-15 km from a single vertical hole (see Figure 2).

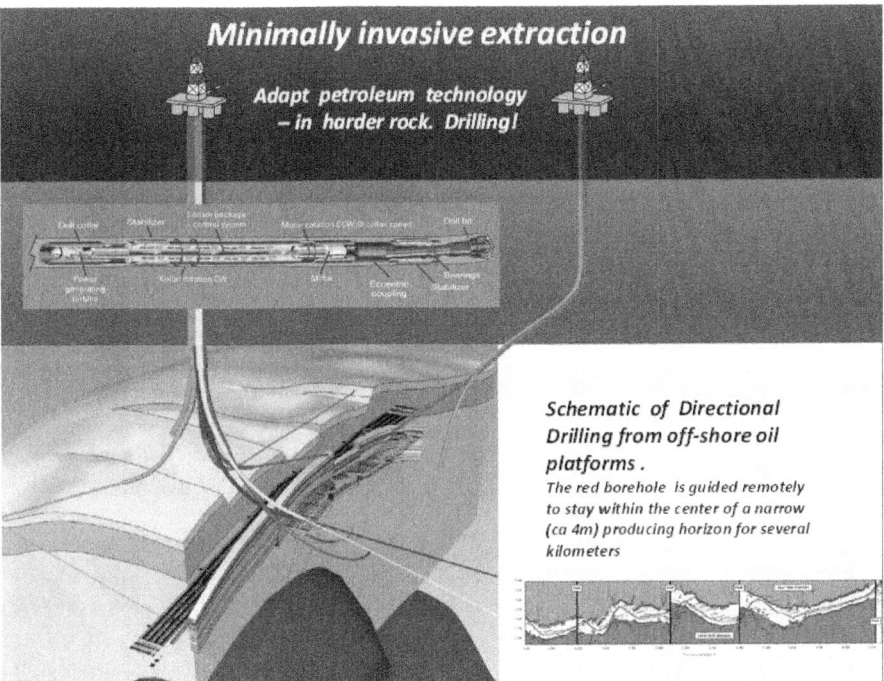

Figure A1-2: Schematic illustration of directional drilling for petroleum production.

These and related developments are stimulating interest in application of borehole technologies to other areas of subsurface engineering, including the development of less-invasive mining technologies, i.e., borehole extraction of minerals. Some applications, e.g., where crystalline rocks are involved, are contingent on the development of significantly lower-cost drilling technologies. The critical dependence of society on reliable and economic subsurface engineering is illustrated by the fact that currently more than 60% of the world's energy is delivered via a borehole. The Deepwater Horizon accident in the Gulf of Mexico in April 2010 provides a sober example of the consequences of error. In summary, hydraulic fracturing and related stimulation technologies are likely to see application to an increasing range of subsurface engineering challenges. HF2013, the first International Conference for Effective and Sustainable Hydraulic Fracturing, is very timely.

Appendix 2

EFFECT OF CORING IN PRE-STRESSED ROCK

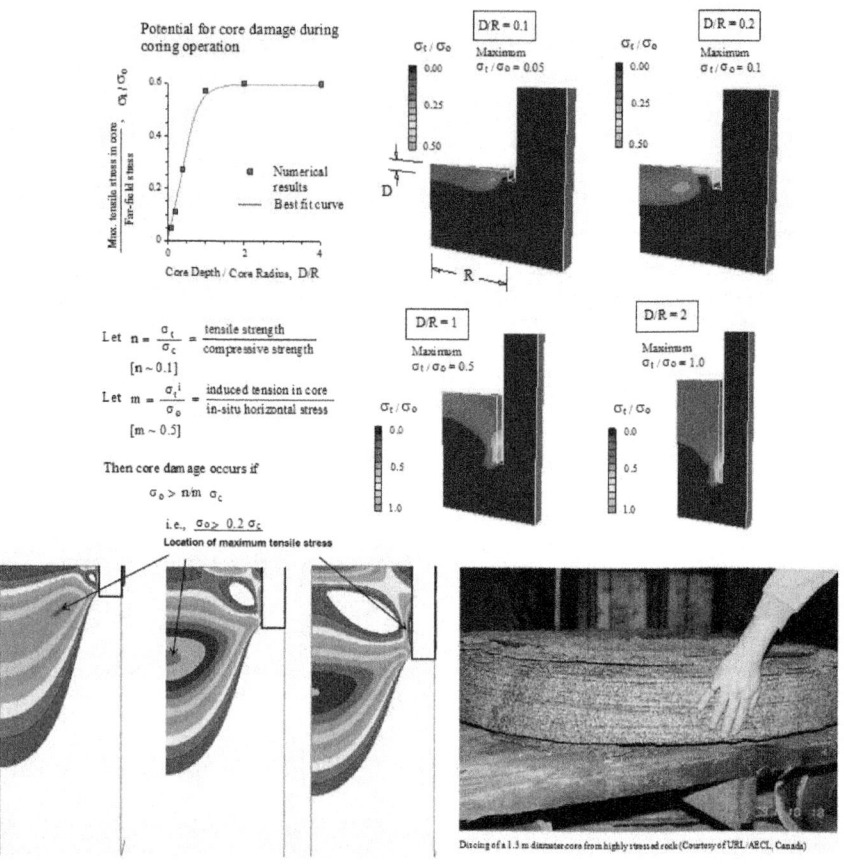

Figure A-2.1: Tensile stress concentrations induced in a brittle rock during coring.

The consequences of disturbing a pre-stressed rock medium are illustrated by examining the rock coring operation. Figure A2-1 shows the stress concentrations in a rock core in a brittle rock. If the in-situ stress normal to the axis of drilling is sufficiently high tensile cracks can develop in the core. Where lateral stresses are very high, then tensile 'spalling' may result, as shown in the photograph of the bottom right of Figure A2-1. Where the rock is more 'ductile' the core may undergo permanent deformation without fracturing. In both cases, the mechanical properties of these cores may differ significantly from those of the rock in situ from which the core was obtained.

ACKNOWLEDGEMENTS

Much of the material and concepts discussed in this paper is the result of work and discussions over many years with colleagues at Itasca Consulting Group, Inc. in Minneapolis and faculty in GeoEngineering at the University of Minnesota, especially in this instance, Professor Emmanuel Detournay. Particular help was received from Itasca colleagues Varun, Branko Damjanac, David Potyondy and Mark Lorig, The influence of numerous stimulating discussions with Professor François Cornet of the Institut de Physique du Globe, Strasbourg, France are clearly evident in the paper. Professors Amos Nur and Mark Zoback, of Stanford University, USA and of Ingrain, Inc., Houston, USA assisted with valuable material, as acknowledged in the text. Dr. Rob Jeffrey and Andrew Bunger of CSIRO, Melbourne, the leaders in arranging HF2013, have provided valuable comments, assistance and understanding throughout. To all, I am very grateful. Such invaluable assistance notwithstanding, I accept full responsibility for the interpretations and views expressed in the paper.

Notes

[1] - 1 exajoule =1018 joules = 1018 watt.seconds.

[2] - Future of Geothermal Energy (2005) Synopsis and Executive Summaryp.1-4 (2).

[3] - Future of Geothermal Energy (2005)Synopsis and Executive Summaryp.1-5 (5).

[4] - A fractured rock mass is typically about two orders of magnitude lower in strength than the strength of a laboratory specimen taken from the rock mass [Cundall (2008); Cundall et al, (2008)].

[5] - Hydraulic fractures generated in classical petroleum applications typically extend (2b) of the order of 25m ~ 50m from a wellbore. The fracture aperture (2a) at the wellbore then will be typically of the order of 0.01 m. Thus, the tensile stress concentration at the tip is very high of the order of 103.

[6] - In his second paper, Griffith (1924), demonstrated that tensile stresses also developed around similar cracks loaded in compression, provided the cracks were inclined to the direction of the major principal (compressive) stress.(He also assumed that the cracks did not close under the compression.) For the optimum crack inclination, an applied compressive stress of eight times the magnitude of the tensile strength was required to develop a tensile stress on the crack boundary (close to, but not at the apex of the crack) equal to the limiting value in the tensile test. He concluded that the uniaxial compressive strength of a brittle material should be eight times greater than the tensile strength.

Interestingly, he did not invoke his second (minimum potential energy) criterion. It was later determined that although a tensile crack could initiate in a compressive stress regime as predicted by Griffith (1924), the crack was stable (i.e., did not satisfy the minimum potential energy criterion). The compressive/tensile strength ratio is greater than 8 (see Hoek and Bieniawski, 1966).

[7] - Tension is assumed to be positive in Figure 3.

[8] - A typical hydraulic fracture may have a length (2a) of the order of 50m and a maximum aperture (2b) of 5mm, so that the stress concentration will be of the order of 2000:1.

[9] - Commonwealth Scientific and Industrial Research Organization.

[10] - Typically, computer tests indicate the unconfined strength of a Synthetic Rock Mass of the order of 50-m to 100-m side length, to be a few percent of the unconfined strength of the laboratory specimen.

[11] - A number of important subsurface engineering problems involve borehole access only. This often means difficulty in establishing reliable, realistic DFN's. In such cases there is no recourse, at least at the start of the project, other than to try to infer fracture networks from borehole observations, perhaps supplemented by local observations of structural geological features . The DFN for Northparkes was available and convenient to use in the example shown in Figure 5.

[12] - Adapted from Fairhurst (1971).

[13] - See also footnote 17 –Appendix 1.

[14] - The URL at Bure was developed in order to determine the suitability of the Calllovo-Oxfordien Argillite formation for permanent storage of high–level nuclear waste.

[15] - DAFSAM -Drilling Active Faults in South African Mines.

[16] - http://www.icdp-online.org/front_content.php?idcat=460

REFERENCES

1.   T. L Anderson, 2005Fracture Mechanics: Fundamentals and Applications. 3rd edition, CRC Press (0-84931-656-1

2.   E. V Artyushkov, 1973Stresses in the Lithosphere Caused by Crustal Thickness

3.   J Inhomogeneities, Geophy.Res. 7832November 10, 1973

4.   A. P Bunger, X Zhang, and R. G Jeffrey, 2012Parameters Affecting the Interaction Among Closely Spaced Hydraulic Fractures" SPE Journal March 2012, 292306

5.   M Calo, C Dorbath, F. H Cornet, and N Cuenot, 2011Large scale aseismic motion identified through 4D P-wave tomography; Geophys. J. Int. 18612951314

6.   P. A Cundall, 2008An Approach to Rock Mass Modelling," in From Rock Mass to Rock Model-CD Workshop Presentations (15 September, 2008)-SHIRMS 2008 (Proc. 1st Southern Hemisphere International Rock Symposium, Perth, Western Australia, September 2008) Y. Potvin et al., Eds. Nedlands, Western Australia: Australian Centre for Geomechanics.

7.   T. T Cladouhos, M Clyne, M Nichols, S Petty, W. L Osborn, and L Nofziger, 2011Newberry Volcano EGS Demonstration Stimulation Modeling" GRC Transactions, 35317322

8.   F. H Cornet, 2012The relationship between seismic and aseismic motions induced by forced fluid injections." Hydrogeology Journal (2012) 20: 1463-1466

9.   F. H Cornet, and T Röckel, 2012Vertical stress profiles and the significance of "stress decoupling". Tectonophysics 5812012193205

10. P. A Cundall, M. E Pierce, and D. Mas Ivars. (2008Quantifying the Size Effect of Rock Mass Strength" in SHIRMS 2008 (op.cit.) 2315

11. B Damjanac, and C Fairhurst, 2010Evidence for a Long-Term Strength Threshold in Crystalline Rock,‖ Rock Mech. Rock Eng., 43, 513-531 (2010).

12. B Damjanac, C Detournay, P. A Cundall, and Varun, (2013Three-Dimensional Numerical Model of Hydraulic Fracturing in Fractured Rock Masses" Proc. HF 2013The International Conference for Effective and Sustainable Hydraulic Fracturing, Brisbane, May 20-22, 2013

13. B Damjanac, and C Fairhurst, Evidence for a Long-Term Strength Threshold in Crystalline Rock,‖ Rock Mech. Rock Eng., 43, 513-531 (2010Duchane, D and D. Brown, (2002) "Hot Dry Rock (HDR) Geothermal Energy Research and Development at Fenton Hill, New Mexico" GHC (Geo-Heat Center) Bulletin, December. 2002 1319

14. C Fairhurst, and C Carranza-torres, 2002Closing the Circle- Some Comments on Design Procedures for Tunnel Supports in Rock," in Proceedings of the University of Minnesota 50th Annual Geotechnical Conference (February 2002), 2184J. F. Labuz and J. G. Bentler, Eds. Minneapolis: University of Minnesota, 2002. [available at www.itascacg. comgo to 'About'and Fairhurst Files]

15. C Fairhurst, 1971Fundamental Considerations Relating to the Strength of Rock. Colloquium on Rock Fracture, Ruhr University, Bochum, Germany, April 1971. (see http://www.itascacg.com/about/ff.php)Revised and

published in Report of the Workshop on Extreme Ground Motions at Yucca Mountain, August 23-25, 2004, U.S. Geological Survey, USGS Open-File Report 20061277T. C. Hanks et al., Eds. Reston, Virginia: USGS, 2006.

16. J Geertsma, and F De Klerk, 1969A Rapid Method of Predicting Width and Extent of Hydraulic Induced Fractures. J Pet Technol 211215711581SPE-2458-PA. http://dx.doi.org/10.2118/2458-PA

17. J. F Geyer, and S Nemat-nasser, 1982Experimental Investigation of Thermally induced Interacting Cracks in Brittle Solids Int. J. Solids and Structures 184349356

18. A. A Griffith, 1921The Phenomena of Rupture and Flow in Solids Phil. Trans. R. Soc. Lond. A 1921,, 221, 163-198 doi:rsta.1921.0006

19. A. A Griffith, 1924Theory of Rupture. Proc. First Int. Cong. Applied Mech (eds Bienzo and Burgers). 5563Delft: Technische Boekhandel and Drukkerij. 1924

20. G Grünthal, and R Wahlström, 2003An Mw-Based Earthquake Catalogue for Central, Northern and Northwestern Europe using a Hierarchy of Magnitude Conversions. J. Seismol. 7, 507-531 (Available at http://seismohazard.gfzpotsdam.de/projects/en/eq_cat/menue_e"q_cat_e.html)

21. E Hoek, and Z. T Bieniawski, 1966Fracture Propagation Mechanism in Hard Rock," in Proceedings of the First Congress of the International Society of Rock Mechanics. Lisbon, September-October, 1243249J. G. Zeitlen, Ed. Lisbon: LNEC.

22. E Hoek, and E. T Brown, 1980Underground Excavations in Rock." Inst'n of Mining and Metallurgy (London) Revised 1982, 164

23. G. C Howard, and C. R Fast, 1970Hydraulic Fracturing" SPE Monograph 2. Henry L.Doherty Series 203 pp. SPE 30402

24. C. E Inglis, 1913Stresses in a Plate Due to the Presence of Cracks and Sharp Corners," Trans. Inst. Naval Arch., London, 55(1), 219141

25. R. G Jeffrey, et al2009Measuring Hydraulic Fracture Growth in Naturally Fractured Rock. SPE 124919; SPE Annual Technical Conference and Exhibition, New Orleans, Louisiana, USA, 47October 2009

26. J. F Knott, 1973Fundamentals of fracture mechanics, Wiley (0-47049-565-0

27. National Research Council2012Induced Seismicity Potential in Energy Technologies." Washington, DC: The National Academies Press, 2012. (300p.) View online at http://www.nap.edu/catalog.php?record_id=13355

28. J. M Nocquet, and E Calais, 2004Geodetic Measurements of Crustal

Deformation in the Western Mediterranean and Europe " Pure Appl. Geophy., 161; 661668

29. R. P Nordgren, 1972Propagation of a Vertical Hydraulic Fracture. SPE J. 124306314SPE-3009-PA. http://dx.doi.org/10.2118/3009-PA.

30. R. B Peck, 1969Advantages and limitations of the observational method in applied soil mechanics. Geotechnique, 192171187

31. T. K Perkins, and L. R Kern, 1961Widths of Hydraulic Fractures. J Pet Technol 139937949SPE-89-PA. http://dx.doi.org/10.2118/89-PA.

32. W. S Pettitt, J. F Hazzard, B Damjanac, Y Han, M Pierce, T Katsaga, and P. A Cundall, Microseismic Imaging and Hydrofracture Numerical Simulations," in Proceedings, 21st Canadian Rock Mechanics Symposium (Alberta, Canada, May 5-9, 2012

33. M Pierce, 2011Discrete Fracture Network Simulation" DFN training session LOP (Large Open Pit). [ppt slides available on request. Itasca Consulting Group: www.itascacg.com]

34. A Riahi, and B Damjanac, 2013Numerical Study of Interaction between Hydraulic Fractures and Discrete Fracture Networks" Proc. HF 2013The International Conference for Effective and Sustainable Hydraulic Fracturing, Brisbane, May 20-22, 2013

35. Royal Society and Royal Academy of Engineering (2012Junep. "Shale Gas Extraction in the UK: a review of hydraulic fracturing" Issued: June 2012, DES2597] View report online at: royalsociety.org/policy/projects/ shale-gas-extraction and raeng.org.uk/shale

36. O Scotti, and F. H Cornet, 1994In-Situ Evidence for Fluid-Induced Asesismic Slip Events along Fault Zones. Int. J. Rock Mech Min.Sci. &Geomech. Abstr. 314347258Control in Mines (1965). South African Institute of Mining and Metallurgy, Johannesburg, 606p

37. A. M Starfield, and P. A Cundall, 1988Towards a Methodology for Rock Mechanics Modelling" Int. J. Rock Mech. Min. Sci.& Geomech. Abstr. 25 (3) 99106

38. J. F Tester, et al2006The Future of Geothermal Energy"-Impact of Enhanced Geothermal

39. Systems (EGS) on the United States in the 21st CenturyMIT Press.

40. J. D Wallis, J Devito, and E Diaz, 2012Digital Rock Physics- A New Approach to Shale Reservoir Evaluation" Oilfield Technology, March 2012 [http://www.ingrainrocks.com/articles/a-new-approach-to-shale-reservoir-evaluation/]

41.   X Zhang, and R. G Jeffrey, 2008Re-initiation or termination of fluid-driven fractures at frictional bedding interfaces" JGR, 113BO 8416, doi:10.1029/2007JB005327,

Chapter 3

HYDRAULIC AND SLEEVE FRACTURING LABORATORY EXPERIMENTS ON 6 ROCK TYPES

Sebastian Brenne[1], Michael Molenda[1], Ferdinand Stöckhert[1] and Michael Alber[1]

[1] Ruhr-University Bochum, Germany

INTRODUCTION

Hydraulic tensile strength is a crucial value for planning reservoir stimulation and stress measurements. It is used in the classical breakdown pressure (P_b) relation by Hubbert & Willis [1], where P_b is a function of major and minor principal horizontal stresses S_H and S_h, hydraulic tensile strength σ_T and pore pressure P_0:

$$P_b = 3S_h - S_H + \sigma_T - P_0 \tag{1}$$

Options

For hydraulic fracturing laboratory experiments (MiniFrac – MF) under isostatic confining pressure P_m this might be reduced to:

$$P_b = cP_m + \sigma_T - P_0 \tag{2}$$

Options

The coefficient cc should be equal to two when porepressure is neglected. However, many laboratory experiments [2,3] resulted in values of about 1 for c, which might be explained by poroelastic effects.

Thus, when poroelasticity is excluded in the experiments by taking dry samples and sealing off the central borehole by an impermeable membrane (like a polymer tube), one would expect that c equals two and σ_T will be in the range of the tensile strength as determined by other tensile strength tests.

However, experiments with jacketed boreholes (sleeve MiniFrac – SMF) yield remarkable high values for cc (about 6 to 8) and also for σ_T (about 3 to 5 times the tensile strength of the material) [4]. As a consequence we use a linear elastic fracture mechanics approach to evaluate our experiments.

Theory of Hydraulic And Sleeve Fracturing On Hollow Cylinders

Fracture mechanics deal with stress concentrations around fractures and the definition of propagation criteria for fractures. The theory is essentially based on the works of Griffith [5] and Irwin [6], which led to the introduction of the stress intensity factor K.

$$K = \sigma\sqrt{\pi a}$$

(3)

KK represents the magnitude of the elastic stress singularity at the tip of a fracture of the length 2a subjected to a uniform stress σ. With this concept, it is possible to formulate a simple fracture propagation criterion $K=K_C$. The fracture propagates when K reaches a critical value K_C (fracture toughness) with the fracture toughness assumed to be a property of the rock.

Mode I stress intensity factors (K_I) for arbitrary tractions $(\sigma(x))$ applied to the surface of a fracture of the length 2a may be computed by following formula [7,8]:

$$K_I = \frac{1}{\pi\sqrt{a}} \int_{-a}^{a} \sigma(x) \left(\frac{a+x}{a-x}\right)^{\frac{1}{2}} dx$$

(4)

The direction of propagation is the x-axis and the stresses are applied perpendicular to the fracture. As can be seen from equation (4), K_I increases with growing fracture length. A simple, 2-dimensional model was assumed for determination of stress intensity factors at the crack tips of the hydraulically induced fractures in MF and SMF tests.

Two fractures of length a are radially emanating from a circular hole of radius r in an infinite plate subjected to a compressive far field stress of the magnitude P_m. A fluid pressure P_{inj} is acting on the borehole wall and the pressure inside the fractures is either zero (SMF) or equal to the pressure in the borehole (MF: $P_{frac} = P_{inj}$). Stress intensities on the fracture tips can be determined by superposition of stress intensity factors resulting from each loading type [2,3]:

$$K_{I-MF} = K_I\left(P_m\right) + K_I\left(P_{inj}\right) + K_I\left(P_{frac}\right)$$

(5)

$$K_{I-SMF} = K_I\left(P_m\right) + K_I\left(P_{inj}\right)$$

$$(6)$$

$K_{I-MF/SMF}$ are not only dependent on the fracture length a (cf. Equations (3) and (4)) but also on the borehole radius r

K_{I-MF} (full pressure in the fracture) gives an upper bound for stress intensities in this geom- etry (actual K_{I-MF} might be lower due to a negative pressure gradient inside the fracture), while K_{I-MF} is only induced by the pressure in the borehole and far-field stresses and is therefore substantially lower than K_{I-MF} (Figure 1).

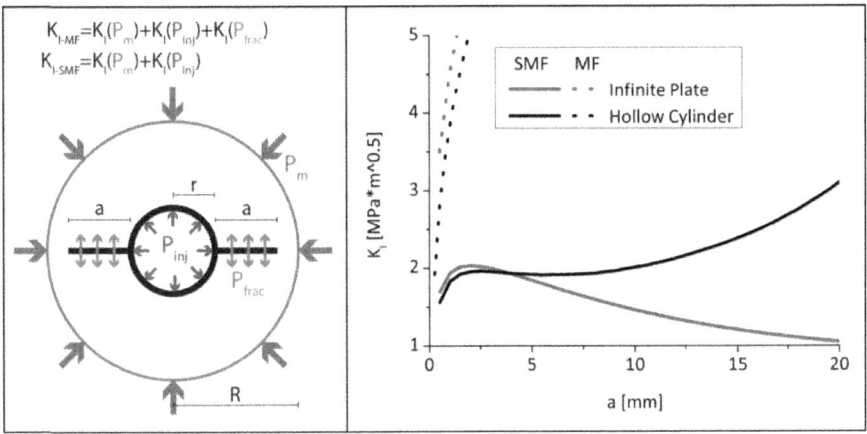

Figure 1. Left side: superposition of stress intensities by each loading type. Right side: stress intensity factor versus fracture length from analytical (infinite plate) and numerical (hollow cylinder) calculations for r =3 mm, P_{inj} =50MPa, P_m =0 and an outer radius R of the hollow cylinder = 30 mm.

As an analytical solution for $K_I(P_m)$ and $K_I(P_{inj})$ for the ring geometry (corresponding to the hollow cylinder) is quite complex, we used the simpler solutions for a circular hole in an infinite plate as described by Rummel and Winter [2,3]. We compared the results of numerical simulations for the ring geometry with analytical solutions for the infinite plate. These results indicate that the simplification might be valid for fracture lengths smaller then $a \approx \frac{R-r}{10}$ with $R = 10r$ (R is the outer radius of the ring geometry (cf. Figure 1).

Solving K_{I-MF} and K_{I-SMF} for P_{inj} and setting $K_{I-MF} = K_{I-SMF} = K_{IC}$ yields a critical injection pressure (P_C (a)) for each crack length a. If P_{inj} reaches P_C (a), the fracture will propagate. From Figure 2 it can be seen, that P_c (a) is very large for very small crack lengths. In consequence, the presence of microcracks is required for the formation of macroscopic fractures.

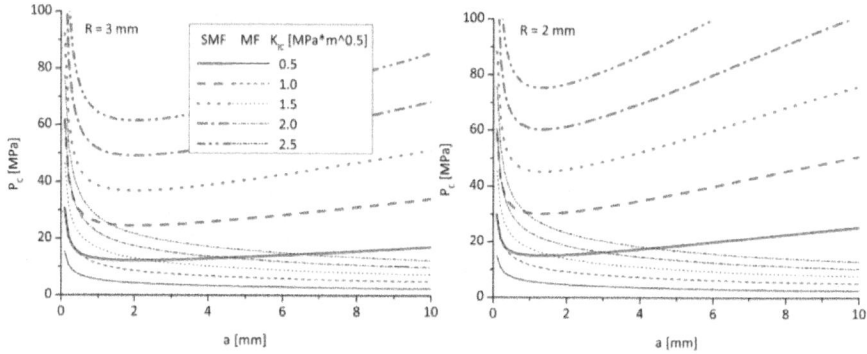

Figure 2. Critical injection pressure for fracture propagation P_C depending on fracture length a for P_m =0. Borehole radii r =3 mm (left), r =2mm (right).

MF-equation (Equation 13) with full injection pressure in the fracture yields unstable fracture propagation at constant injection pressures as soon as microcracks start to propagate. On the other hand, the SMF-equations (Equation 14) show a minimum. Thus, after a fracture reaches the crack length corresponding to the minimum critical injection pressure, stable fracture propagation (i.e. to propagate the fracture, the injection pressure has to be increased) could be expected.

To calculate the coefficient c from Equation 2, we assume the presence of microcracks of a fixed length a_0 in the sample. The corresponding P_C (a_0) versus Pm for the MF case (pressure in fracture = injection pressure) yields a coefficient c =1, which is independent of a_0. P_C (a_0) while for SMF the c value depends strongly on the assumed microcrack length a_0 and gives c >2 (increasing a_0 yield higher c).

SAMPLE PREPARATION AND ROCK TESTING

The core specimens are drilled either with 40 mm or 62 mm water cooled diamond core drills. Core end planes are cut with a water flushed diamond saw blade and ground coplanar to a maximum deviation of ± 0.02 mm. The length and diameter ratio is chosen between 1.5:1 and 2.25:1. After sample preparation core specimens were dried for two days at a temperature of 105°C. For calculations of porosity Φ, measurements of bulk density ρ_d and of grain density ρ_s via pycnometer were done. Static geomechanical parameters were determined by uniaxial and triaxial compressive as well as Brazilian disc tensile strength test series according to ISRM and DGGT suggested methods [9,10]. Mode I fracture toughness was determined using the Chevron notched three-point bending test accord- ing to [9]. Furthermore, a dynamic rock parameter, the compressional ultra-sonic wave velocity (v_p) was measured. For MF/SMF specimens a central axial borehole was drilled into cores, using a water flushed diamond hollow drill with an outer diameter of 4 mm or 6 mm.

rock type	era & period	quarry localization	Microstructure
marble	Triassic Upper	Carrara Italy	coarse monocrystalline polygonal fabric
limestone	Jurassic upper Malm	Treuchtlingen South Germany	micritic limestone with abundant fossils and stylolites
sandstone	Carboniferous Mississippian	Dortmund/Hagen West Germany	fine-grained arcose
andesite D	Permian Rotliegend	Doenstedt N German Basin	porphyric fine-grained partly altered and pre-fractured
rhyolite	Permian Rotliegend	Flechtingen N German Basin	porphyric fine-grained partly pre-fractured and sealed joints
andesite R	Permian Rotliegend	Thuringian Forest Rotkopf	porphyric coarse-grained and pre-fractured

Table 1: Rock types used in our experiments.

Stress Field and Injection

Figure 3 shows schematically the components of the MF and SMF experimental set-up. The stress field is induced by a hydraulic ram (capacity 4500 kN) through a servo controlled MTS Test Star II system with a Hoek triaxial cell which is pressurized using a hand pump to achieve simultaneous pressure increase of confining pressure and axial load. In all tests axial stress is set to be 2.5 MPa higher than P_m to prevent leakage. Distilled water is pumped into borehole as the injection fluid (MF) or into a polymer tube inside the borehole (SMF). A servo controlled pressure intensifier with a maximum injection pressure of 105 MPa was used to perform a constant pumping rate of 0.1 ml/s. With this apparatus also steadystate flow tests were conducted to obtain rock permeability values (according to the procedure described in [11]).

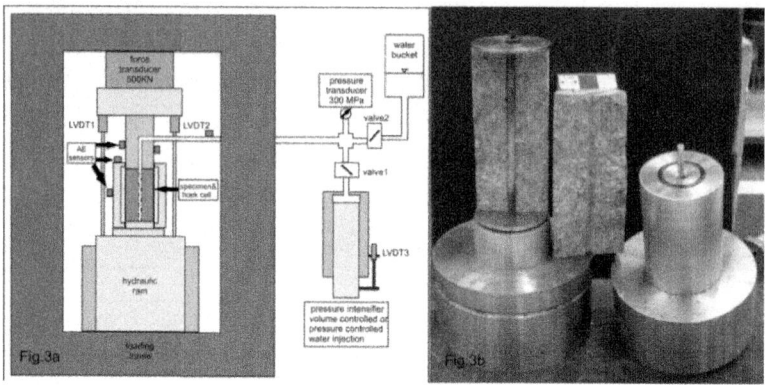

Figure 3: (a) Sketch of MF experimental set up including AE monitoring sensors (not shown are the pressure transduc- er and hand pump system to regulate confining pressure in the Hoek cell). (b) Typical specimen after SMF experiment.

Acoustic Emission Monitoring

Acoustic Emission (AE) signals are acquired with an AMSY5 Acoustic Emission Measurement System (Vallen Systeme GmbH, Germany) equipped with 5 Sensors of type VS150-M. The VS150-M Sensors operate over a frequency range of 100-450 kHz with a resonance frequency at 150 kHz. Due to machine noise in the range below 100 kHz incoming signals are filtered by a digital bandpass-filter that passes a frequency range of 95-850 kHz. AE data are sampled with a sampling rate of 10 MHz. The sensors are fixed using hot-melt adhesive to ensure best coupling characteristics. Pencil-break tests (Hsu-Nielsen source [12]) and sensor pulsing runs (active acoustic emission by one sensor) are used to test the actual sensor coupling on the sample.

RESULTS

Petrophysical and Mechanical Parameters

An overview of the rock properties is given in Table 2. A wide range of low porosity/perme- ability rocks with K_{IC} from 1 to 2 MPam $\sqrt{m}$ were tested.

Table 2. Mean values and standard deviations of petrophysical and mechanical parameters of tested rocks: dry bulk density ρ_d , porosity Φ, permeability k, compressive wave velocity v_p , fracture toughness from Chevron notched threepoint bending tests K_{IC} , cohesion C and friction angle φ from a Mohr-Coulomb fit, Young's modulus Estat , σ_T as determined by Brazilian disc tensile strength tests

rock type	ρ_d [g/cm³]	Φ [-]	k [m²]	v_p [m/s]	KIC [MPa·√m]	C/φ [MPa]/[°]	E_{stat} [GPa]	σ_T [MPa]
marble	2.71	0.40	1E-19	5.67	1.57	29/22	36.0	6.4
	±0.002	±0.08		±0.06	±0.11 (N=3)		±1.0	±1.5
limestone	2.56	5.64	1E-18	5.59	1.19	27/53	32.2	8.2
	±0.008	±0.04		±0.05	±0.14 (N=8)		±1.6	±2.2
sandstone	2.57	4.39	8E-18	4.61	1.54	36/50	29.4	13.2
	±0.006	±0.06		±0.13	±0.13 (N=4)		±1.6	±2.1
rhyolite	2.63	1.02	9E-19	5.39	2.16	20...36/55	30.2	15.8
	±0.015	±0.12		±0.34	±0.10 (N=4)		±1.9	±3.2
andesite D	2.72	0.51	6E-19	5.26	1.90	20...41/50	28.7	14.6
	±0.023	±0.09		±0.28	±0.08 (N=2)		±3.1	±4.5
andesite R	2.60	1.70	4E-20	4.35	1.63	31/46	21.3	11.4
	±0.013	±0.08		±0.27	±0.24 (N=4)		±0.9	±2.8

MF and SMF Experiments

A schematic example of typical experiment data for MF and SMF tests is shown in Figure 4. Acoustic emission recordings are used to identify fracture processes in the test speci- mens. AE counts (threshold crossings per time interval – corresponding to AE activity) can directly be linked to localized fracture propagation [4]. The pressure at which the AE count rate raises rapidly is defined as P_{AE}, which is further used as initial fracture propagation pressure. P_{AE} is picked where the AE count rate permanently exceeds 1 / 10 of the test's average (see Figure 4).

In MF experiments, there is almost no AE activity prior to failure. Failure occurs in a very short time span just before sample breakdown (which occurs at maximum injection pressure $P_{inj\,max} = P_b)$, therefore in MF experiments $P_{AE} \approx P_b$. In contrast, SMF experiments show an exponential increase in AE activity at injection pressures that are substantially lower than the actual breakdown pressure ($P_{AE} < P_b$), but much higher than P_{AE} in MF experiments. Therefore, it is possible to interrupt the experiment after AE activity started but before sample breakdown. The latter occurs in SMF experiments when the sample is completely splitted into two parts, which results in a tube breakdown and therefore in an injection pressure drop. Thin sections of specimens, where the experiment was interrupted, show macroscopic fractures emanating several millimeters into the sample but without any connection to the outer surface.

Noteworthy is the discrepancy between the MF and SMF initial fracture propagation pressures P_{AE} at zero confining pressure. This result would imply different hydraulic tensile strength values for the same rock type when using equation (2). Furthermore there is a significant difference between the values of coefficient c calculated for MF and SMF experiments. This

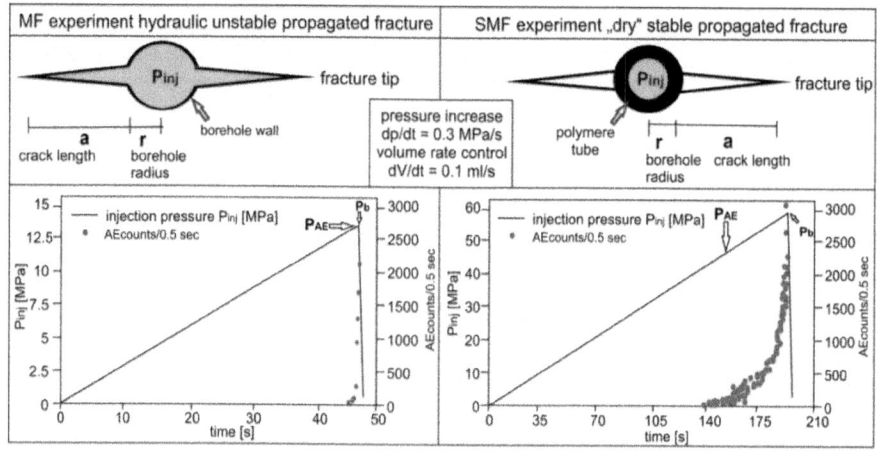

Figure 4: Schematic differences between MF (left) and SMF (right) experiments.

can be seen clearly in Figure 5. Scale effects in P_{AE} (Figure 2) with borehole radius are not evident for the 2 and 3 mm borehole radius samples due to data scattering. One single SMF test of a sandstone with a 6.35 mm borehole radius showed a significantly lower P_{AE} as can be seen in Figure 5.

Table 3: Results of all MF and SMF rock type test series in form of P_{AE0} and coefficient c (see equation (2)). N gives the number of tested samples per lithology and borehole diameter.

rock type	Borehole diam.	MF			SMF		
		N	PAE0 [MPa]	c	N	PAE0 [MPa]	c
marble	4 mm	8	7.7	1.03	6	31.7	6.97
	6 mm	8	9.4	0.96	4	19.6	8.54
limestone	4 mm	9	10.3	1.00	6	26.7	6.06
	6 mm	8	8.2	1.01	7	29.1	5.79
sandstone	4 mm	8	18.2	1.13	5	41.7	6.29
	6 mm	8	18.5	1.14	4	40.5	7.26
rhyolite	4 mm	11	18.2	0.89	4	51.6	6.04
	6 mm	8	16.0	0.85	5	50.9	5.88
andesite D	4 mm	9	16.1	1.00	3	64.2	4.17
	6 mm	6	10.9	0.87	4	48.1	4.83
andesite R	4 mm	10	10.0	1.17	4	47.4	6.26
	6 mm	6	8.2	1.17	5	29.7	7.44
		$\sum$ 93	-	$\varnothing$ 1.02	$\sum$ 57	-	$\varnothing$ 6.33

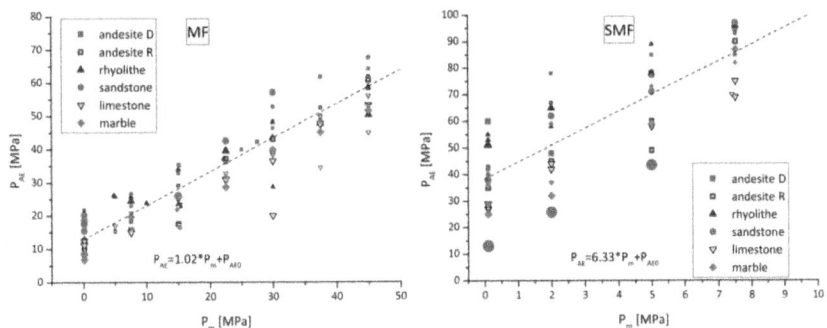

Figure 5: Experimental results of MF (left) and SMF (right) initial fracture propagation pressures for different confining pressures. Dashed line – linear regression of test data. Symbol size refers to borehole radius r (small – r =2mm; inter- mediate – r =3mm; large – r =6.35mm)

CONCLUSION

With SMF tests, stable fracture propagation was achieved over a wide range of injection pressure. Fracture initiation can be confidently linked to the AE count rates. This can be concluded from experiments that were interrupted after P_{AE} but below breakdown pressure. Physical examination revealed the presence of distinct fractures in these speci- mens (see Figure 6).

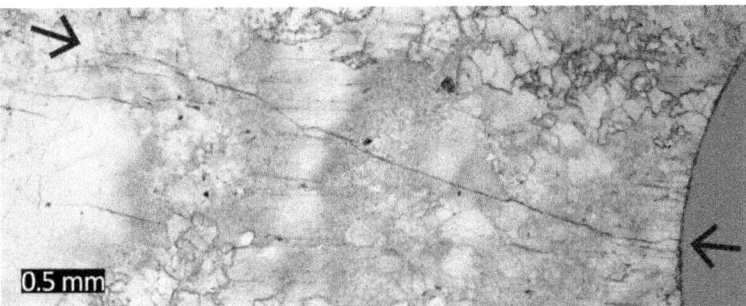

Figure 6: Thin-section of a marble specimen (r =2mm) after SMF test. Clearly visible is a "dry" fracture (indicated by arrows) emanating radially from the borehole (at the right side of the picture). The experiment was interrupted before specimen breakdown. The fracture did apparently not propagate to the outer wall of the specimen.

Due to high data scatter, the theoretical scale effect (critical injection pressure P_c is higher for smaller borehole radii) cannot be resolved by our data. However, tests with a larger (r =6.35 mm) borehole give some support to the notion.

The simple fracture mechanics model is able to explain the higher P_{AE} in SMF experiments. Equations 5 and 6 include the influence of fractures (with or without pressure inside), which is omitted in the classical approach (Equation 1). The high coefficient c in SMF test can only be explained by assuming high microcrack lengths $(a_0 \approx 6 \text{ mm})$.

We excluded poroelastic effects in our analysis due to the use of initially dry rocks with low permeabilities.

Superposition of stress intensity factors for two radial cracks of length a emanating from an internally pressurized (P_{inj} injection pressure in the borehole, P_{frac} pressure inside the fracture) circular hole of radius r in an infinite plate subjected to an isostatic far-field stress Pm as described by [2] and [3] :

$$K_I(P_m) = P_m \sqrt{r} * f_{Pm}(a,r)$$

(7)

$$K_I(P_{inj}) = P_{inj} \sqrt{r} * f_{P_{inj}}(a,r)$$

(8)

$$K_I(P_{frac}) = P_{frac} \sqrt{r} * f_{P_{frac}}(a,r)$$

(9)

$$f_{Pm}(a,r) = 2\left(1+\frac{a}{r}\right)^2 \left(\frac{\left(1+\frac{a}{r}\right)^2 - 1}{\left(\pi\left(1+\frac{a}{r}\right)\right)^7}\right)^{\frac{1}{2}} + \left(\pi\left(1+\frac{a}{r}\right)\right)^{\frac{1}{2}}\left(1-\frac{2}{\pi}\sin^{-1}\left(\frac{1}{1+\frac{a}{r}}\right)\right)$$

(10)

$$f_{P_{inj}}(a,r) = \left(1.3\frac{\frac{u}{r}}{1+\left(1+\frac{a}{r}\right)^{\frac{3}{2}}} + \frac{7.8\left(\sin\left(\frac{2a}{r}\right)\right)}{2\left(1+\frac{a}{r}\right)^{\frac{5}{2}} - 1.7}\right)$$

(11)

$$f_{P_{frac}}(a,r) = \left(\pi\left(1 + \frac{a}{r} \right) \right)^{\frac{1}{2}} \left(1 - \frac{2}{\pi}\sin^{-1}\left(\frac{1}{1 + \frac{a}{r}} \right) \right)$$

(12)

Note: In equations 10 and 12 the borehole was excluded from the integration of stresses (cf. equation 4). The critical fracture propagation pressure at a given fracture length a, borehole radius r and mode I fracture toughness K_{IC} for the unjacketed $(P_{c\text{-}MF})$ and the jacketed $(P_{c\text{-}SMF})$ case:

ACKNOWLEDGEMENTS

The authors wish to thank the German Federal Ministry for the Environment, Nature Conservation and Nuclear Safety for financing our project (FKZ 0325279B). Many core specimens were prepared and analyzed by our student staff: T. Hoferichter, J. Braun, S. Hönig, K. Bartmann and A. Kraft. A great praise to the precision mechanics workshop guys for the construction of the fine working pressure intensifier system. We appreciate fruitful discussions with geomecon GmbH, Potsdam

REFERENCES

1. Hubbert M, Willis D. Mechanics of hydraulic fracturing. Petroleum Transactions. 1957;210:153–68.

2. Rummel F. Fracture Mechanics Approach to Hydraulic Fracturing Stress Measure- ments. In: Atkinson BK, editor. Fracture mechanics of rock. Academic Press geology series. London

3. .u.a..: Academic Pr; 1987. p. 217–39.

4. Winter R. Bruchmechanische Gesteinsuntersuchungen mit dem Bezug zu hydrauli- schen Frac-Versuchen in Tiefbohrungen. Berichte des Instituts für Geophysik der Ruhr-Universität Bochum: Reihe A. Bochum; 1983.

5. Ito T, Hayashi K. Physical background to the breakdown pressure in hydraulic frac- turing tectonic stress measurements. International Journal of Rock Mechanics and Mining Sciences & Geomechanics Abstracts. 1991;28:285–93.

6. Griffith AA. The Phenomena of Rupture and Flow in Solids. Philosophical Transac- tions of the Royal Society of London. Series A, Containing Papers of a Mathematical or Physical Character. 1921;221:163–98.

7. Irwin GR. Analysis of stresses and strains near the end of a crack

traversing a plate. Journal of Applied Mechanics. 1957;24:361–64.

8.   Sih GC. Handbook of stress-intensity factors: Stress-intensity factor solutions and formulars for reference. Bethlehem, Pa: Lehigh Univ., Inst. of Fracture and Solid Me- chanics; 1973.

9.   Tada H, Paris PC, Irwin GR. The stress analysis of cracks handbook. 3rd ed. New York: ASME Press; 2000.Ulusaihutsen

10.   Ulusay R, Hudson JA, editors. The complete ISRM suggested methods for rock char- acterization, testing and monitoring: 1974-2006. 2007th ed. Ankara: Commission on Testing Methods, International Society of Rock Mechanics; 2007.

11.   Mutschler T. Neufassung der Empfehlung Nr. 1 des Arbeitskreises "Versuchstechnik Fels" der Deutschen Gesellschaft für Geotechnik e. V.: Einaxiale Druckversuche an zylindrischen Gesteinsprüfkörpern. Bautechnik. 2004;81:825–34.

12.   Selvadurai APS, Jenner L. Radial Flow Permeability Testing of an Argillaceous Lime- stone. Ground Water. 2013;51:100–07.

13.   ASTM E976. Standard guide for determining the reproducibility of acoustic emsis- sion sensor response. American Society for Testing and Materials. 1994;386-391.

Chapter 4

THREE-DIMENSIONAL NUMERICAL MODEL OF HYDRAULIC FRACTURING IN FRACTURED ROCK MASSES

B. Damjanac[1], C. Detournay[1], P.A. Cundall[1] and Varun[1]

[1]Itasca Consulting Group, Inc., Minneapolis, Minnesota, USA

ABSTRACT

Conventional methods for simulation of hydraulic fracturing are based on assumptions of continuous, isotropic and homogeneous media. These assumptions are not valid for most rock mass formations, particularly shale gas reservoirs, as these typically consist of a large volume of naturally fractured rock in which propagation of a hydraulic fracture (HF) involves both fracturing of intact rock and opening or slip of pre-existing discontinuities (joints). The pre-existing joints can significantly affect the HF trajectory, the pressure required to propagate the fracture and also the leak-off from the fracture into the surrounding formation. None of these effects can be simulated using conventional methods.

HF Simulator is a new three-dimensional numerical code that can simulate propagation of hydraulic fracture in naturally fractured reservoirs, accounting for the interaction between the hydraulic fracture and pre-existing joints. In *HF Simulator*, fracture propagation occurs as a combination of intact-rock failure in tension, and slip and opening of joints. The code uses a lattice representation of brittle rock consisting of point masses (nodes) connected by springs. The pre-existing joints are derived from a user-specified discrete fracture network (DFN).

HF Simulator can model fluid injection or production from one or multiple boreholes each with one or multiple clusters. Non-steady, hydro-mechanically coupled fluid flow and pressure within the network of joint segments and the rock matrix are considered.

An outline of the code hydro-mechanical formulation is presented and examples are provided to illustrate the code capabilities.

INTRODUCTION

A new generation tool that uses the bonded particle model (BPM) [1] and the synthetic rock mass (SRM) concept [2] has been developed to model hydraulic fracture (HF) propagation in naturally fractured reservoirs (NFRs).

Most rock mass formations, and shale gas reservoirs in particular, consist of a large volume of fractured rock in which propagation of an HF involves both fracturing of intact rock and opening or slip of pre-existing discontinuities (joints). The pre-existing joints can significantly affect the HF trajectory, the pressure required to propagate the fracture, but also the leak-off from the fracture into the surrounding formation. None of these effects can be simulated using conventional hydraulic fracturing simulation methods, based on assumptions of continuous, isotropic and homogeneous media.

To address this challenge, a numerical approach called SRM method [2] has been developed recently based on the distinct element method. SRM method usually is realized as a bonded-particle assembly representing brittle rock containing multiple joints, each one consisting of a planar array of bonds that obey a special model, namely the smooth joint model (SJM). The SJM allows slip and separation at particle contacts, while respecting the given joint orientation rather than local contact orientations. Overall fracture of a synthetic rock mass depends on both fracture of intact material (bond breaks), as well as yield of joint segments.

Previous SRM models have used the general-purpose codes *PFC2D* and *PFC3D* [3,4], which employ assemblies of circular/spherical particles bonded together. Much greater efficiency can be realized if a "lattice," consisting of point masses (nodes) connected by springs, replaces the balls and contacts (respectively) of *PFC3D*. The lattice model still allows fracture through the breakage of springs along with joint slip, using a modified version of the SJM. The new 3D program, *HF Simulator* described in this paper, is based on such a lattice representation of brittle rock. *HF Simulator* overcomes all main limitations of the conventional methods for simulation of hydraulic fracturing in jointed rock masses and is computationally more efficient than *PFC*-based implementations of the SRM method.

The formulation of the code is described in this paper. The examples of code verification and application are also presented.

MODEL DESCRIPTION

Background: Synthetic Rock Mass Approach

Over past years, the SRM has been developed [2] as a more realistic representation of mechanical behavior of the fractured rock mass compared to conventional numerical models. The SRM consists of two components: (1) the bonded particle model (BPM) of deformation and fracturing of intact rock, and (2) the smooth joint model (SJM) of mechanical behavior of discontinuities.

The BPM, originally implemented in *PFC*, is created when the contacts between the particles (disks in 2D and spheres in 3D) are assigned certain bond strength (both in tension and shear). It was found that BPM quite well approximates mechanical behavior of the brittle rocks [1]. The elastic properties of the contacts (i.e., contact shear and normal stiffness) can be calibrated to match the desired elastic properties (e.g., Young's modulus and Poisson's ratio) of the assembly of the particles. Similarly, the tensile and shear contact strengths can be adjusted to match the macroscopic strengths under different loading conditions (e.g., direct tension, unconfined and confined compression).

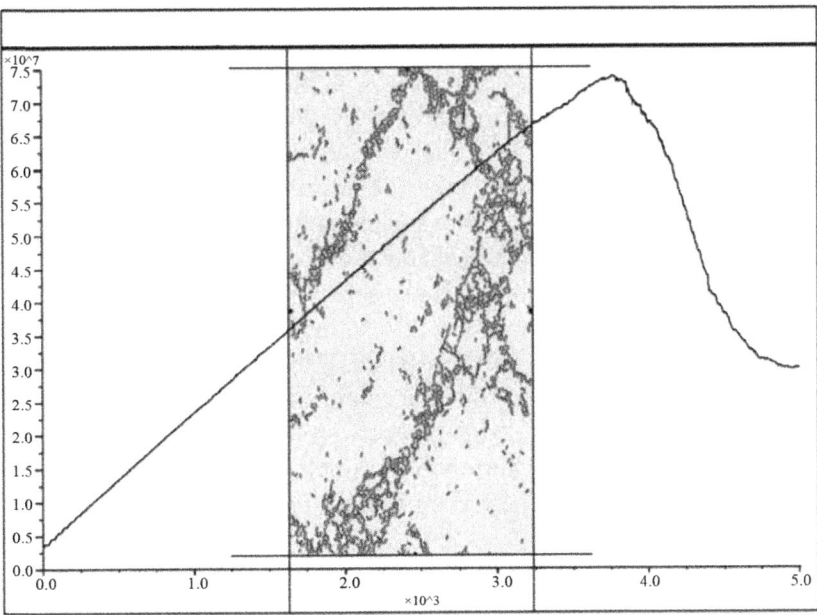

Figure 1: Example of unconfined compressive test using bonded particle model (BPM).

In the BPM, the contact behavior is perfectly brittle. Breakage of the bond, a function of the forces in the contact and the bond strength, corresponds to formation of a microcrack. An example of unconfined compression test conducted using *PFC2D* is illustrated in Figure 1, which shows recorded axial stress-strain response and the model configuration with generated microcracks. The shear microcracks are black; the tensile microcracks are red. Shown is the state when the sample is loaded beyond its peak strength. The stress-strain curve exhibits characteristics typical of brittle rock response. For the load levels less than ~80% of the peak strength, the stress-strain response is linearly elastic, with the slope of the line equal to the Young's modulus. Some microcracks, randomly distributed within the sample, start developing at the load levels greater than ~40% of the peak strength. Significant non-linearity develops as the load exceeds 80% of the peak strength. In this phase, the microcracks begin to coalesce, forming fractures on the scale of the sample. After the peak strength is reached, the material starts to soften (i.e., to lose the strength). At this stage, as shown in Figure 1, the failure mechanism and the "shear bands" are well developed. It is interesting that in the unconfined compression test, the majority of cracks are tensile (red lines in Figure 1). The "shear bands" on the scale of the sample are formed by coalescence of a large number of tensile microcracks.

In order to model a typical rock mass in the BPM, it is also necessary to represent pre-existing joints (discontinuities). A straightforward approach is to simply break or weaken the bonds (in the contacts between the particles) intersected by the pre-existing joints. The created discontinuity will have roughness with the amplitude and wavelength related to the resolution, or the particle size of the BPM. The mechanical behavior of discontinuities is very much affected by their roughness. The problem is that the selected particle size (or resolution) typically is not related to actual roughness of the pre-existing joints. The SJM overcomes this limitation. The contacts in the BPM model are oriented in the direction of the line connecting the centers of the particles involved in the contact. The SJM contacts are oriented perpendicular to the fracture plane irrespective of the relative position of the particles. Consequently, the particles can slide relative to each other in the plane of the fracture as if it is perfectly smooth.

The SRM and its components are shown in Figure 2. The BPM represents the intact rock, its deformation and damage. The pre-existing joints are represented explicitly, using the SJM. They can be treated deterministically, by specifying each discontinuity by its position and orientation as mapped in the field. However, typically, for practical reasons, it is not possible to treat the DFN deterministically. Instead, fracturing in the rock mass is characterized statistically. The synthetic DFNs that are statistically equivalent (i.e., fracture spacing, orientation and size) to fracturing of the rock mass are generated and imported into the SRM using SJM (Figure 2). Very often a reasonable compromise is to represent few dominant structures (faults) with their deterministic position and orientation and the rest of the fracturing in the rock mass (smaller structures) using a synthetic DFN.

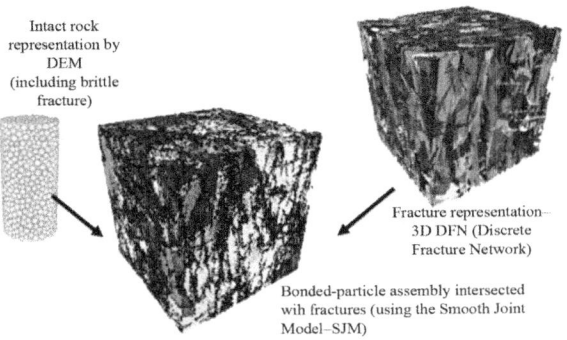

Figure 2: Synthetic rock mass (SRM).

One of the advantages of the SRM is that the components, the intact rock and the joints, can be mechanically characterized by standard laboratory tests. The mechanical response of the rock mass and the size effect are the model results, functions of the model size, DFN characteristics and mechanical properties of the components. Thus, it is not necessary to rely on empirical relations to estimate the rock mass properties and to account for the size effect considering the size of the samples tested in the laboratory and the scale of interest in the model.

The new code, *HF Simulator*, is based on implementation of the SRM in the lattice, which is a simplified, but also a computationally more efficient version of particle flow code (*PFC*). Despite simplifications, the lattice approach represents all physics important for simulation of hydraulic fracturing.

Lattice

The lattice is a quasi-random array of nodes (with given masses) in 3D connected by springs. It is formulated in small strain. The lattice nodes are connected by two springs, one representing the normal and the other shear contact stiffness. The springs represent elasticity of the rock mass. In *HF Simulator*, the calibration factors for spring stiffness are built-in and the user may specify typical macroscopic elastic properties as it is done for other conventional numerical models. The tensile and shear strengths of the springs control the macroscopic strength of the lattice. As for elastic constants, calibration factors are built-in for the strength parameters.

The model simulation is carried out by solving an equation of motions (three translations and three rotations) for all nodes in the model using an explicit numerical method. The following is the central difference equation for the translational degrees of freedom:

$$\dot{u}_i^{(t+\Delta t/2)} = \dot{u}_i^{(t-\Delta t/2)} + \Sigma F_i^{(t)} \, \Delta t \, / \, m$$

$$u_i^{(t+\Delta t)} = u_i^{(t)} + \dot{u}_i^{(t+\Delta t/2)} \Delta t$$

$$(1)$$

where $\dot{u}_i^{(t)}$ and $u_i^{(t)}$ are the velocity and position (respectively) of component i(i=1,3)i(i=1,3) at time t, ΣF_i is the sum of all force-components i, acting on the node of mass m, with time step Δt. The relative displacements of the nodes are used to calculate the force change in the springs:

$$F^N \leftarrow F^N + \dot{u}^N k^N \Delta t$$

$$F_i^S \leftarrow F_i^S + \dot{u}_i^S k^S \Delta t$$

$$(2)$$

where "N" denotes "normal," "S" denotes shear, k is spring stiffness and F is the spring force. If the force exceeds the calibrated spring strength, the spring breaks and the microcrack is formed. In other words, if $F^N > F^{Nmax}$, then $F^N = 0$, $F_i^S = 0$, , and a "fracture flag" is set.

Fluid Flow

Fluid-flow model and hydro-mechanical coupling are essential parts of *HF Simulator*, as a code for simulation of hydraulic fracturing. The fluid flow occurs through the network of pipes that connect fluid elements, located at the centers of either broken springs or springs that represent pre-existing joints (i.e., springs intersected by the surfaces of pre-existing joints). (The code also can simulate the porous medium flow through unfractured blocks as a way to represent the leakoff. This capability is not discussed further in this paper.) The flow pipe network is dynamic and automatically updated by connecting newly formed microcracks to the existing flow network. The model uses the lubrication equation to approximate the flow within a fracture as a function of aperture. The flow rate along a pipe, from fluid node "A" to node "B," is calculated based on the following relation:

$$q = \beta k_r \frac{a^3}{12\mu} \left[p^A - p^B + \rho_w g \left(z^A - z^B \right) \right]$$

(3)

where a is hydraulic aperture, $\mu\mu$ is viscosity of the fluid, p^A and p^B are fluid pressures at nodes "A" and "B", respectively, z^A and z^B are elevations of nodes "A" and "B", respectively, and ρ_w is fluid density. The relative permeability, k_r, is a function of saturation, s:

$$k_r = s^2(3 - 2s)$$

(4)

Clearly, when the pipe is saturated, $s=1s=1$ and the relative permeability is 1. The dimensionless number $\beta\beta$ is a calibration parameter, a function of resolution, used to match conductivity of a pipe network to the conductivity of a joint represented by parallel plates with aperture aa. The calibrated relation between $\beta\beta$ and the resolution is built into the code.

Hydro-Mechanical Coupling

In *HF Simulator*, the mechanical and flow models are fully coupled.

- Fracture permeability depends on aperture, or on the deformation of the solid model.
- Fluid pressure affects both deformation and the strength of the solid model. The effective stress calculations are carried out.
- The deformation of the solid model affects the fluid pressures. In particular, the code can predict changes in fluid pressure under undrained conditions.

A new coupling scheme, in which the relaxation parameter is proportional to $K_R a/R$, where K_R is rock bulk modulus and R is the lattice resolution, is implemented in *HF Simulator*, allowing larger explicit time steps and faster simulation times compared to conventional methods that use fluid bulk modulus as a relaxation parameter.

VERIFICATION TEST: PENNY-SHAPED CRACK PROPAGATION IN MEDIUM WITH ZERO TOUGHNESS

The non-steady response of rock to injection of fluid depends on fracture toughness, the viscosity of the fluid and the rate of leak-off. In the case of zero fracture toughness and no leak-off, the response is viscosity-dominated, which corresponds to the "M-asymptote" identified by [5]. This condition is used for verification of *HF Simulator*.

In the simulated example, fluid is injected at a constant rate into a penny-shaped crack of low initial aperture (10^{-5}m). The crack has zero normal strength, and the in-situ stresses are also zero. Thus, the test conditions approximate those of the analytical solution for the no-lag case (i.e., no fluid pressure tension cut-off) provided by [5]. The injection rate is 0.01 m^3/s; the dynamic viscosity is 0.001 Pa×s. The mechanical properties of the rock are characterized by Young's modulus of 7×10^7Pa and Poisson's ratio of 0.22. Figure 3 provides a visualization of the state of the model at 10 s of elapsed time. Note that pressures are negative in the outer annulus of the flow disk.

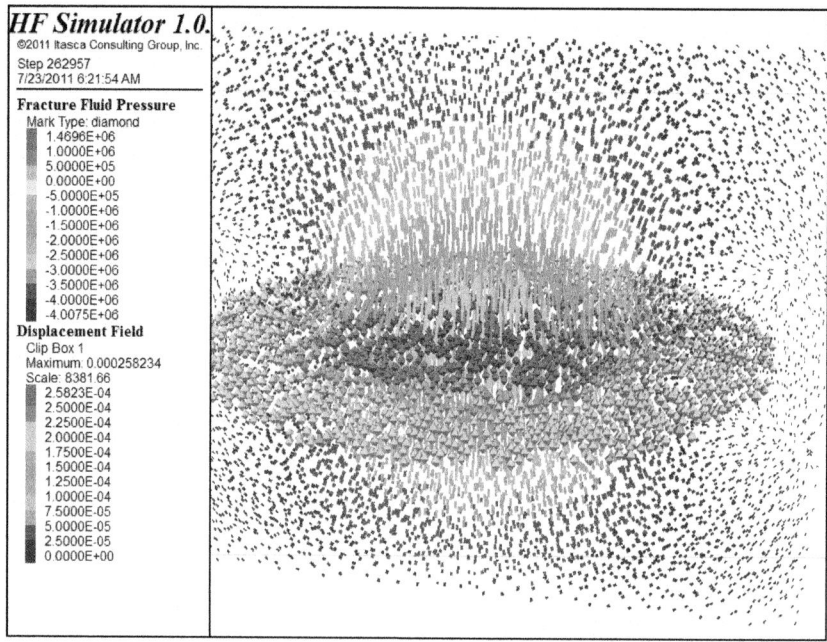

Figure 3: View of pressure (Pa) field (icons, colored according to magnitude) and cross-section of displacement (m) field (vectors, colored according to magnitude).

Figure 4 shows the aperture profiles at three times during the simulation — averaged numerical results (for 30 radial distances), together with asymptotic solutions (derived from the equations of [5]). Figure 5 shows the pressure profile at 10 s, together with the asymptotic solution. Note that there is a lack of match at small and large radial distances: at small distances, the numerical source is a finite volume, rather than a point source (which is assumed in the exact solution); at large distances, the finite initial aperture allows seepage (compared to zero seepage in the exact solution, which assumes zero initial aperture).

EXAMPLE APPLICATION

Two example problems are discussed in this section. Fracture propagation in a homogeneous (unfractured) and fractured media is analyzed. These two problems involve a horizontal borehole segment with two injection clusters with centers at 4.8 m distance (Figure 6). The model domain is 18 m × 18 m × 18 m, and the lattice resolution was set to 0.5 m. Fluid is injected into the clusters at rate of 0.01 m³/s. The assumed stress state is anisotropic with σ_{xx}=1MPa, σ_{yy}=12MPa and σ_{zz}=10MPa. The least principal stress is aligned with the horizontal section of the borehole. This stress state favors crack propagation in the direction normal to the horizontal section of the borehole. In order to initiate the fluid calculation, fluid-filled joints have been placed at the center of each cluster; these joints are slightly larger than the cluster size. The initial apertures in these joints have been set to 0.1 mm. Both example problems use this model configuration. The example shown on the left in Figure 6 simulates the response of an unfractured medium to fluid injection. Three discrete joints that interact with the induced fractures are introduced in the example on the right in Figure 6.

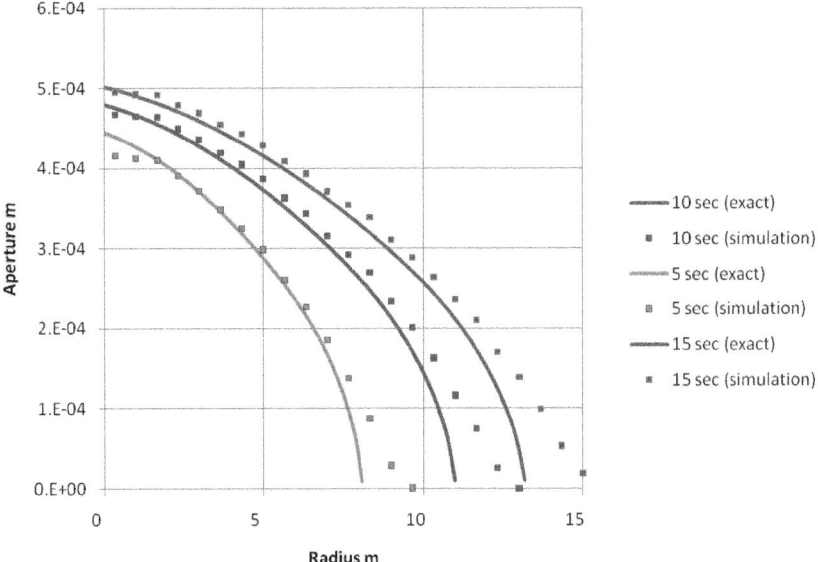

Figure 4: Aperture profiles for three times.

The induced microcracks in the homogeneous model after 15 s of injection are shown in Figure 7. The microcracks form two roughly circular (penny-shape) hydraulic fractures. In this example, the fractures are not parallel. There is a slight trend of fractures curving away from each other as a result of stress interaction.

In the second example, the HF propagation is clearly affected by the pre-existing joints, as shown inFigure 8. When the HF intersects the pre-existing joint, the fluid is diverted into the pre-existing joints. (In general case, the HF can cross or be diverted into the pre-existing joint, depending on a number of parameters, including stress state, strength and permeability of the pre-existing joint.) The propagation continues by reinitiation along the edges of pre-existing joints.

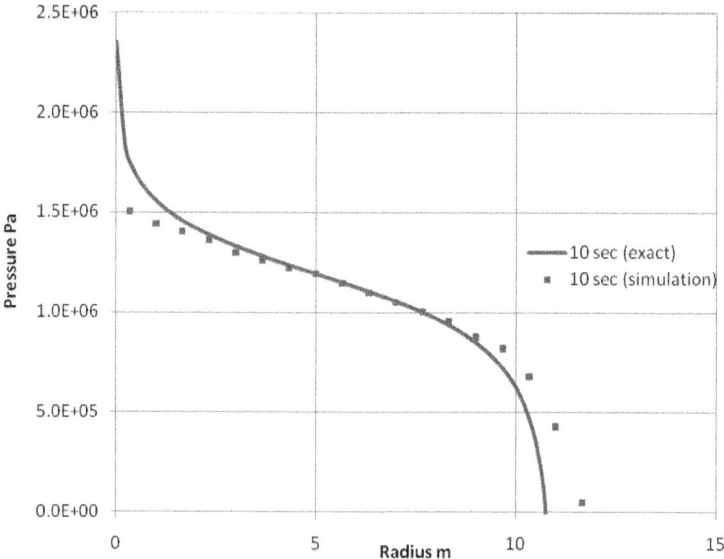

Figure 5: Pressure profile at 10 s.

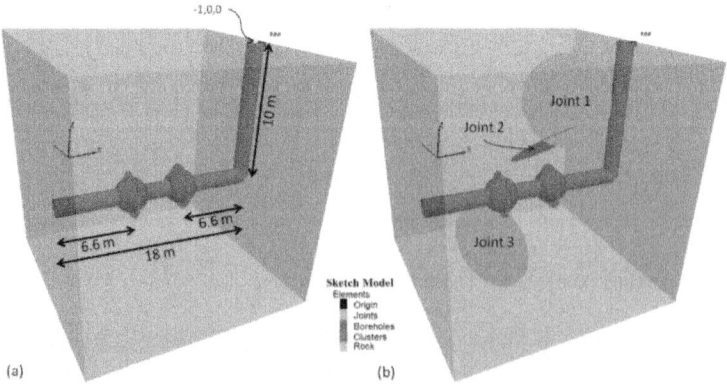

Figure 6: Geometry of two example problems.

CONCLUSION

HF Simulator is a powerful 3D simulator for hydraulic fracturing in jointed rock mass that allows the main mechanisms (nonlinear mechanical response, fluid flow in joints and coupled fluid-mechanical interaction) to be reproduced. The formulation of *HF Simulator* is based on a quasi-random lattice of nodes and springs.

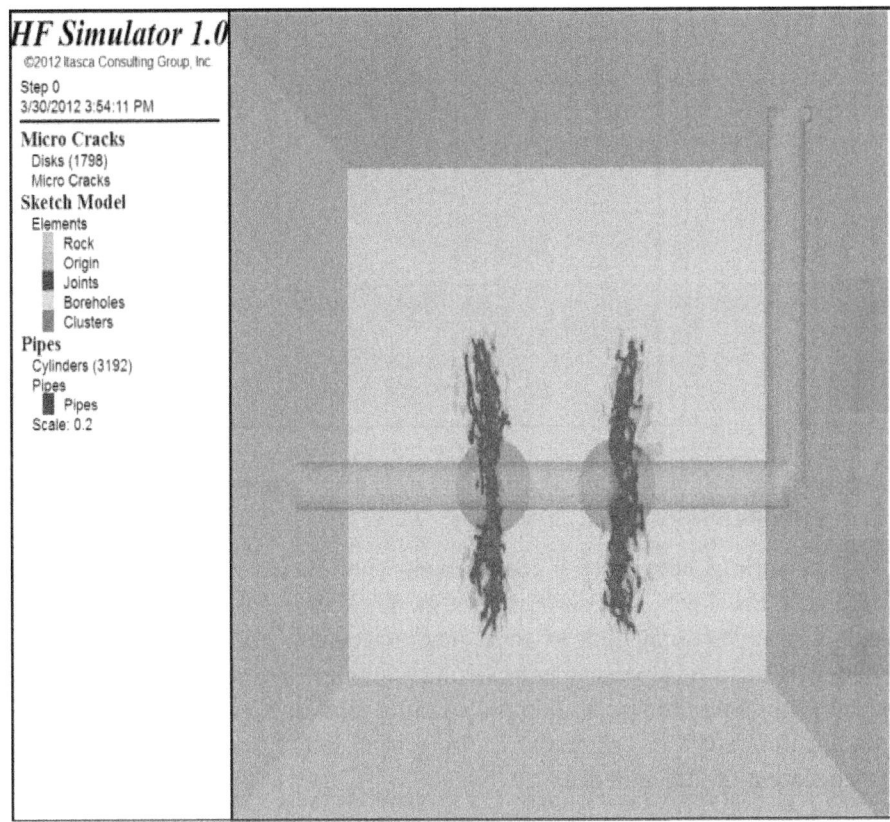

Figure 7: Hydraulic fractures generated in a homogeneous medium (dark blue disks are microcracks).

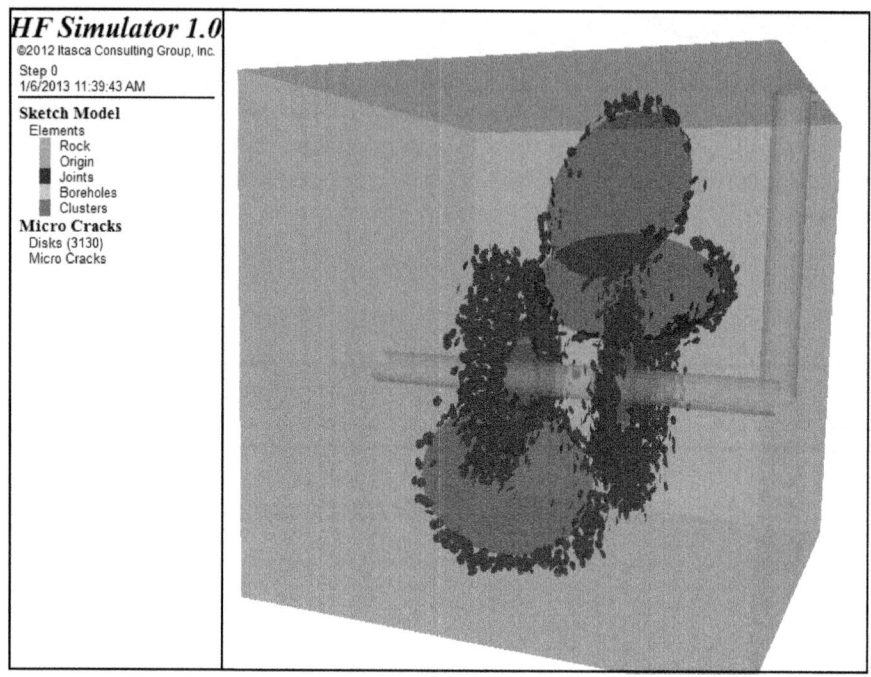

Figure 8: Hydraulic fractures generated in a medium with three pre-existing joints (blue disks are microcracks).

The springs between the nodes break when their strength (in tension) is exceeded. Breaking of the springs corresponds to the formation of microcracks, and microcracks may link to form macrofractures. The SJM (smooth joint model) is used to represent pre-existing joints in the model. Thus, the SJM allows simulation of sliding of a pre-existing joint in the model, unaffected by the apparent surface roughness resulting from lattice resolution and random arrangement of lattice nodes.

The model is fully coupled hydro-mechanically. There are several ways in which fluid interacts with the rock matrix. First, fluid pressures may induce opening or sliding of the fractures. Second, mechanical deformation of fractures causes changes in joint pressures. Third, the mechanical deformation changes the permeability of the rock mass as the joint apertures change.

The new code is a promising tool for simulation and understanding of complex processes, including propagation of HF and its interaction with DFN, during stimulation of unconventional reservoirs.

ACKNOWLEDGEMENTS

The development of the numerical code described in this paper was funded by BP America. The authors would like to thank BP America for their support. Matt Purvance, Jim Hazzard and Maurilio Torres of Itasca Consulting Group, Inc. are thanked for their valuable work on HF Simulator.

REFERENCES

1. D. O Potyondy, P. A Cundall, A Bonded-Particle Model of Rock. Int. J. Rock Mech. & Min. Sci., 41132913642004

2. M Pierce, Mas Ivars D., Cundall P.A., Potyondy D.O. "A Synthetic Rock Mass Model for Jointed Rock," in Rock Mechanics: Meeting Society's Challenges and Demands (1st Canada-U.S. Rock Mechanics Symposium, Vancouver, May 2007), 1Fundamentals, New Technologies & New Ideas, 341349E. Eberhardt et al., Eds. London: Taylor & Francis Group; 2007

3. Itasca Consulting GroupInc. PFC2D (Particle Flow Code in 2 Dimensions), Version 4.0. Minneapolis: Itasca; 2008

4. Itasca Consulting GroupInc. PFC3D (Particle Flow Code in 3 Dimensions), Version 4.0. Minneapolis: Itasca; 2008

5. A Peirce, E Detournay, An Implicit Set Method for Modeling Hydraulically Driven Fractures, Comput. Methods Appl. Mech. Engrg., 197285828852008

Chapter 5

MODEL TEST OF ANCHORING EFFECT ON ZONAL DISINTEGRATION IN DEEP SURROUNDING ROCK MASSES

Xu-Guang Chen[1, 2, 3] Qiang-Yong Zhang[3] Yuan Wang[1] De-Jun Liu[3] and Ning Zhang[3]

[1]Institute of Tunnel and Urban Railway Engineering, Hohai University, Nanjing 210098, Key Laboratory of Ministry of Education for Geomechanics and Embankment Engineering, Hohai Univ., Nanjing 210098, China

[2]State Key Laboratory for GeoMechanics and Deep Underground Engineering, China University of Mining & Technology, Xuzhou 221000, China

[3]Research Center of Geotechnical and Structural Engineering, Shandong University, Jinan 250061, China

ABSTRACT

The deep rock masses show a different mechanical behavior compared with the shallow rock masses. They are classified into alternating fractured and intact zones during the excavation, which is known as zonal disintegration. Such phenomenon is a great disaster and will induce the different excavation and anchoring methodology. In this study, a 3D geomechanics model test was conducted to research the anchoring effect of zonal disintegration. The model was constructed with anchoring in a half and nonanchoring in the other half, to compare with each other. The optical extensometer and optical sensor were adopted to measure the displacement and strain changing law in the model test. The displacement laws of the deep surrounding rocks were obtained and found to be nonmonotonic versus the distance to the periphery. Zonal disintegration occurs in the area without anchoring and did not occur in the model under anchoring condition. By contrasting the phenomenon, the anchor effect of restraining zonal disintegration was revealed. And the formation condition of zonal disintegration was decided. In the procedure of tunnel excavation, the anchor strain was found to be alternation in tension and compression. It indicates that anchor will show the nonmonotonic law during suppressing the zonal disintegration.

INTRODUCTION

Shallow-buried resources have been decreasing with the rapid progress in global economy. Thus, the exploitation of deeply buried resources has drawn interest from a number of countries. South Africa, Russia, India, and China have recently conducted a series of exploitations of deep mines with embedded depths of more than 1,000 m. In China, the Jinchuan nickel mine, Tongling copper mine, and Dingji coal mine are more than 1,000 m deep, whereas the Jinping II Hydropower Station is 2,600 m deep. The Kolar gold mine in India is 2,400 m deep, and the deepest gold mine worldwide residing in South Africa is 3,700 m deep. This series of new phenomena in underground engineering with increasing embedded depth has caused the emergence of a failure phenomenon called zonal disintegration. Shemyakin et al. [1] defined zonal disintegration as the alternated regions of fractured and relatively intact rock masses appearing around or in front of the working slope during the excavation of tunnels in the deep rock mass. This phenomenon has not been observed in shallow rock engineering. Moreover, zonal disintegration presents a serious hazard to the stability of surrounding rocks (Qian) [2].

Zonal disintegration in many deep tunnels has been monitored using the physical probe method. Adams and Jager [3] were the first to observe such phenomenon by using a bore periscope at an embedded depth of 2,000 m to 3,000 m in the Witwatersrand gold mine in South Africa. He reported that zonal disintegration occurred when the tunnel was excavated either by drilling and blasting or the mechanized method. However, explosion was eliminated as the result of zonal disintegration after Shemyakin [1, 4–6] explored zonal disintegration in Taimyrskii deep mine in Russia by using a resistivity meter.

Zonal disintegration phenomenon differs from the engineering response in shallow rock excavation and as such cannot be explained perfectly under the framework of traditional rock theory. In accordance with the concepts of traditional continuum mechanics, the enclosing rock mass around a deep tunnel is divided into fractured, plastic, and elastic regions from the periphery of the tunnel to infinity. Zonal disintegration, a characteristic of deep rock masses, has been the focus of recent investigations.

A number of experts have used various methods to explain zonal disintegration. Sellers and Klerck [7] indicated that the discontinued surface could be one of the derivations of zonal disintegration. Malan and Spottiswoode [8] analyzed the relationship between the shock bump and the zonal disintegration of a top plate in the surrounding rocks of a mining field. Zhou et al. [9] investigated the dynamic excavation of a deep tunnel to determine the residual strength and the forming time of fractured zones. Gu et al. [10] conducted a compression test on cylinder specimen and regarded axial stress as an important factor for zonal disintegration. Other studies on zonal disintegration have applied different techniques such as a series of compression tests Pan [11, 12], nonequilibrium thermodynamics (Metlov et al.) [13], Hamiltonian time-domain variation (Li et al.) [14], and the non-Euclidean model (Guzev and Paroshin) [15]. In addition, some elastic-plastic theories have been adopted to analyze the forming mechanism of zonal disintegration (Wang et al. [16, 17]; He et al., [18]; Zhou et al. [19–24]; Reva and Tropp, [25]; Tan et al., [26]; Wu et al., [27]; Odintsev, [28]). A zonal disintegration phenomenon is shown in Figure 1.

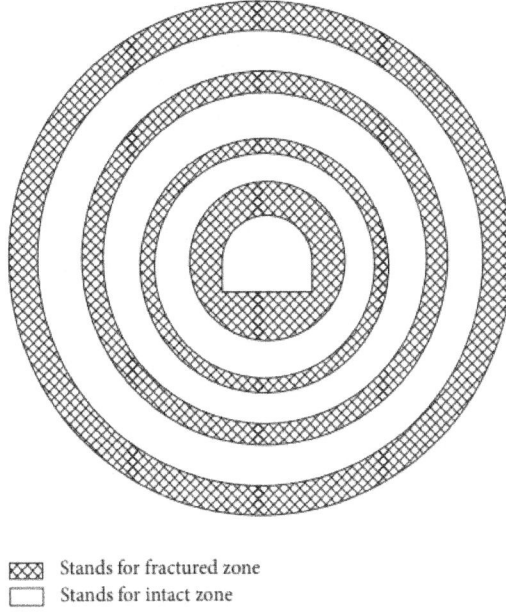

▨	Stands for fractured zone
▢	Stands for intact zone

Figure 1: Sketch of the zonal disintegration phenomena in deep tunnel.

Zonal disintegration is a unique failure phenomenon posing a large-scale disaster during excavation of deep rock masses (Laptev and Potekhin) [29]. It threatens the stability of deep tunnel and will cause large collapse of rock mass which induces a great loss. It is of great importance to know the anchoring effect on zonal disintegration and the mechanical behavior under anchoring condition in deep rock masses, for the stability of deep tunnel. To the authors' knowledge, anchoring effect on zonal disintegration phenomenon in deep rock masses is not investigated previously.

In this paper, the Huainan coal mine in which zonal disintegration occurs in China was taken as the engineering background. The model tests on zonal disintegration were carried on in the condition of anchoring and without anchoring, in separate. The model was built using an independently developed barites-iron-sand cementation analogical (BISA) material. Through the analogical model test, the damage pattern with and without anchoring was observed. The nonlinear deformation changing laws were clarified by using a precise optical apparatus. Based on this, the anchoring effect and forming condition of zonal disintegration in deep rock masses is revealed.

SIMILARITY THEORY AND ANALOGICAL MATERIAL

The geomechanical model test is an important scientific research method. Similar to prototype engineering, the model was designed based on the similarity principle. An optical measuring apparatus was used in the geomechanical model test. The stress and displacement changing rules of the model and strain of the anchor were monitored to determine the deformation laws of prototype engineering. The model test exhibits an advantage in studying the failure mechanism of underground cavities over in situ observation, which relies on auditory-visual perception and is time-consuming.

The geomechanical model test is an effective reduced scale method used for investigating special engineering problems based on the similarity principle. The changing laws of stress, strain, and displacement can be monitored by designing the model similar to that of prototype engineering. Following the similarity principle, the data observed from the model test can be used to reveal the stress distribution laws and the mechanism in prototype engineering, thereby solving actual problems.

Similarity Theory

The geomechanical model test requires a suitable similar material that can reflect the mechanical behavior of a rock type. The similar material and its prototype must comply with the similarity principle. The theory requires several similarity coefficients defined as ratios of prototype parameters to model parameters to be constant (Fumagalli) [28, 30]:

$$C_\sigma = C_\gamma C_L, \tag{1a}$$

$$C_\delta = C_\varepsilon C_L, \tag{1b}$$

$$C_\sigma = C_\varepsilon C_E, \tag{1c}$$

$$C_\varepsilon = 1, \qquad C_f = 1,$$

$$C_\phi = 1, \qquad C_\mu = 1, \tag{1d}$$

where $C_\sigma, C_\varepsilon, C_E, C_L, C_\gamma, C_f, C_\phi,$ and C_μ indicate the similarity ratios of stress, strain, MOE, geometry, volume weight, COF, internal friction angle, and passion ratio, respectively.

The analogical material should have the properties of high volume-weight, low deformation module, and changeable inner friction angle. No crude material can fulfill all these demands, and thus, the similar material should be assembled artificially. According to the similarity theory, the mechanical parameters of the model can be readily obtained through the prototype.

Proportion of the Similar Material

There are several Institutes researching on the similar material, such as ISMES (Institute of Experimental Models and Structures) in Italy, LNEC (National Laboratory for Civil Engineering) in Portugal, and Tsinghua University in China [30, 31]. Their work shows that whether the model test can reflect the prototype engineering's mechanical response depends on the chosen materials. The suitable material should reflect the mechanical behavior of prototype engineering. The proportion of each component is important for model simulation.

The barite powder, iron powder, and quartz sand are selected to form the aggregate, whereas the alcoholic solution of rosin is used as the mucilage glue (Figure 2). The proportion of the aggregates and the concentration of the alcohol solution of rosin decide the mechanical behavior of the BISA material. The barite-iron-sand (BISA) material was developed through hundreds of groups of proportioning tests. The specimens of similar material were built by pouring the material into a mould and compressing it (Figure 3). The material exhibits the following advantages: stability in performance, widely variable mechanical parameters, low price, high volume-weight, easy processing, and no toxicity or side effects. The BISA material, which can be used for modeling a tunnel or underground powerhouse, has obtained a patent in China. The material can be used to simulate all kinds of rocks, including hard and soft rocks. The proportion of the material composition for surrounding rocks in the Dingji coal mine was determined via the physical-mechanical parameters test on the analogical material. The mechanical parameters of the material were tested in the proportion of each component in the material (Figure 3).

Figure 2: Proportion of the analogical material. The barites powder, iron powder and quarts sand are mixed together to make the aggregate; the solution of rosin and alcohol make the glue.

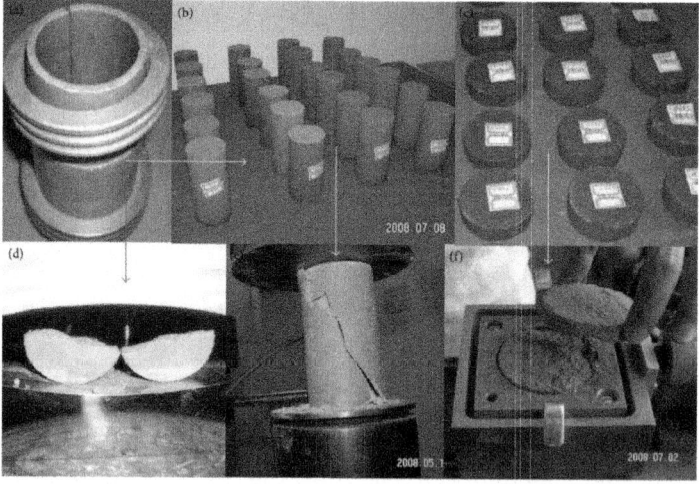

Figure 3: Mechanical test on analogical material. (a) is the specimen mould. (b) is the specimens for UCS. (c) is the specimens for direct shear test. (d) is the Brazilian test. (e) is the uniaxial test. (f) is the direct shear test.

The laws between the mechanical parameters of material and components proportion were derived by mechanical testing on hundreds of specimens.

The medium sandstone in the bed stratum from the Dingji coal mine and the Huainan mining area was processed into a specimen. The physical-mechanical parameters of equivalent anchors were tested. The mechanical parameters of the medium sandstone consist of the following: unconstrained compressive strength (UCS) of 88.55 MPa, tensile strength of 14.01 Mpa, and deformation modulus of elasticity (Edef) of 12.97 GPa. The similarity ratio of volume weight for the analogical material is set to 1 : 1, whereas the similarity ratio of geostress is set to 1 : 50. Thus, according to the similarity principle, the mechanical parameters of the prototype and similar material are as follows (Table 1).

Table 1: Physical-mechanical parameters of the prototype rock and model material

Material type	Volume weight/KN·m⁻³	Edef/MPa	Cohesion/MPa	φ	UCS/MPa	TS/MPa	NUXY
Prototype	2.62	12970	10	43	88.55	14.01	0.268
Model	2.62	51.88	0.4	43	3.54	0.56	0.268

According to the curves between the mechanical parameters and the material proportion, the proportion of each component for medium sandstone is as follows (Table 2).

Table 2: Proportion of the analogical material

I : B : S	Portion of the gypsum	Concentration of the solution	Portion of the solution
1 : 1.1 : 0.42	2.5%	7.5%	5.0%

Equivalent Anchors

The parameters of anchor adopted in engineering are: ϕ 20@800 × 800 mm, L = 2.2 m. According to the similar principle, the parameters of model anchor can be got from the prototype anchor (Table 3).

Table 3: Physical-mechanical parameters of prototype and equivalent anchor

Anchor	Edef/GPa	TS/MPa	Yield strength/MPa	length/cm
Prototype	210	510	345	220
Model	4.2	10.2	6.90	4.4

After the mechanical test on the serious metal materials (Figure 4), the aluminum wire is selected as the equivalent anchor.

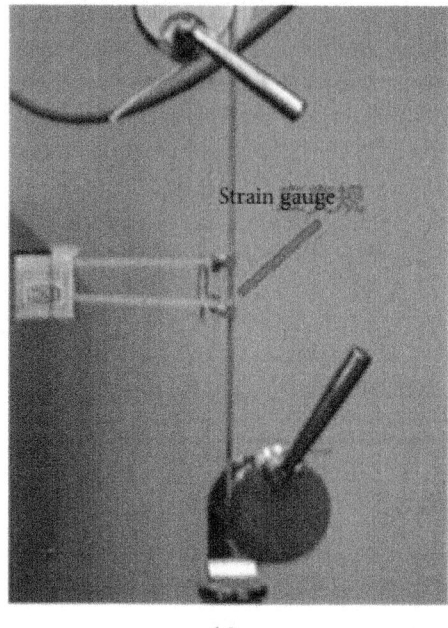

(a)

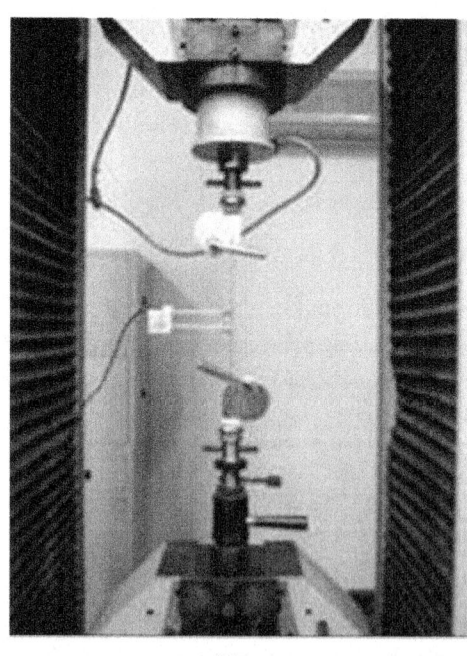

(b)

Figure 4: Mechanical test on equivalent anchor.

DEVELOPMENT OF THE MODEL TEST SYSTEM

Development of the Triaxial Model Test System

In order to simulate the 3 D geostress state of tunnel precisely, the high geostress-triaxial loading model test system was developed independently. Figure 5 is the design sketch and the photo of the system.

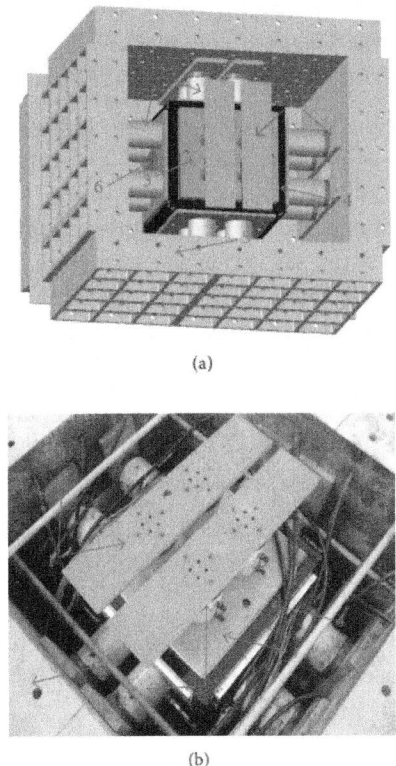

(a)

(b)

Figure 5: Sketch and photos of the true 3D model test system. The model was set inside the frame, and the platen was set into the frame at 0.5 cm depth. Then, each direction of the model has the reserved displacement of 9 cm. Notes: (1) junction plate, (2) hydraulic jacks, (3) loading platen, (4) oriented frame, and (5) combination reaction frame for loading (6) excavation guiding platen.

The model test system is comprised of the high geostress loading system, the computerized numerical control (CNC) hydraulic system, and the support reaction apparatus. The high geostress loading system consists of jacks and loading platens, which are used to apply high geostress. It is set inside the support reaction apparatus. One end of the jacks was disposed with loading platen, while the other was fixed to the support reaction apparatus with bolts. There are 6 platens in all, and one contacts each surface of the model, which can apply loading at 3 directions. The dimension of each platen is $0.6 \times 0.6 \times 0.04$ m, which is equal to the size of model surface. There arrange 4 hydraulic jets on each platen, and the capacity of each jack is 40 tons. The loading capacity of each platen is 1600 KN. The high geostress loading system is connected with the CNC hydraulic system. The CNC hydraulic system is used to regulate and control loading value. The support reaction apparatus is composed of box cast steel components, which can be assembled flexibly. It is solid enough to undertake support reaction to 5000 KN high.

During the triaxial compression on 3D model, two adjacent loading platens will interrupt each other because of the volume contraction. So, an oriented frame was developed to solve the problem. The frame consists of 12 steel poles which are connected with each other. Each pole was 5 cm thick and 70 cm long. The model was set inside the frame, and the platen was set into the frame at 0.5 cm depth. Then, each direction of the model has the reserved displacement of 9 cm. A circle platen whose size is similar to the tunnel was fixed into the two platens in front and back of the model. And the platens will be fixed during the loading on the model and taken off when excavation of the tunnel (Figure 5). So, the excavation in 3D model can be solved.

The advantages of the system are shown as follows.

(1) High geostress can be applied to the model in all 3 directions independently and synchronously. And the geostress can be remained stable for long term.

(2) The problem of two adjacent loading platens that interrupted each other which was caused by the triaxial compression was solved. And the excavation in 3D model was realized.

(3) The loading capacity is large enough to 2000 KN, which can simulate the cavern deep to 5000 m. And the loading precision reaches up to 0.5%.

Development of the High Precision Optical Measuring System

A micrometer is necessary because the multi-point displacement meters are too huge for the model. Thus, a micrograting ruler multipoint displacement measuring System (GRDS) was developed to monitor the displacement in the rock mass of the model (Figure 6). The Moiré fringe is used to enlarge the displacement in the model. The GRDS is composed of rack, grating rulers, steel wires, balance weight, fixed end, signal translating system (STS), and data acquisition system (DAS). Several fixed ends were buried at the measuring spots inside the model. The fixed ends were connected to the grating ruler outside the model with the steel wire. The steel wire was made up of special sterepsinema without any axial deformation, and thus, it is very flexible and can be bent arbitrarily. The steel wire was wrapped with Teflon pipe to eliminate friction between the steel wire and the model. The GRDS transformed the displacement of the model to the grating ruler through the mechanical method. Then, the real displacement was transferred to the optical signal by using an optical ruler. The STS was connected to the grating ruler with the guide line, and the optical signal was transformed to digital signal and then transferred to the DAS. The displacement data of the model was displayed and stored instantaneously. The high precision (1 µm) GRDS can be used to measure the displacement in any direction as it can be bent at any angle.

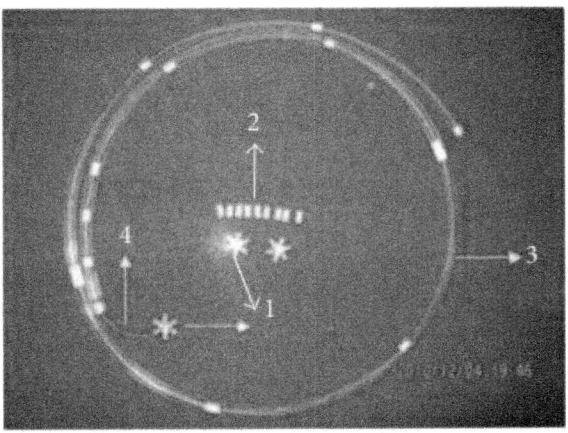

(a) The disassembled part

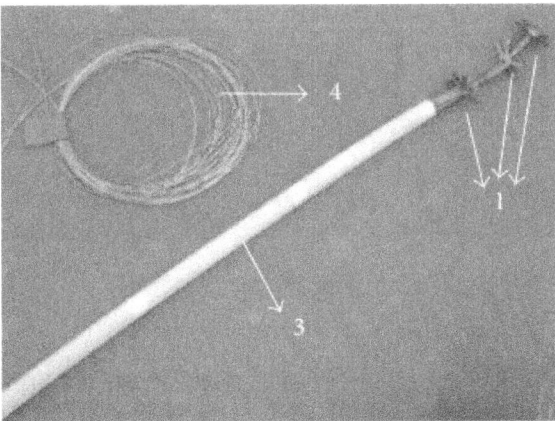

(b) The assembled device

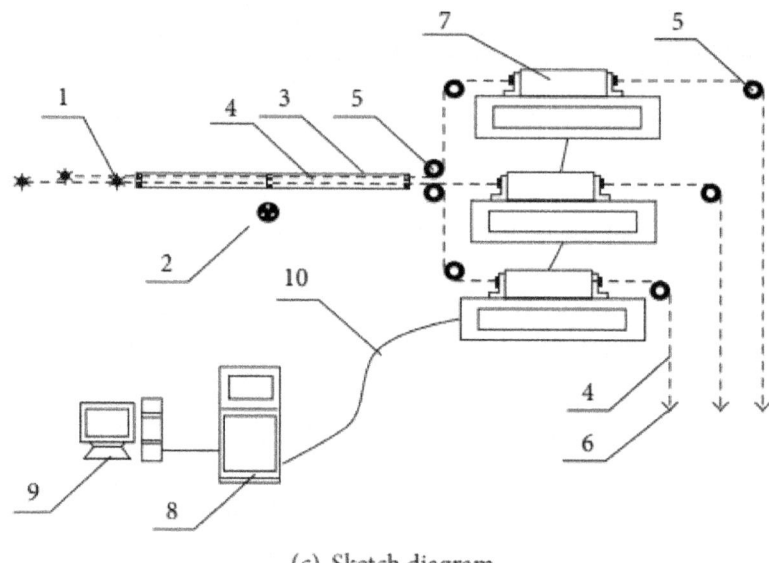

(c) Sketch diagram

(d) The optical ruler

Figure 6: Grating ruler multipoints displacement measuring system (GRDS). (1) Fixed end, (2) guding frame, (3) soft pipe, (4) flexible steel wire, (5) displacement transfer roller, (6) balance weight, (7) optical ruler, (8) STS, (9) DAS, (10) cable.

The fiber bragg grating (FBG) was stuck to the analogical material blocks to make the grating strain sensor. And it can be used to measure the strain in each direction inside the model (Figure 7).

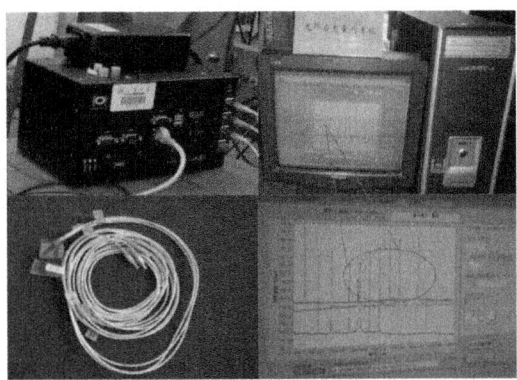

Figure 7: Grating strain sensor measuring system.

The FBG can measure the strain changing outside through the centre wavelength mobile. The relationship between the centre wavelength λ_B and the effective refraction index n_{eff} of grating and the period of grating Λ is

$$\lambda_B = 2n_{\text{eff}}\Lambda, \tag{2}$$

where λ_B is centre wavelength of the Bragg grating; Λ is the period of grating; n_{eff} is the effective refraction index.

When the fiber is stretched, Λ and n_{eff} will change, and then the centre wavelength λ_B will drift. The wavelength gets larger when the fiber is stretched, while it gets smaller when the fiber is compressed. The linear relationship will be satisfied:

$$\Delta\lambda_B = \lambda_B\left(1 - P_e\right)\varepsilon = K_e\varepsilon, \tag{3}$$

where, $\Delta\lambda_B$ is the variation of wavelength; P_e is the effective optical coefficient (0.22, generally); ε is the axial strain of fiber; K_e is the sensitivity of strain measuring.

When the grating generates strain in the stress condition, the pitch of the grating will change to $\Delta\Lambda$, which will make the wavelength change to $\Delta\lambda$, so the strain can be got:

$$\varepsilon = \frac{\Delta l}{l} = \frac{\Delta\lambda}{\lambda}. \tag{4}$$

CONSTRUCTION OF THE MODEL

To study the anchoring effect on zonal disintegration, a model was built and divided to two halves to make compassion: one half is anchored and the other half is nonanchoring.

Simulation Range of the Model

The model size is limited by the reasonable size of the steel frame. Within the frame, if the similarity constant for the geometry C_L is smaller than 1 : 50, then the model tunnel will be too small to be excavated. If the C_L is very large, then the relevant monitoring devices will be too difficult to install and the monitoring data will be inaccurate. Considering these prior factors, 100 is taken as the optimal similarity coefficient C_L.

The simulation range of the prototype is 30 m × 30 m × 30 m. According to the geometry similar scale 1/50, the dimension of the model is determined to 0.6 m × 0.6 m × 0.6 m and the model tunnel is 100 mm × 77.6 mm.

Flow of Model Construction

The model was made delaminating. Each layer is 10 cm high, so 7 times are needed in all to finish the entire model. The measuring components were set up when the material reached the designed height (see Figure 8).

(a) Charge mixture   (b) Compaction   (c) One layer finished

Figure 8: Procedure of the model building.

Burying of the Measuring Components

The measuring components were disposed in the model, including the optical multipoint displacement meter and the grating Bragg optical strain sensor, which were used to monitor the displacement and strain changing in the surrounding rocks, respectively. Figures 9–10 show the section sketch and burying procedure of the measuring components, in separate.

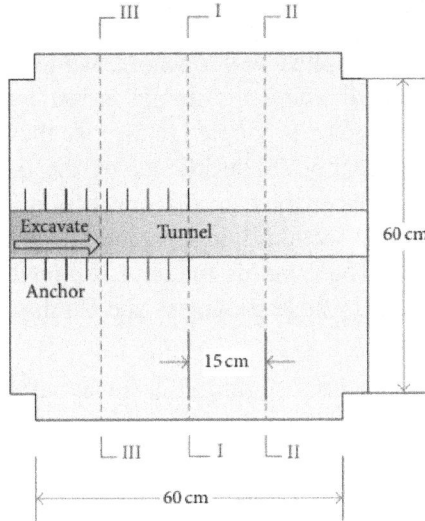

Figure 9: Layout sketch of monitoring sections in the model. The model is divided to two parts: anchored area and the area without anchoring. The measuring sections were set to 3, which divides the model to 4 parts evenly. The II measuring section was chosen to be the middle part of the model, while the I and III sections are quarter and three-quarters of the model. Then, the measuring section can be used to monitor the deformation inside the model.

(a) Burying of optical Bragg fiber strain sensor

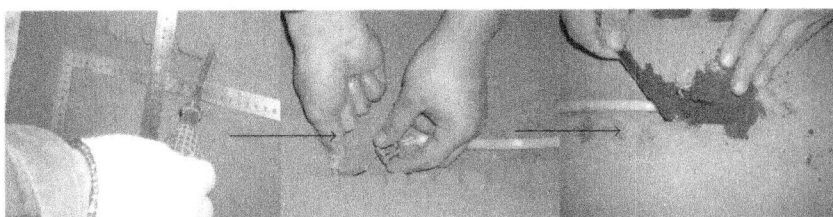

(b) Burying of GRDS

Figure 10: Burying procedure of the optical measurement sensor. The sensor is Fiber Optical Bragg grating sensor, where the precision is $10^{-3} \, \mu\varepsilon$. The spacing between those sensors is 1 cm.

Distribution and Burying of the Anchor

The anchored area was applied with anchors, while the other half was not. Because the anchor surrounding the tunnel is too intensive, an equivalent principle of pulling resistance is adopted to dispose the anchor. A thick anchor is replaced by 4 thin anchors. Considering of the quincuncial disposal of anchor, it is not equal between the two adjacent sections. There are 8 anchors in section A and 7 anchors in section B, in separate. The two sections are arranged alternately (Figure 11). The grouting material is mixed by the high thickness alcohol solution of rosin, whose stickiness and fluidity are both well to meet the grouting demand.

(1)   Arrangement of anchors. Figure 11 is the layout sketch of the equivalent anchor. As the figure shows, the equivalent anchor is arranged according to the quincunx. There are 8 anchors in section A and 7 anchors in section B. The inter-row spacings are both 1.6 m. The two sections are arranged alternately along the axial direction of tunnel.

(2)   Burying of the anchors. The embedded method and grouting burying method are adopted together. The embedded method is adopted for the anchor of the middle of model. The grouting burying method is adopted for the entrance of the tunnel.

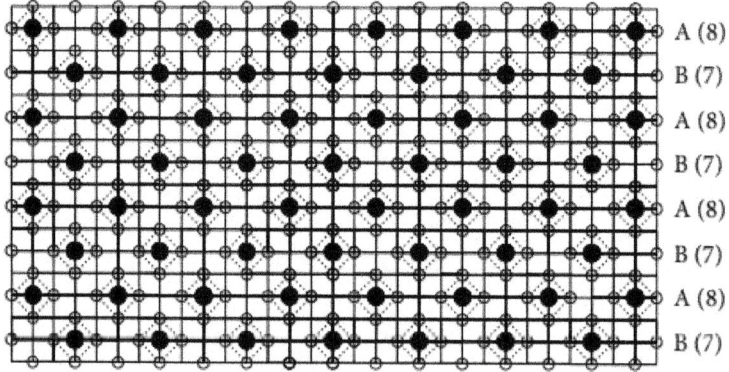

○ Stands for rock type anchor, 0.8 m × 0.8 m
● Stands for equivalent anchor, 1.6 m × 1.6 m

(a) Distribution sketch along the axial direction

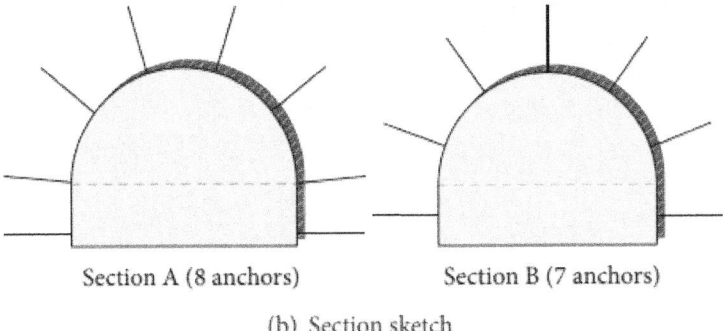

Section A (8 anchors)   Section B (7 anchors)

(b) Section sketch

Figure 11: Layout sketch of the equivalent anchor.

Figure 12 is the burying procedure of the embedded anchor, where (a) is for the side wall and (b) is for the arc crown. After burying, the material is backfilled and tamped, till the completion of the entire model.

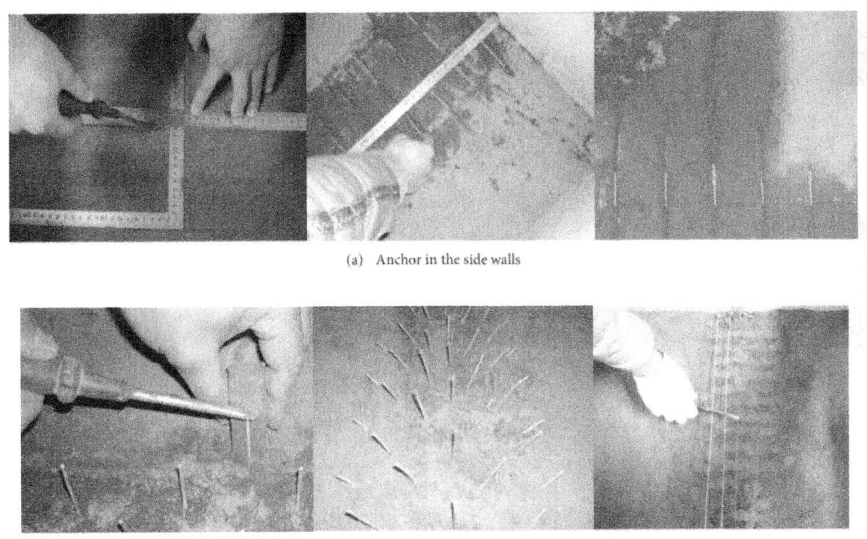

(a) Anchor in the side walls

(b) Anchor in the arc crown

Figure 12: Burying procedure of analogical anchor.

The anchor burying of two sections near the cavern was carried during the procedure of tunnel excavation. The details are shown in Figures 14(b)–14(d).

Tunnel Excavation and Model Testing

The deep tunnel is in the 3D geostress state. In order to simulate the geostress fields accurately, the model was loading in true 3D state (Figure 13). During the excavation, the displacement and strain changing laws were monitored and recorded (Figure 13).

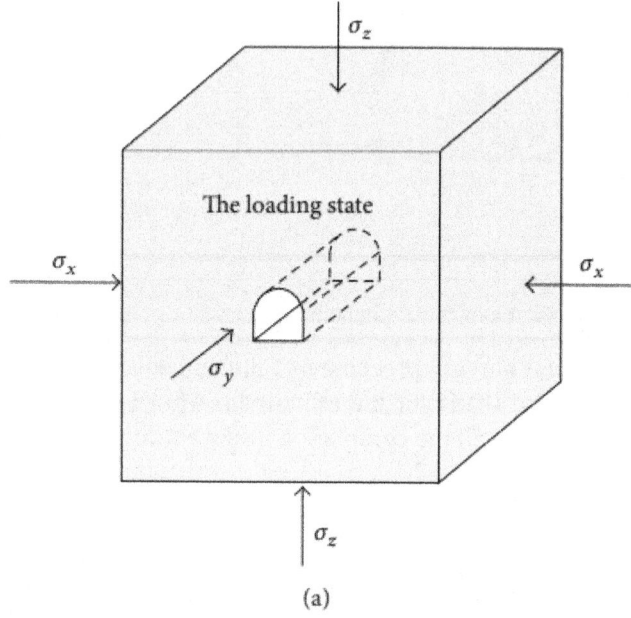

(a)

(b)

Figure 13: Loading state and test procedure of the model.

(a) Excavation

(b) Applying anchor

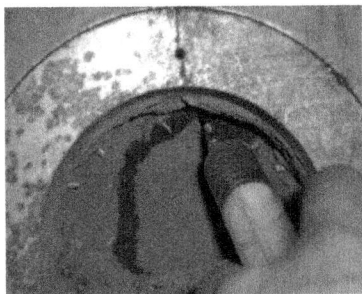

(c) Grouting

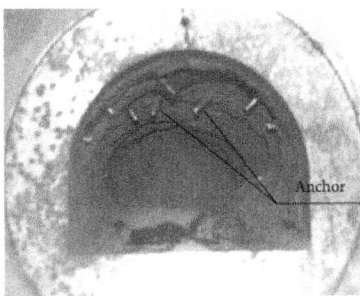

(d) Anchor finished

(e) Next excavation step

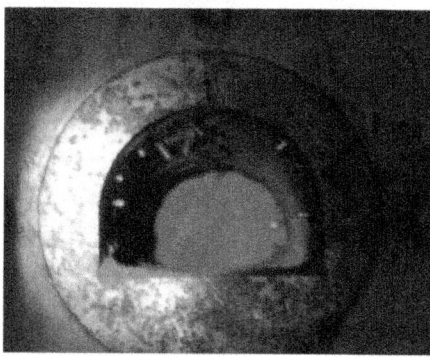

(f) Breakthrough

Figure 14: Excavation procedure of model tunnels.

The self-weight stress is γh. The loading which is perpendicular to the tunnel axial is 1.5 γh (coefficient of horizontal pressure is 1.5). Where γ is the volume weight, h is the embedded depth of the tunnel. The loading value is listed in Table 4.

Table 4: The loading value of the model

Loading direction	Self-weight stress	Perpendicular to the axial	Parallel to the axial
Loading value/Mpa	0.5	0.75	0.75
Loading direction	Self-weight stress	Perpendicular to the axial	Parallel to the axial

The excavation procedure of the model is shown as follows: (Figure 14).

RESULTS OF MODEL TEST AND DISCUSSION

Displacement from the Grating Extensometer

Figure 15 shows the drawing sketch of grating extensometer. The displacement of surrounding rocks is labeled and connected using the smooth curve after excavation. Then, the changing laws of displacement can be got. The displacement changing sketch is placed together to compare. It is known from the sketch of displacement surrounding the tunnel.

(1)   The displacement shows a very different changing law between the anchored model and nonanchoring model. In the anchored model, the displacement decreases monotonously as the distance to the tunnel wall increases. It is similar to the shallow embedded tunnel. While in the nonanchoring model, the displacement presents the undulate changing status, wherein the wave crest and the trough are arranged alternately. It completely differs from the shallow embedded tunnel.

(2)   Compared with the dissembled model, the area of larger displacement is the severely damaged region. In the anchored model, the displacements of no. 1 spot near the periphery are larger than the other area. It indicates that this area is the damage zone in the traditional perception. This is in accordance with the phenomenon of periphery damage seriously during the excavation. The displacement in arc crown is larger than the side walls. This is corresponding to the phenomenon of serious damage and collapse in arc crown. In the model without anchoring, the wave crest region with a larger displacement is the fractured zone, whereas the trough region with a smaller displacement is the intact zone. The additional displacement in the wave crest area, which is caused by the circular fracture, increases the total displacement.

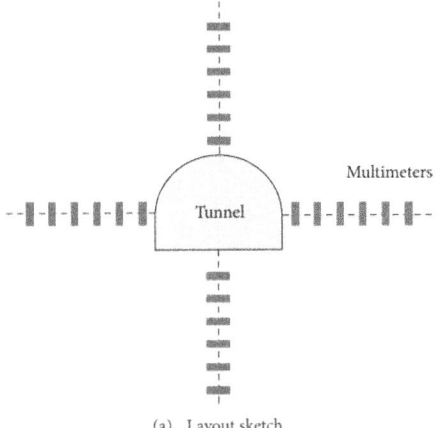

(a) Layout sketch

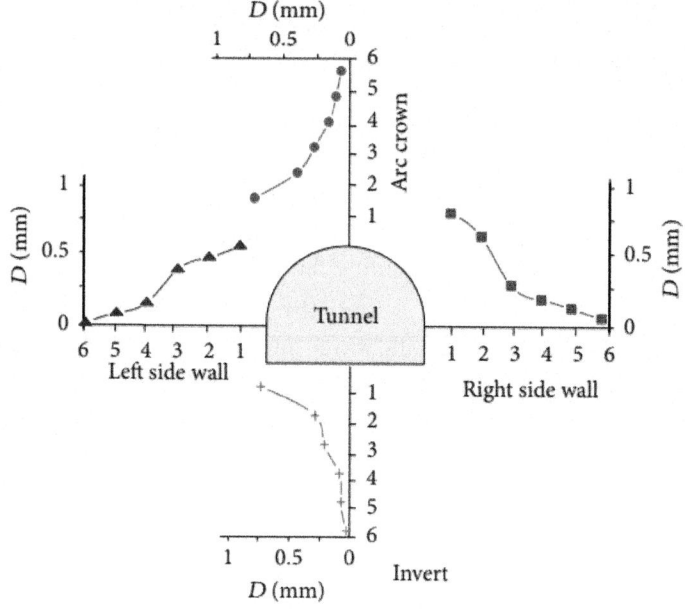

(b) Data of anchored model

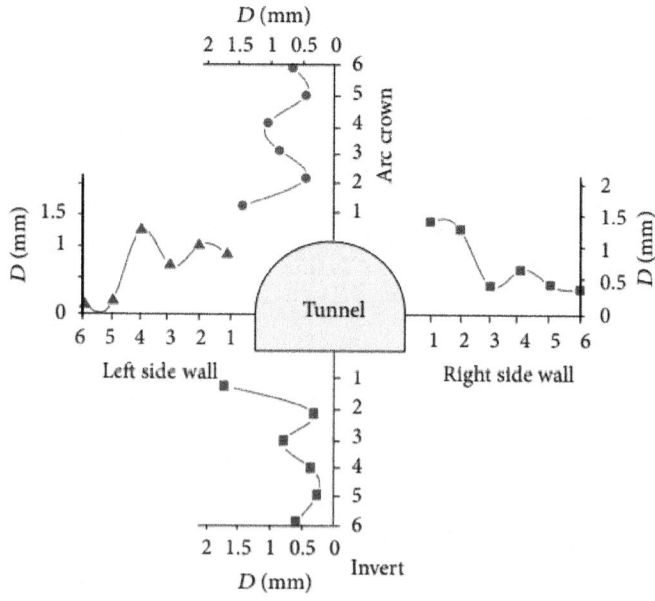

(c) Data of model without anchoring

Figure 15: Displacement surrounding the model tunnel.

Strains from FBG Strain Sensor

The strain results show that both the radial and tangential strains were negative before the excavation. This indicates that the surrounding rocks were in compressive state. When it is excavated to the strain monitoring section, the radial strain ε_r turns to positive, which indicates that the tensile strain appear in the radial direction. The phenomenon was in accordance with the elastic-plastic strain field analysis of the ideal circular tunnel.

Figure 16 is the diagram sketch of FBG strain sensor. The radial strain is labeled in the surrounding rocks nearby the tunnel after the tunnel excavation and measurement.

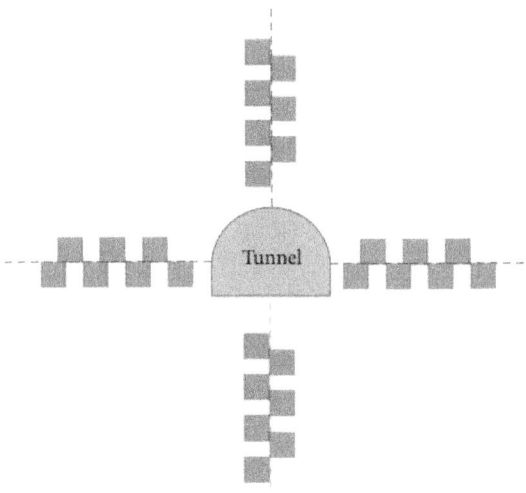

(a) Layout sketch of optical fiber strain sensor

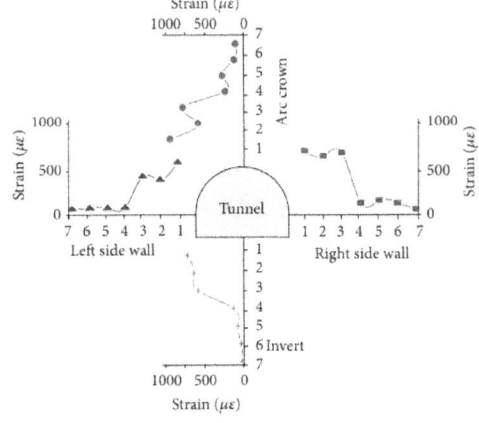

(b) Strain of anchored model

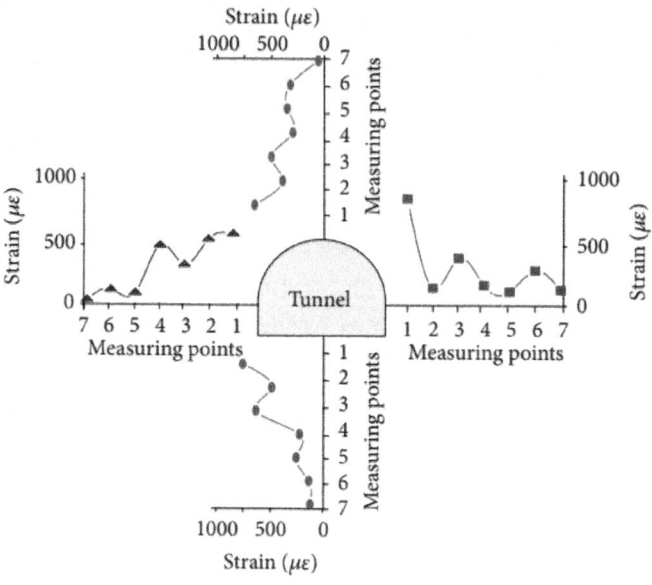

(c) Strain of model without anchoring

Figure 16: Radial strain around tunnel measured by optical fiber strain sensor.

As is shown in Figure 16, the radial strain in the surrounding rocks both in the anchored model and nonanchoring model shows the same changing law: it presents the fluctuate distribution with the distance increasing from the periphery. The wave crest and the trough distribute alternately. This law is totally different from that in the shallow embedded tunnel. The phenomenon indicates that under the high geostress loading, the radial tensile strain is the highest in a certain area of the surrounding rocks. This area is almost circular and concentrated in the tunnel, and identified as the elastoplastic zone of the surrounding rock.

Zonal disintegration phenomenon does not occur in the side walls and arc crowns of anchored model. It is because the reinforcement effect improves the capacity of fracture-resistance of surrounding rocks. There is no reinforcing effect in the nonanchoring area, such as the invert of anchored half model and the half model without anchoring. Zonal disintegration phenomenon occurs in these areas.

Anchor Strain in the Middle Section of the Model

Figure 17 is the changing laws of anchor strain in the middle section of the model measured by the strain gages.

(1)   The anchor strain is minus at the beginning of tunnel excavation.
       It indicates that the anchors are in compressive state in high loading
       condition. It turns to positive at the excavation of step 6 which is near
       the middle of the model (there are 14 excavation steps in all). It indicates
       that anchors turn into tensile state and work.

(2)   Most anchors (except nos. 1, 6, and 11) turn to maximum state at
       excavation of 7 or 8. This is because the anchor works at the largest
       effect when excavated in the middle of the model, which is the section
       of anchor strain measurement.

(3)   A mount of anchors (nos. 1, 4, 5, 8, 10, 13, and 14) shows the phenomenon
       of positive and negative alternation during the excavation process. It
       indicates the tension and compression alternating of anchor, which is
       first observed in the geomechnical model test.

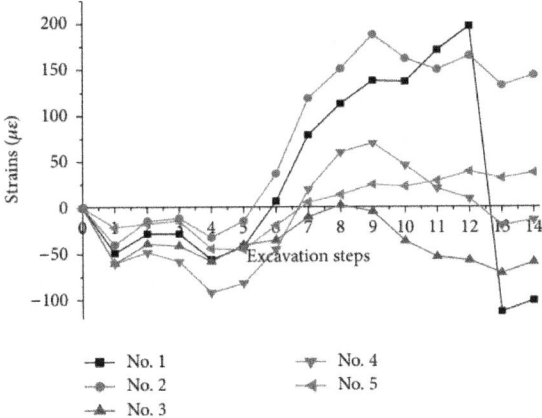

(a) Nos. 1–5 anchor strain changing with excavation steps

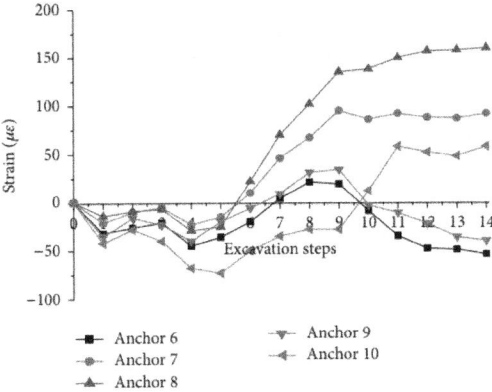

(b) Nos. 6–10 anchor strain changing with excavation steps

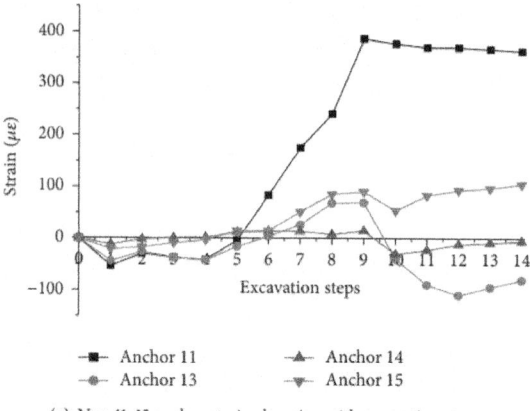

(c) Nos. 11–15 anchor strain changing with excavation steps

Figure 17: Anchor strain changing with the excavation steps. Note: No. 12 anchor was damaged during the test and the data could not be recorded.

Fang [32] used to observe the similar special phenomenon of tension and compression alternating in the deep tunnel of Jinchuan nickel mine, 1150 m deep. He believes that it is the self-organized phenomenon and the inherent characteristic of deep tunnel.

Fracture Shape inside the Model

Figure 18 is the photos of dissembled model which show the damage pattern.

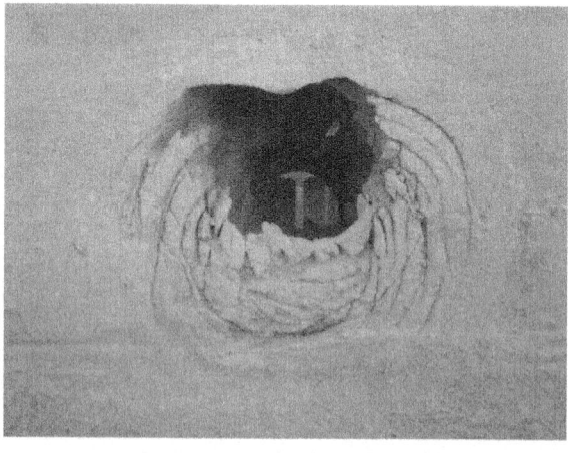

(a) The non-anchoring half model

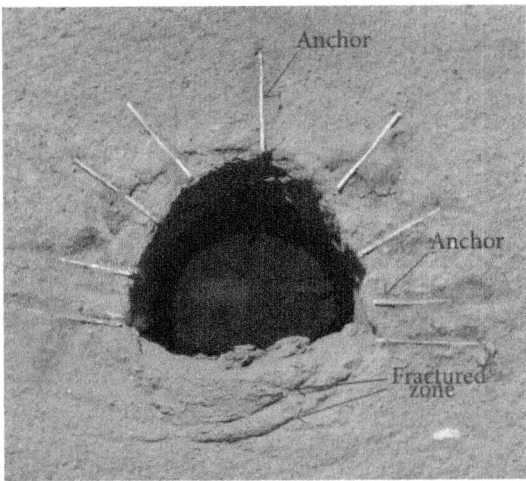

(b) The anchored half model

Figure 18: Failure distribution of the surrounding rock mass after tunnel excavation.

The zonal disintegration phenomenon of alternating fractured zone and intact zone occurs in nonanchoring model. There are 4 fractured zones and 4 intact zones arranging in space which are in accordance with the fractured shape of the in-situ observation in DINGJI coal mine. There are 2 obvious fracture lines in the bottom of the anchored model. Zonal disintegration also occurs in the invert of the model. But there is not any fracture line occurs in the arc crown and side walls; that is, zonal disintegration phenomenon did not occur in the anchored part of model.

After comparing with the model of anchored and nonanchoring, it indicates that the anchoring effect reinforces the anchoring area of model to be a unity. This impact makes zonal disintegration difficult to occur.

Analysis of the Anchoring Effect on Zonal Disintegration

Under high geostress conditions, the surrounding rocks near the cavern wall yield to plasticity. The principal stress field in the plastic and the elastic zones is as follows (Figure 19). As is shown, the tangential stress is the maximum principal stress. At the location of $r=R_p$, the tangential stress is the summit value.

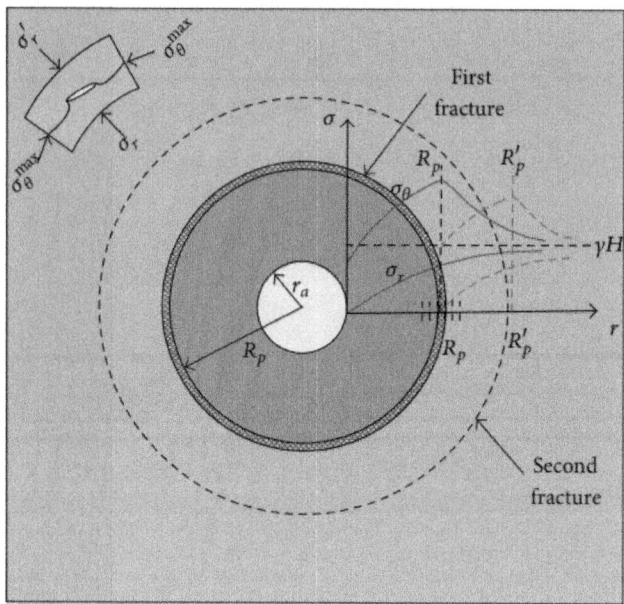

Figure 19: Elastoplastic geostress fields in surrounding rock mass and forming process of zonal disintegration.

According to Griffith's criterion, a fracture occurs when the UCS of rocks under pressure reaches the threshold value. Fairhurst and Cook (1966) [33] indicated that microcracks would initiate and extend in the direction of the maximum principal stress when compressive stress reaches Griffith's strength σ_s. This finding explains the longitudinal splitting of the rock specimen and the slabbing of the surrounding rocks in the rectangular openings, which also holds true for the circular cavern. Given a unit rock in the system of the polar coordinates (Figure 19), the second circular fracture extends towards the direction of the maximum principal stress (i.e., the tangential direction) when stress fulfills its relationship with the mechanical parameters of the surrounding rocks. The fracture is expected to transfix and form the circular fracture (i.e., the first fracture of the zonal disintegration) when compressive loading is sufficiently large. Fracture formation causes geostress redistribution in the surrounding rocks, inducing the formation of other ultimate equilibrium plastic zones. The second circular fracture occurs when the summit tangential stress is sufficiently large. Zonal disintegration occurs during the process cycle (Figure 19).

In the condition of high axial geostress, the surrounding rock mass has the trend of zonal disintegration. When model is anchored, it is reinforced and the threshold value of fracture is enlarged. So, at the same geostress, the zonal disintegration did not occur. The model test indicates the trend of zonal disintegration phenomenon existing under a condition of high axial geostress. And the reinforcement of anchor suppresses the zonal disintegration in the anchored area. During the anchor working, the trend of zonal disintegration which induces it shows the tension and compression alternation.

CONCLUSION

(1)   Under high axial geostress, zonal disintegration phenomena occur in the nonanchored area, while it did not occur in the fully anchored area. It indicates that anchoring suppresses the zonal disintegration obviously.

(2)   The radial strain of surrounding rocks displays the fluctuation state where wave crest and rough arrange in interval, both in anchored and nonanchoring area. It indicates that the deep rock masses have the trend of zonal disintegration under the high axial geostress.

(3)   The tension and compression alternation phenomenon of anchor is observed during the tunnel excavation procedure. This special phenomenon is different from the shallow buried tunnel. It indicates that the trend of zonal disintegration exists in the deep tunnel. The reinforcement of anchor suppresses the occurrence of zonal disintegration. During the work of anchor, it shows the alternation of tension and compression. It indicates that the mechanical behavior of the surrounding rocks is nonmonotonic.

The anchor suppresses the growth of zonal disintegration, so the optimization of parameters and disposal of anchor are of great importance to the stability of deep tunnel. The further study is needed to conduct in the future.

ACKNOWLEDGMENTS

This work was supported by the National Natural Science Foundations of China (Grant nos. 51209074 and 41172268), China Postdoctoral Science Foundation (Grant nos. 2012M511189 and 2013T60494), and the Fundamental Research Funds for the Central Universities (Grant no. 2012B02714), supported by State Key Laboratory for Geomechanics and Deep Underground Engineering, China University of Mining & Technology, under Grant SKLGDUEK1206, the Open Research Fund of State Key Laboratory of Geomechanics and Geotechnical Engineering, and the Institute of Rock and Soil Mechanics, Chinese Academy of Sciences, under Grant no. Z012008, and funded by CRSRI Open Research Program CKWV2012306/KY, by the Key laboratory of coal-based CO_2 capture, and geological storage Open Research Program 2012KF08. The authors are deeply grateful for these supports.

REFERENCES

1. E. I. Shemyakin, G. L. Fisenko, M. V. Kurlenya et al., "Zonal disintegration of rocks around underground workings, part II: rock fracture simulated in equivalent materials," Soviet Mining Science, vol. 22, no. 4, pp. 223–232, 1986.

2. Q. H. Qian, "The characteristic scientific phenomena of engineering response to deep rock mass and the implication of deepness," Journal of East China Institute of Technology, vol. 27, no. 1, pp. 1–5, 2004.

3. G. R. Adams and A. J. Jager, "Etroscopic observations of rock fracturing ahead of the stope faces in deep-level gold mines," Journal of the South Africa Institute of Mining and Metallurgy, vol. 21, no. 2, pp. 115–127, 1980.

4. E. I. Shemyakin, G. L. Fisenko, M. V. Kurlenya et al., "Zonal disintegration of rocks around underground workings, part 1: data of in situ observations," Soviet Mining Science, vol. 22, no. 3, pp. 157–168, 1986.

5. E. I. Shemyakin, G. L. Fisenko, M. V. Kurlenya et al., "Zonal disintegration of rocks around underground mines, part III: theoretical concepts," Soviet Mining Science, vol. 23, no. 1, pp. 1–6, 1987.

6. E. I. Shemyakin, M. V. Kurlenya, V. N. Oparin et al., "Zonal disintegration of rocks around underground workings. IV. Practical applications," Soviet Mining Science, vol. 25, no. 4, pp. 297–302, 1989.

7. E. J. Sellers and P. Klerck, "Modelling of the effect of discontinuities on the extent of the fracture zone surrounding deep tunnels," Tunnelling and Underground Space Technology, vol. 15, no. 4, pp. 463–469, 2000.

8. D. F. Malan and S. M. Spottiswoode, "Time-dependent fracture zone behavior and seismicity surrounding deep level stopping operations," in Rockbursts and Seismicity in Mines, S. J. Gibowicz and S. Lasocki, Eds., pp. 173 177, A. A. Balkema, Rotterdam, The Netherlands, 1997.

9. X. Zhou, Q. Qian, and B. Zhang, "Zonal disintegration mechanism of deep crack-weakened rock masses under dynamic unloading," Acta Mechanica Solida Sinica, vol. 22, no. 3, pp. 240–250, 2009.

10. J. Gu, L. Gu, A. Chen, J. Xu, and W. Chen, "Model test study on mechanism of layered fracture within surrounding rock of tunnels in deep stratum," Chinese Journal of Rock Mechanics and Engineering, vol. 27, no. 3, pp. 433–438, 2008.

11. Y. Pan, Y. Li, X. Tang, and Z. Zhang, "Study on zonal disintegration of rock," Chinese Journal of Rock Mechanics and Engineering, vol. 26,

supplement 1, pp. 3335–3341, 2007.

12. X. Tang, Y. S. Pan, and M. T. Zhang, "Mechanism analysis of zonal disintegration in deep level tunnel,"Journal of Geological Hazards and Environment Preservation, vol. 17, no. 4, pp. 80–84, 2006.

13. L. S. Metlov, A. F. Morozov, and M. P. Zborshchik, "Physical foundations of mechanism of zonal rock failure in the vicinity of mine working," Journal of Mining Science, vol. 38, no. 2, pp. 150–155, 2002.

14. S. Li, Q. Qian, D. Zhang, and S. Li, "Analysis of dynamic and fractured phenomena for excavation process of deep tunnel," Chinese Journal of Rock Mechanics and Engineering, vol. 28, no. 10, pp. 2104–2112, 2009. ·

15. M. A. Guzev and A. A. Paroshin, "Non-Euclidean model of the zonal disintegration of rocks around an underground working," Journal of Applied Mechanics and Technical Physics, vol. 42, no. 1, pp. 131–139, 2001.

16. M. Wang, H. Song, D. Zheng, and S. Chen, "On mechanism of zonal disintegration within rock mass around deep tunnel and definition of 'deep rock engineering'," Chinese Journal of Rock Mechanics and Engineering, vol. 25, no. 9, pp. 1771–1776, 2006.

17. M.-Y. Wang, P.-X. Fan, and Z. K. Guo, "Elastoplastic model for discontinuous shear deformation of deep rock mass," Journal of Central South University of Technology, vol. 18, no. 3, pp. 866–873, 2011.

18. Y. He, B. Jiang, L. Han, P. Shao, and H. Zhang, "Study of intermittent zonal fracturing of surrounding rock in deep roadways," Journal of China University of Mining and Technology, vol. 37, no. 3, pp. 300–304, 2008. ·

19. X. Zhou and Q. Qian, "Zonal fracturing mechanism in deep tunnel," Chinese Journal of Rock Mechanics and Engineering, vol. 26, no. 5, pp. 877–885, 2007. ·

20. X. Zhou, Q. Qian, and H. Yang, "Effect of loading rate on fracture characteristics of rock," Journal of Central South University of Technology, vol. 17, no. 1, pp. 150–155, 2010.

21. X. P. Zhou, H. F. Song, and Q. H. Qian, "Zonal disintegration of deep crack-weakened rock masses: a non-Euclidean model," Theoretical and Applied Fracture Mechanics, vol. 55, no. 3, pp. 227–236, 2011.

22. Q. Qian, X. Zhou, and E. Xia, "Effects of the axial in situ stresses on the zonal disintegration phenomenon in the surrounding rock masses around a deep circular tunnel," Journal of Mining Science, vol. 48, no. 2, pp. 276–285, 2012.

23. X. P. Zhou, G. Chen, and Q. H. Qian, "Zonal disintegration mechanism of cross-anisotropic rock masses around a deep circular tunnel," Theoretical and Applied Fracture Mechanics, vol. 57, no. 1, pp. 49–54, 2012.

24. X. P. Zhou and J. Bi, "Zonal disintegration mechanism of cross-anisotropic rock mass around a deep circular tunnel under dynamic unloading," Theoretical and Applied Fracture Mechanics, vol. 60, no. 1, pp. 15–22, 2012.

25. V. N. Reva and E. A. Tropp, "Elastoplastic model of the zonal disintegration of the neighborhood of an underground working," in Physics and Mechanics of Rock Fracture as Applied to Prediction of Dynamic Phenomena (Collected Scientific Papers), pp. 125–130, Mine Surveying Institute, Saint Petersburg, Russia, 1995.

26. Y. L. Tan, J. G. Ning, and H. T. Li, "In situ explorations on zonal disintegration of roof strata in deep coalmines," International Journal of Rock Mechanics and Mining Sciences, vol. 49, pp. 113–124, 2012.

27. H. Wu, Z. Guo, Q. Fang, Y. Zhang, and J. Liu, "Mechanism of zonal disintegration phenomenon in enclosing rock mass around deep tunnels," Journal of Central South University of Technology, vol. 16, no. 2, pp. 303–311, 2009.

28. V. N. Odintsev, "Mechanism of the zonal disintegration of a rock mass in the vicinity of deep-level workings," Journal of Mining Science, vol. 30, no. 4, pp. 334–343, 1994.

29. B. V. Laptev and R. P. Potekhin, "Burst triggering by zonal disintegration of evaporites," Soviet Mining Science, vol. 24, no. 3, pp. 238–241, 1988.

30. E. Fumagalli, Statical and Geomechanical Models, Springer, New York, NY, USA, 1973.

31. E. Fumagalli, "Geomechanical models of dam foundation," in Proceedings of the International Colloquium on Physical and Geomechanical Models, Bergamo, Italy, March 1979.

32. Z. Fang, "Support principles for roadway in soft rock and its controlling measures," in Soft Rock Tunnel Support in Chinese Mines: Theory and Practice, H. E. Manchao, Ed., pp. 64–70, Coal Industry Publishing Press, Beijing, China, 1996 (Chinese).

33. C. Fairhurst and N. G. W. Cook, "The phenomenon of rock splitting parallel to the direction of maximum compression in the neighborhood of a surface," in Proceedings of the 1st Congress of the International Society for Rock Mechanics, pp. 687–692, Lisbon, Portugal, September 1966.

Chapter 6

PETROLOGICAL AND GEOCHEMICAL CHARACTERISTICS OF MAFIC GRANULITES ASSOCIATED WITH ALKALINE ROCKS IN THE PAN-AFRICAN DAHOMEYIDE SUTURE ZONE, SOUTHEASTERN GHANA

Prosper M. Nude[1], Kodjopa Attoh[2], John W. Shervais[3] and Gordon Foli[4]

[1]Department of Earth Science, University of Ghana, Legon-Accra, Ghana

[2]Department of Earth &Atmospheric Sciences, Cornell University, Ithaca, NY 14853, USA

[3]Department of Geology, Utah State University, Logan UT 84322, USA

[4]Department of Earth and Environmental sciences, University for Development studies, Navrongo Campus, Ghana

INTRODUCTION

Most alkaline complexes are characterized by the presence of a distinctive zone where alkaline emanations appear to affect the wall rocks and the contact zones with country rocks (Winter, 2001). Such alkaline solutions and magmas may be effective agents for transporting trace elements and modifying the compositions of the host rocks (Wallace & Green, 1988, Rudnick et al., 1993). As a result the primary minerals can be replaced by alkaline minerals such as nepheline and feldspar. In this way nepheline-bearing rocks and other metasomatic derivatives of variable compositions can arise (Dawson et al. 1990). In the Pan-African Dahomeyide suture zone in southeastern Ghana, variably deformed alkaline rocks, comprising nepheline syenite and carbonatitic rocks, referred to as the Kpong complex (KC), occur in tectonic contact with high-pressure (HP) mafic granulite rocks of garnet-pyroxeneamphibole composition (Nude et al., 2009). The Dahomeyide mafic granulites have been found to preserve geochemical imprints of island arc theoleiitic (IAT) basalts as well as rocks with N-MORB-like affinities (Agbossoumonde et al., 2001, Attoh & Morgan 2004). Thus the mafic granulites possess distinct geochemical signatures that differ significantly from the alkaline rocks. This paper presents petrological and geochemical data on the nepheline-bearing mafic rocks previously referred to as mafic nepheline gneiss (Holm, 1974) at the contact zone between the HP mafic granulites and the KC rocks.

The data are used to evaluate the distinctive mineralogical and trace element contents of the nepheline-bearing mafic rocks, and also infer the interactions of the alkaline magma with the mafic granulites at the contact zone.

REGIONAL GEOLOGICAL SETTING

The Dahomeyide orogen in southeastern Ghana and adjoining parts of Togo and Benin is the southern segment of the Pan-African Trans-Saharan belt (TSB). The TSB defines the eastern margin of the West African craton (WAC) and extends for over 2500 km from the Sahara to the Gulf of Guinea (Caby, 1987). The Pan-African orogen resulted in the assembly of northwest Gondwana (Hoffman, 1991; Cordani et al., 2003; Tohver et al., 2006). In southeastern Ghana and adjoining parts of Togo and Benin the Dahomeyide is interpreted to have resulted from easterly subduction after resorbtion of oceanic lithosphere at rifted margin of WAC (Affatton et al., 1991; Agbosoumonde et al., 2004, Attoh & Nude, 2008) with a preserved suture. These rocks are also exposed in the Amalaoulaou complex to the north in the Gourma fold and thrust belt in Mali (Berger et al., 2011) and shares comparable geochemical, metamorphic and tectonic evolution to the rocks of the Dahomeyides to the south in Benin, Togo and Ghana.

Figure1 is a geologic map of the Dahomeyide orogen in southeastern Ghana, and adjoining parts of southern Togo and Benin (Sylvain et al., 1986, Castaing et al., 1993; Attoh et al., 1997) showing the principal lithologies of the orogen. From the west is the deformed margin of the WAC that include 2.1 Ga granitoids (Agyei et al., 1987; Agbossoumonde et al., 2007), known as Ho gneisses, now deformed into proto-mylonites, and its cover rocks (Atacora nappes) occurring on the rifted passive margin. These are bounded to the east by distinctive high-pressure (HP) mafic granulite and eclogite facies rocks known locally as the Shai-Hill gneisses that form the suture zone unit (Attoh, 1998; Agbossoumonde et al., 2001; Attoh & Morgan, 2004) and mark the zone of collision of WAC with presumed exotic blocks to the east. Granitoids to the east of the suture zone comprise migmatites and dioritic gneisses which represent the arc terrane that is postulated to have formed during the subduction and accompanying oceanic closure.

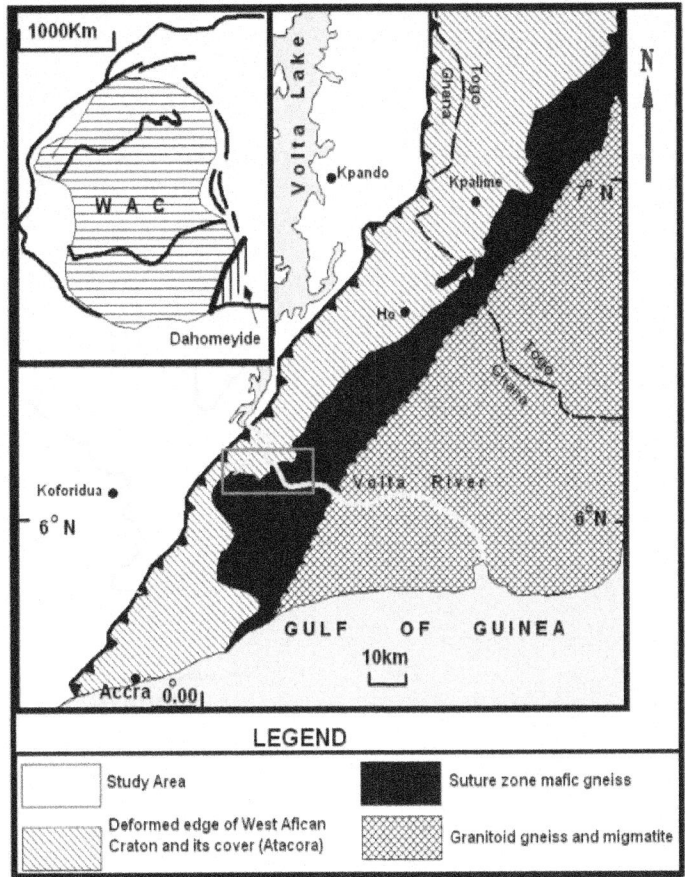

Figure 1: Tectonic map of the Dahomeyides in southeastern Ghana and its northern extension (After Attoh, 1998) showing the study area.

LITHOLOGICAL DISTRIBUTIONS AND PREVIOUS GEOCHRONOLOGICAL WORK

Lithological Distribution

The lithological distributions of the alkaline rocks in relation to the mafic granulite gneiss and other lithological units have been described by several workers including Holm (1974), Attoh et al. (2007), Nude et al. (2009), and the geology is shown in Figure 2. The alkaline rocks comprise alternating layers and interfolded units of nepheline syenite gneiss and carbonatite along the inferred sole thrust of the suture that separates the mafic granulite gneiss from rocks of the deformed edge of the WAC.

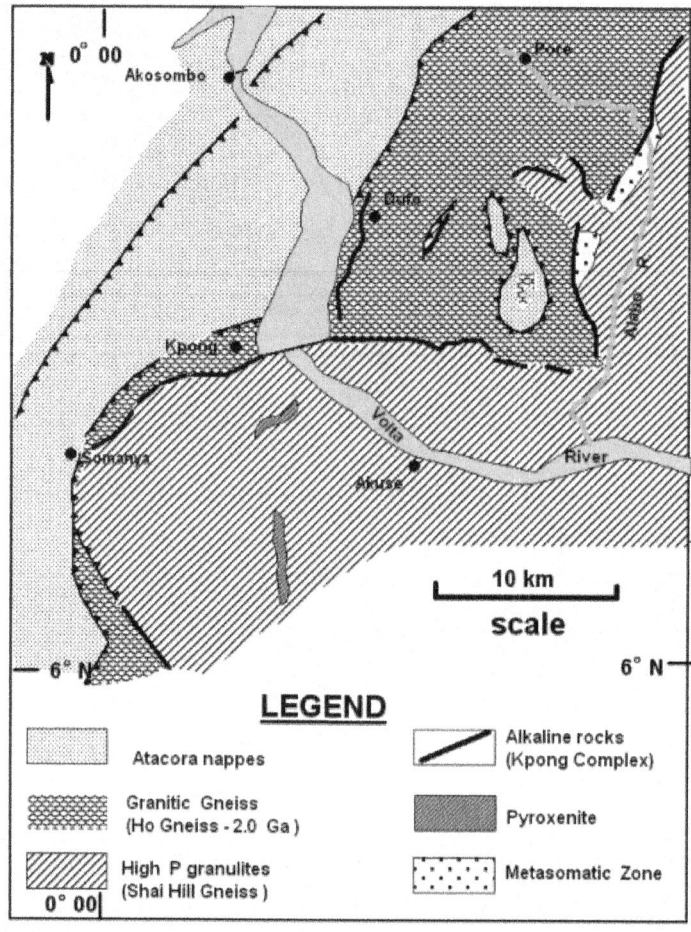

Figure 2: Geological map of the study area showing the lithological relatioships and the metasomatic zone where the samples were taken.

The nepheline-bearing mafic granulite which forms the basis of this study is a garnet-bearing rock that is restricted to the contact zone with the alkaline rocks and the Shai Hills gneisses. It occurs in isolated outcrops in the northeast of the area (Fig. 2) where it is typically folded with steep axial surfaces, subvertical hinge zones and asymmetrical limbs. Attoh et al. (1997) interpreted the structure of the suture zone to have resulted from early east–west compression, which produced the north– south imbricate thrust slices followed by NNW-directed thrusting.

Previous Geochronological Work

Geochronological studies of the suture zone mafic granulite gneisses (Shai Hill gneisses) and the alkaline rocks provide constraints on the chronology of the tectonic record of the area. U-Pb zircon ages determined from the mafic granulites from the suture zone in Ghana by Attoh et al. (1991) and interpreted as peak metamorphic age was 610 ± 2 Ma. Also Hirdes and Davis (2002) reported U-Pb zircon ages of 603 ± 5 Ma from the mafic granulites from the Shai Hills area which confirm the timing of peak metamorphism in the suture zone. Similar age of 613 ± 1 Ma from zircon evaporation ($^{207}Pb/^{206}Pb$) was reported by Affaton et al. (2000) for the suture zone rocks in northern Togo. Hornblende separates from the mafic granulites yielded $^{40}Ar/^{39}Ar$ ages between 587 and 567 Ma, interpreted as the time of exhumation of the nappes (Attoh et al., 1997). Thus taken together high pressure metamorphism of the suture zone rocks occurred around 603-613 Ma and exhumation through the hornblende ages around 580-570 Ma (Attoh et al., 2007). U-Pb ages on zircon separates determined by Bernard-Grifiths et al. (1991) from eclogite facies rocks from the suture zone in southern Togo had a discordant lower intercept of ~640 ± 53 Ma and Nd model ages (T_{DM}) of 1150 Ma (Bernard-Grifiths et al., 1991). Nd model age of 940 Ma was obtained by Attoh and Schmitz (1991) in the HP mafic granulites from the Shai Hills area in Ghana. The model ages suggest that the mantle derivation of the protoliths of these rocks may have occurred earlier. In the Amalaoulaou arc in Mali, the magmatic activity was found to have occurred at least c. 793 - 660 Ma followed by UHP metamorphism at c. 623 Ma (Berger et al., 2011).

Analyses of zircons separates from the carbonatite and the nepheline syenite gneiss samples yielded ages of 592-594 ± 4 Ma interpreted as the time of intrusion of the alkaline rocks (Nude et al., 2006). So the available age data suggest the emplacement of the alkaline rocks during syn-orogenic rifting, but this occurred after peak granulite metamorphism (Attoh et al., 2007). Overall therefore the alkaline rocks appear to have been emplaced later than the mafic granulites.

PETROGRAPHIC AND GEOCHEMICAL CHARACTERISTICS OF THE MAFIC GRANULITES IN THE SUTURE ZONE

The mafic granulites (Shai Hills gneiss) are variably sheared and deformed, and have a streaky appearance. The rocks are characterized by prominent modal layering consisting of alternating but discontinuous garnet-rich and hornblende—rich zones that are cut by veins of all sizes and orientations. The microstructural features of the Shai Hills gneisses have been described by Attoh and Nude (2008). Generally the rocks are composed of variable proportions of garnet, diopside pyroxene and scapolite. The following petrographic types have been identified by Attoh (1998): a) hornblende-rich granulite with typical modal compositions of 42% hornblende, 38 % plagioclase, 9 % garnet, 4% diopside and 5% quartz, and b) garnet-rich granulites that have similar mineral assemblage but with different mineral proportions of 29% garnet, 26% plagioclase, 20% diopside, 9% hornblende, 10% quartz and 2% scapolite. Geochemical features determined by Attoh and Morgan (2004) suggest that the mafic granulites have predominantly island arc tholeiite imprints with subordinate N-MORB signatures and trace element patterns that are very similar to lower crust compositions.

PETROGRAPHIC AND GEOCHEMICAL CHARACTERISTICS OF THE ALKALINE ROCKS

The alkaline rocks consist of nepheline syenite gneiss and carbonatite, and their petrographic features have been described by Holm (1974), Nude et al. (2009). The nepheline syenite gneiss is composed of nepheline (20–30%) which sometimes shows replacement by cancrinite, Other major phases are sodic feldspar (An0–An4, 30–50%), perthitic microcline and/or orthoclase (15–30%), annitic biotite (5–15%). Titanite is a widespread accessory constituent. Minor accessories include fine grained calcite, zircon, apatite, and muscovite. More syenitic varieties occur locally consisting essentially of albite, microcline, accessory biotite and nepheline. Modally, the carbonatite consists of coarse-grained mosaics of subhedral to euhedral equant calcite (35–50%) and annitic biotite (25–40%), with feldspar (albite and microcline/orthoclase, 5–20%) and nepheline (2–20%) and rare zircon. Common mineral phases such as calcite, nepheline, feldspar and biotite in the nepheline syenite gneiss and the carbonatite have similar compositions (Attoh & Nude, 2008; Nude et al., 2009). The calcites show homogeneous compositions; CaO concentrations fall within 49.07–57.36 wt% and they are enriched in Sr with SrO values up to 1.4 wt%. Nepheline in both rock suites is generally similar in composition; it is relatively sodium rich, and compositions fall within $Na_{6.0-8.1}K_{0.4-1.7}Al_{7.3-7.9}Si_{8.0-8.2}O_{32}$. K-feldspar in the rocks is almost pure orthoclase with over 94 mol% Or in the nepheline syenite. Plagioclase is essentially albite, and common in almost all samples with compositions from 78 to 99 mol% Ab in the carbonatite, 94–98 mol% Ab in some nepheline syenite gneiss samples, confirming the compositional similarities in both rock suites. Biotite from the rocks is generally annitic with the composition falling within $K_{1.8-1.9}Fe_{3.1-3.5}Mg_{1.2-1.4}Si_{5.2-5.3}Al_{3.1-3.4}O_{20}(OH,F_{0.1-0.4})$. Geochemically the alkaline rocks are characteristically enriched in alkalis ($Na_2O + K_2O$ is up to 16.4 wt %), Ba (3389-4665 ppm), Sr (3891-5481 ppm), Nb (78-135 ppm). The rocks show strong LREE fractionations and large deletions of Zr and Hf relative to primitive mantle (Nude et al., 2009). Most carbonatite and related rocks worldwide are known to have these geochemical features (Potter, 1996; Nelson et al., 1988; Woolley & Kemp, 1989; HornigKjarsgaard, 1998; Bell & Tilton, 2001; Thompson et al., 2002; Chakhmouradian et al., 2007).

PETROGRAPHY OF THE MAFIC GRANULITES IN THE METASOMATIC ZONE

Representative samples of the mafic granulites analyzed in this study were taken from the metasomatic zone (Fig. 2). Generally these rocks which were previously mapped as mafic nepheline gneiss (Holm, 1974, Kesse, 1985) are found in isolated outcrops as a dense, foliated rock close to the alkaline rocks. The dark colour, coarse texture and significant modal content of garnet and pyriboles make the mafic granulite gneiss conspicuous in the bluish-gray nepheline gneiss and the dark-grey carbonatite. The rock contains feldspar and nepheline rich veinlets in the shear zone. Major modal compositions are variable and are composed of garnet (10-25 vol. %), sodic plagioclase (~30 vol. %), microcline (~15 vol. %), nepheline (~20 vol. %), aegirine–augite (~35 vol. %), ferro-pargastite amphibole (10-30 vol. %), coarse titanite (~5 vol. %). The feldspars are generally coarse but in some of the crystals they occur as equigranular, granoblastic and interstitial grains. Accessory constituents include calcite, mostly found in cleavage cracks, zircon and rare kaersutite.

Composition of Common Mineral Phases in the Mafic Granulites from the Metasomatic Zone and the Alkaline Rocks

The common mineral phases in the mafic granulites from the metasomatic zone and the alkaline rocks are calcite, nepheline, and feldspar. The compositions of these mineral phases were determined from representative samples of the mafic granulites with the objective of comparing their chemical contents with those from the alkaline rocks determined from previous studies by Attoh and Nude (2008) and then Nude et al. (2009). This will provide an insight into the extent of similarities in these common phases in the adjacent rocks. Two representative samples PN32A and PN56 which represent the variability of the compositional phases were selected for phase chemistry analysis. The mineral chemistry analysis was done using a Cameca SX-50 electron microprobe at the University of Utah. The minerals were tentatively identified using energy dispersive spectrometry (EDS). Table 1 lists the results of the microprobe analysis.

Calcite

Calcite is the only carbonate in the rocks; CaO contents range from 51.0 – 53.8 wt %. The totals of the major element concentrations are limited and fall within 55-58 wt % excluding volatiles and. The mineral is characteristically Sr-rich, with values within 1.3- 1.5 wt %.

Nepheline

Nepheline compositions in the rocks are variable, but a key feature is that it is Na-rich, and the variable compositions fall within $Na_{2.9-6.0}K_{0.0-1.7}Al_{4.2-8.2}Si_{8.0-11.8}O_{32}$. Two varieties of the nepheline have been recognized from the samples (Table 1b). The first variety is relatively SiO_2-rich and Al_2O_3-poor. This type is also relatively low in alkalis especially K_2O. The second type is relatively poor in SiO2, but has high contents of Al_2O_3 and Na (Table 1b).

Feldspar

Feldspar compositions are also variable within the samples. The mineral is present as twofeldspar components, comprising albite and orthoclase in some samples (PN 32A, Table 1c), with representative compositions of 21-32 mol% Ab and 67-78 mol% Or, or as single feldspar comprising almost pure albite with composition of 96-99 mol% Ab (PN 56, Table 1c).

Table 1: Representative compositions of calcite, nepheline and feldspar in the mafic granulites from the metasomatic zone

Sample:	PN 32A		PN 56	
Analyses no:	1	2	3	4
(a) Calcite				
FeO	0.23	0.26	0.21	0.28
MnO	0.39	0.3	0.45	4.58
MgO	0.02	0.06	0.03	0.01
CaO	53.82	53.3	53.3	51.96
SrO	1.28	1.5	1.33	1.39
Total	55.74	55.42	55.32	58.22
Mg#	15.5	28.3	20.9	4.8

Sample:	PN 32A		PN 56		
Analyses no:	1	2	1	2	3
(b) Nepheline					
SiO_2	67.9	65.33	42.69	42.17	42.31
Al_2O_3	20.77	20.37	34.94	34.75	35.17
FeO	0	0.02	0.15	0.04	0.18
CaO	0.54	0.41	0.46	0.69	0.61
Na_2O	11.33	8.5	15.7	15.5	16.06
K_2O	0.08	3.26	6.76	6.51	6.41
Total	100.62	97.89	100.7	99.66	100.74
Si	11.8	11.782	8.165	8.143	8.095
Al	4.255	4.329	7.878	7.908	7.93
Fe	0	0.003	0.024	0.007	0.028
Ca	0.101	0.079	0.094	0.142	0.125
Na	3.819	2.973	5.821	5.801	5.958
K	0.017	0.749	1.65	1.603	1.564
Total	19.992	19.915	23.632	23.604	23.7

Sample:	PN 32A		PN 56	
Analyses no:	1	2	1	2
(c) Feldspar				
SiO_2	67.07	68.05	62.16	61.29
Al_2O3	20.92	20.23	20.38	20.42
FeO	0	0.21	0	0.16
CaO	0.61	0.15	0.1	0.05
Na_2O	11.35	11.79	3.24	2.18
K_2O	0.12	0.07	10.43	12.12
BaO	0	0	3.51	3.87
Total	100.07	100.5	99.82	100.09
Si	2.934	2.964	2.897	2.88
Al	1.079	1.039	1.119	1.131
Fe	0	0.008	0	0.006
Ca	0.029	0.007	0.005	0.003
Na	0.963	0.995	0.293	0.199
K	0.006	0.004	0.62	0.727
Ba	0	0	0.064	0.071
Total	5.011	5.017	4.998	5.017
Mol% An	2.9058	0.6958	0.5447	0.3229
Mol% Ab	96.493	98.9066	31.9172	21.4209
Mol% Or	0.6012	0.3976	67.5381	78.2562

Total Fe as FeO

A notable feature in the mafic granulites from this study is that calcite, nepheline and feldspars are similar in their compositions to those from the alkaline rocks, with nepheline and feldspars showing similar variability as in the alkaline rocks (Nude et. al., 2009). These comparable features suggest mineralogical influence of the alkaline rocks on the mafic granulites.

GEOCHEMISTRY

Analytical Methods

Whole rock samples were analyzed from representative samples for 10 major elements (SiO_2, TiO_2, Al_2O_3, total Fe as Fe_2O_3*, MnO, MgO, CaO, Na_2O, K_2O,

P_2O_5) and 12 trace elements (Nb, Zr, Y, Sr, Rb, Zn, Cu, Ni, Cr, Sc, V, Ba) at Utah State University, and the analytical techniques have been described by Nude et al. (2009). The analysis was carried on Philips 2400 X-ray fluorescence spectrometer using pressed powders for both major and trace elements, with selected U.S.G.S. and international standards prepared identically to the samples. Accepted concentrations were taken from the compilation of Potts et al. (1992). Matrix corrections were carried out within the Philips SuperX software package, which uses the fundamental parameters approach (Rousseau, 1989) to calculate theoretical alpha coefficients for the range of standards. Replicate analyses of selected standards as unknowns suggest percent relative errors ≈1% for silica, ≈2-4% for less abundant major elements, and ≈1-6% for trace elements. The concentrations of rare earth elements (REE) and other trace elements in whole rock samples were determined using Perkin–Elmer 6000 Inductively Coupled Plasma Mass Spectroscopy (ICP-MS) at Centenary College, Shreveport, Louisiana, with acid digestion techniques. Standard reference samples were used in the quantitative analyses of the elements. Table 2 shows the major and trace elements concentrations in the representative samples.

Major Elements

The representative samples of the mafic granulites have SiO_2 contents in the range of 35.0 and 52.0 wt% while CaO contents are from 8.0 to 24.0 wt%. Al_2O_3 contents range from 12.9 to 17.2 wt%; Fe_2O_3 total values range from 7.9 to 10.5 wt% whereas TiO_2 and P_2O_5 are from 1.2 to 1.8 and 0.5 to 0.7 wt% respectively. The total alkalis ($Na_2O + K_2O$) contents are relatively high, with values ranging from 9.7 to 14.1 wt%. Figure 3 are Harker plots in which selected major elements concentrations and total alkalis compositions in the metasomatic mafic granulites are compared to that of the alkaline rocks. The data for the alkaline rocks are from Nude et al. (2009). Apart from K_2O the other major elements from the mafic granulites display linear trends with those from the alkaline rocks. The deviation of K_2O from this trend is not surprising because it is much more mobile and susceptible to alteration. From the present data the linear trends suggest mechanical mixing of the rocks rather than fractional crystallization which can also show linear trend. Attoh and Morgan (2004) carried out geochemical investigations of the mafic granulites which they sampled from nearby the areas where the present study was carried out, but outside the metasomatic zone, specifically to the east and south of the zone. The following major element ranges (wt %) were reported by these authors: SiO_2 = 42.4-52.0, TiO_2 = 0.9-3.4, Al_2O_3 = 8.6-18.9, Fe_2O_3 total = 6.3- 6.8, MgO = 4.4-11.6, CaO = 7.7–11.1, Na_2O = 1.52- 4.36 + and K_2O = 0.01– 0.57.

Their major element results appear similar to those obtained in the present study; exceptions are Fe2O3total,, CaO, the alkalis, Na$_2$O and K$_2$O, which are relatively enriched in the rocks from the metasomatic zone compared to those obtained by Attoh and Morgan (2004).

Table 2: Major and trace element concentrations in the mafic granulites from the metasomatic zone

	PN-32A	PN-36	PN-39	PN-42	PN-46	PN-55	PN-56	PN-61	PN-63
SiO$_2$	35.98	50.03	49.2	38.2	39.33	43.7	42.46	38.4	50.48
TiO$_2$	1.47	1.37	1.32	1.2	1.31	1.88	1.83	1.32	1.22
Al$_2$O$_3$	12.94	17.07	17.11	14.08	14.93	17.23	18.53	14.34	17.21
Fe$_2$O$_3$	10.86	8.2	9.41	9.54	9.42	12.34	10.52	9.93	7.92
MnO	0.30	0.28	0.35	0.29	0.29	0.36	0.36	0.31	0.27
MgO	3.05	1.01	1.52	3.71	2.43	1.82	1.48	2.49	0.96
CaO	23.85	8.48	8.01	19.59	17.8	10.25	9.55	19.38	8.36
Na$_2$O	4.32	8.53	8.95	6.62	7.49	8.24	10.13	6.94	8.64
K$_2$O	5.25	3.89	2.95	5.14	5.32	2.92	3.97	5.22	3.82
P$_2$O$_5$	0.77	0.55	0.64	0.66	0.64	0.58	0.50	0.63	0.56
Total	98.79	99.41	99.46	99.03	98.96	99.32	99.33	98.96	99.44
Mg#	35.8	19.7	24.2	43.5	33.8	22.6	21.8	33.2	19.3
ppm									
Nb	90	256	261	124	146	157	216	144	245
Zr	339	398	365	295	324	266	257	267	352
Y	37	34	36	28	32	46	40	30	36
Sr	3815	2045	1574	3562	3366	1585	1007	3433	1996
Rb	139	82	53	157	139	77	117	137	83
Sc	38	17	16	32	24	16	13	31	14
V	158	60	115	162	191	155	126	199	53
Cr	57	31	21	46	27	33	1	27	44
Ni	12	3	1	23	6	1	1	7	2
Cu	23	4	5	8	17	4	1	16	6
Zn	96	105	111	101	86	87	77	92	106
Ba	3783	1623	1786	2769	2899	2812	2789	2924	1647
La	141.26	124.45	112.01	140.33	114.2	153.2	138.07	146.92	149.81
Ce	258.3	256.77	230.63	222.26	203.32	290.48	275.09	233.75	280.12
Pr	26.46	29.49	25.35	24.28	18.79	31.57	25.17	22.33	34.11
Nd	84.44	99.03	82.34	76.93	59.1	98.99	75.28	68.09	113.59
Eu	3.93	5.03	3.97	3.41	2.84	4.07	2.71	3.32	5.7
Sm	13.33	15.63	12.81	11.59	9.11	13.78	9.63	10.08	17.44
Gd	10.56	11.46	9.53	8.46	6.83	10.01	7	7.42	13.4
Tb	1.26	1.48	1.23	1	0.82	1.29	0.82	0.92	1.72
Dy	6.65	7.35	6.36	4.86	4.15	6.96	4.23	4.42	8.36
Ho	1.25	1.32	1.2	0.89	0.73	1.35	0.87	0.86	1.5
Er	3.2	3.29	2.95	2.25	1.72	3.71	2.37	2.13	3.84
Tm	0.43	0.43	0.41	0.3	0.25	0.53	0.32	0.3	0.54
Yb	2.57	2.49	2.48	1.77	1.4	3.4	1.88	1.74	3.06
Lu	0.38	0.39	0.41	0.26	0.21	0.51	0.29	0.24	0.47
Hf	3.21	3.2	3.37	1.82	1.89	4.25	4.34	1.81	3.32
Ta	4.48	21.4	16.86	6.98	10.44	12.54	18.16	8.64	20.11
Pb	10.43	0.45	0.59	0.35	6.13	-0.76	-1.35	7.08	0.24
Th	4.17	10.99	6.89	4.88	3.77	13.45	12.11	3.62	12.73
U	1.75	4.45	1.55	1.38	2.07	2.13	2.45	1.54	4.2
Zr/Hf	105.6	124.4	108.3	162.19	171.4	62.6	59.2	147.5	106.0
Nb/Ta	20.1	11.0	15.5	17.8	13.0	12.5	11.9	16.7	12.2

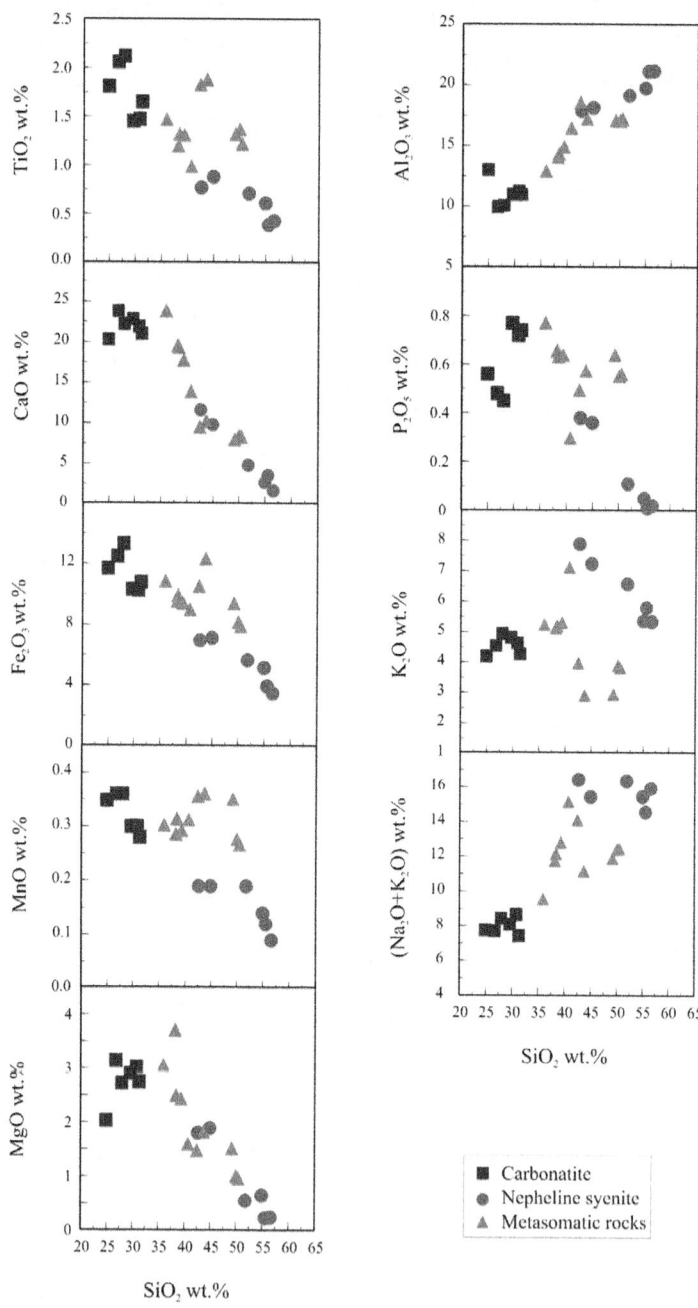

Figure 3: Harker plots comparing selected major element concentrations in the metasomatic mafic granulite rocks with the alkaline rocks (carbonatite and nepheline syenite). Data for the alkaline rocks are from Nude et al. (2009).

For example K$_2$O contents in the metasomatic rocks are several folds enriched (concentrations range from 2.95 to 5.25 wt %, Table 2) compared to the concentrations in the non-metasomatic varieties (0.01 – 0.57 wt %) determined by Attoh and Morgan (2004). Na$_2$O also shows similar enrichment in the metasomatic rocks (4.32-10.13 wt %) compared to the non-metasomatic varieties (1.52- 4.36 wt %). The overall major element contents show that the mafic granulites in the metasomatic zone are particularly alkaline, presumably from the addition of Na- and K-rich fluids from the adjacent alkaline rocks. The rocks are also evolved and contain variable amounts of CaO.

Trace Element Contents and Variations

The trace element contents of the metasomatic zone mafic granulites show high absolute values of Sr (1574-3815 ppm), Ba (1623-3783 ppm), Nb (90-256 ppm). The rocks also have Nb/Ta values ranging from 11-20 and very high Zr/Hf values of 59-171. The Nb/Ta values from the analysed samples compares with chondritic values of 17.6, but the Zr/Hf values are far higher than chondritic values of approximately 36 determined in most reservoirs of the silicate earth (Weyer et al., 2003; Potter, 1996). In Figure 4, Sr and Ba concentrations in the metasomatic mafic granulites are compared with those in the alkaline rocks determined by Nude et al. (2009). From the figure, Sr displays linear trend as is Ba, although 2 samples of the nepheline syenite and a sample of the metasomatic mafic granulites show anomalously high Ba values and deviate from the linear trend. The linear trend has also been shown in the selected major elements in Figure 3, confirming the possible mechanical mixing of those rocks.

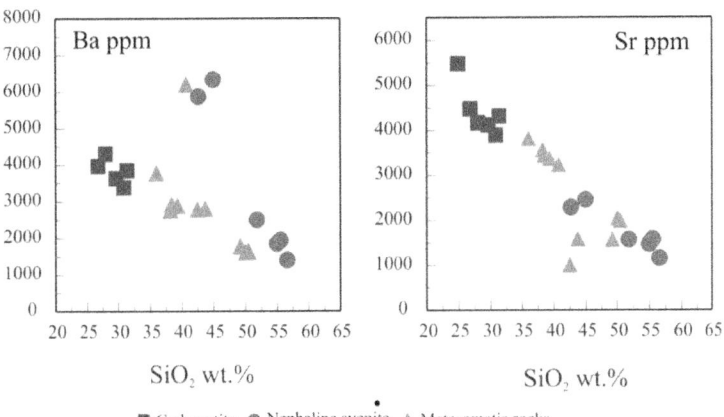

Figure 4: Ba and Sr concentrations in the metasomatic mafic granulite rocks compared with the alkaline rocks (carbonatite and nepheline syenite). Data for the alkaline rocks are from Nude et al. (2009).

Figure 5 is a primitive mantle-normalized incompatible elements plot of the metasomatic zone mafic granulites compared with alkaline rocks. The similarities of the alkaline rocks to the analyzed samples are shown especially in the relative depletions of the HREE, Rb Hf, Ti, Y and enrichment of K, Eu and Sm relative to Primitive Mantle. The analyzed rocks also differ from the alkaline rocks in elevated U, Ta, Nb, and Zr. Another difference is the prominent troughs at Th, U and Hf, Zr shown in the alkaline rocks; a feature not shown in the mafic granulites. Of particular interest is the relative fractionation of Zr from Hf in the analyzed samples.

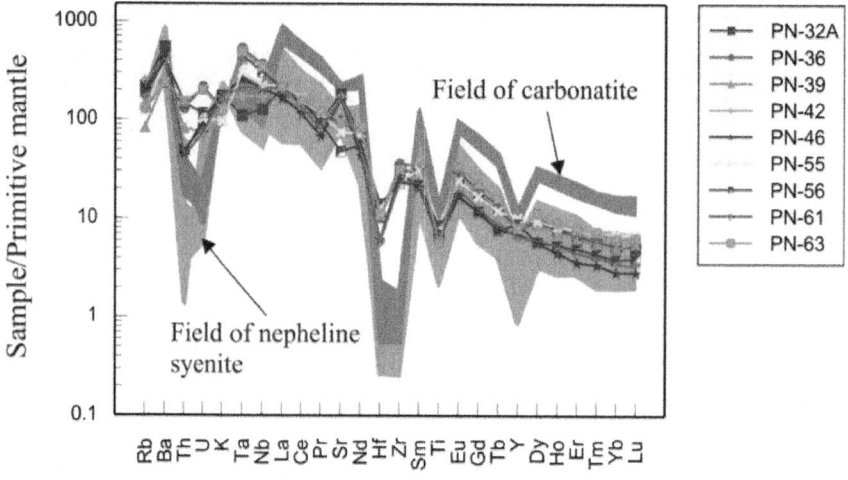

Figure 5: Primitive mantle-normalized concentrations in the mafic granulites from the metasomatic zone compared with the alkaline rocks (nepheline syenite and carbonatite). Data of the alkaline rocks are from Nude et al. (2009). Normalizing values are from McDonough et al. (1991).

In the REE plot (Fig. 6) the rocks show LREE enrichment, slight positive Eu anomaly and spread at the HREE end. However, the slight positive Eu anomaly is not shown in the carbonatite samples. Overall, the REE patterns of the metasomatic mafic granulites are similar to those of the alkaline rocks, particularly the nepheline syenite. Compared to the suture zone mafic granulites analyzed by Attoh and Morgan (2004) the trace element patterns shown by the rocks from the present study differ markedly. First is the LREE fractionation and steep REE pattern, second is the slight positive Eu anomaly and, third is the overall incompatible trace elements enrichments in the metasomatic granulites.

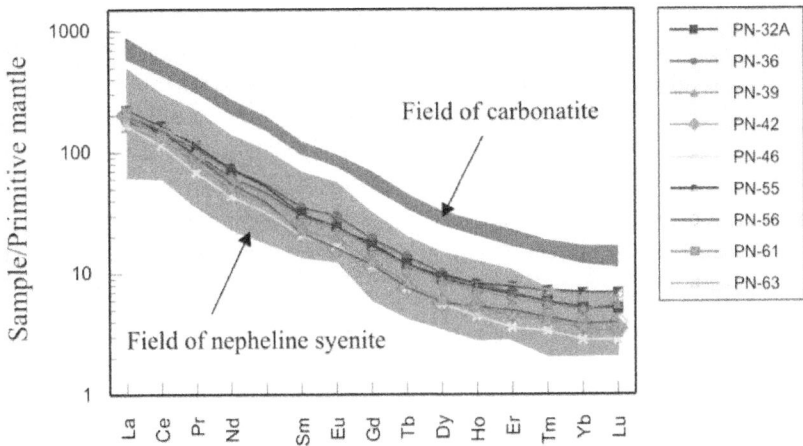

Figure 6: Primitive mantle-normalized REE plot for the mafic granulites from the suture zone compared with the alkaline rocks (nepheline syenite and carbonatite). Data of the alkaline rocks are from Nude et al. (2009). Normalizing values are from McDonough et al. (1991).

DISCUSSION

The new data presented in the present study on the modal contents, mineral chemistry and geochemical compositions of the mafic granulite gneiss at the contact zone between the alkaline rocks and the Shai Hills gneiss show that the rocks possess unique compositions which are suggestive of metasomatic transformation of the mafic granulites and hence the modification of the petrography and geochemistry of the original rocks. For example the ubiquitous presence of modal nepheline, feldspar and to a lesser quantity calcite in the mafic granulites is reflective of the mineralogy of the alkaline rocks. Additionally, the common mineral phases such as calcite, nepheline and feldspars in both rock suites have similar mineral chemistry. Partly, the Ca-Na- and K-rich fluids which formed these phases most likely emanated from the alkaline rocks. This evidence is supported by the comparable high CaO, Na_2O and K_2O contents in the mafic granulites under study to the alkaline rocks nearby, and suggests carbonate-alkali fluid interaction.

As shown from the present results the metasomatic mafic granulites are also enriched in Sr, Ba, Nb and show strong REE fractionation; these features are characteristic of carbonatitic melts. Most incompatible trace element concentration and patterns and the REE plots on the mantle-normalized diagrams also show similarities with the alkaline rocks. The observed very high Zr/Hf values in the rocks from the present study which is indicative of Hf fractionation is often associated with carbonate metasomatism (Dupuy et al., 1992) as the element pairs are expected to behave congruently in both fluids and melts (Jochum et al., 1986). Taken together, the major element compositions, the trace element contents and patterns constitute strong evidences to suggest the influence of the alkaline rocks on the overall modal compositions and bulk rock chemistry of the mafic granulites. The linear trends shown in the Harker plots and the spread at the HREE end of the mantle-normalized plots are indicative of mechanical mixing of the rocks, although this requires further evidence to confirm.

However, the metasomatic mafic granulites from the present study preserve some textural, modal and geochemical features which are similar to the Shai Hill gneisses and which also make them different from the alkaline rocks in the area. These features are their coarse texture and presence of garnet, and pyriboles. On geochemistry the mafic granulites again differ in their relative enrichment of Th and U, and depletion of Hf relative to Zr in the spider plots. Importantly the slight positive Eu anomaly in the mantled-normalized plots is absent in the carbonatite. These features provide compelling evidence to suggest that the mafic granulites are alkaline facies of the Shai hills rocks; the trace element budget are likely to have resulted from alkaline fluid interactions.

Available age data show that the emplacement of the alkaline rocks postdates the formation of the mafic granulites (Attoh et al., 2007), so the alkaline rocks are strong candidates for the source of the trace elements and particularly LREE enrichment in the mafic granulites. Although the mechanism of the interaction of the alkaline fluids is not clear, it is possible that it could have resulted from deformation associated with the emplacement of the alkaline rocks through percolation of alkaline fluids and/or mechanical mixing of the rock suites. This hypothesis requires further testing using isotopic data. However, carbonatitic melts have been shown from experiments to be of low viscosity, and capable of separating from their source at low degree melt fractions, and can percolate wall rocks by low angle dihedral flow (Hunter and Mackenzie, 1989; Hammouda and Laporte, 2000). Thus the alkaline fluids are likely to have emanated from the carbonatite and the nepheline syenite into the mafic granulites.

Tectonic and Petrological Implications

The new geochemical data from this study also provide further constraints on the evolution of the alkaline rocks and associated mafic granulites exposed along the suture zone of the Pan African Dahomeyide orogen in West Africa. Evidence for the Pan African suture and high pressure metamorphism have been provided in the literature by several workers including Berger et al.,

(2011), Agbossoumonde, et al. (2001, 2004), Attoh (1998) and Caby, (1987). These data, together with geochemical data on the Shai Hills granulites (Attoh & Morgan, 2004) infer a lower crust chemical composition for the rocks. The latter authors argued that the Dahomeyide mafic granulites preserve chemical imprints of basaltic rocks with trace element compositions similar to those of lower continental crust. The postulation is that the suture zone mafic granulites represent the roots of Pan African volcanic arc that formed from subduction to great depths, followed by HP granulite facies metamorphism accompanied by partial melting and later thrusting along ductile shear zones to produce crystalline nappes along the margin of the West African craton (Attoh, 1998).

The new geochemical data from this study shows a slight positive Eu anomaly in the mantle-normalized trace element patterns in the metasomatic rocks along the suture zone. This feature is interesting because positive Eu anomaly in the Shai Hills rocks has not been previously reported. But evidence of positive Eu anomaly in mafic granulite terrains that formed from basaltic lower crust has been shown by Rudnick (1992). It therefore appears that the mafic granulites from this study preserve a geochemical characteristic that may be of lower crustal affinity, a feature consistent with the findings of Attoh and Morgan (2004). So its tectonic association with the alkaline rocks along the sole thrust of the suture could be partly responsible for the unique alkalic and trace element compositions, as from available age data the alkaline rocks formed possibly after peak granulite metamorphism related to the Pan African orogeny (Attoh et al., 2007).

CONCLUSIONS

We have provided petrological and geochemical data on the mafic granulites exposed at the contact zone with alkaline rocks of the Kpong complex in the Pan African Dahomeyide suture zone southeastern Ghana. The mafic granulites have been found to have distinct modal and geochemical compositions which are in many ways similar to the alkaline rocks and different from the other mafic granulites outside the contact zone. Some of these are the presence of nepheline-rich veinlets and calcite along cleavage cracks in the mafic granulites. Together with other features such as compositions of modal phases, namely, calcite, nepheline, and feldspar, enrichment of alkalis, Ba, Sr, Nb and LREE, very high Zr/Hf values and overall steep REE patterns observed in the mafic granulites provide compelling evidences which suggest interaction of the alkaline fluids with the mafic granulites at the contact zone.

From the present data the mafic granulites from this study which seem to have a precursor texture and mineralogy identical to the Shai Hill mafic granulite gneisses (Attoh, 1998, Attoh and Morgan 2004), and described by

Holms (1974) and Kesse (1985) as mafic nepheline gneiss, is an alkaline facies of the Shai Hills gneisses. The unique composition resulted from alkali fluid interaction from the carbonatitic and nepheline syenite along the tectonized zone. The degree and style of the alkali matasomatism may have varied because of the variable compositions of some the common mineral phases.

ACKNOWLEDGEMENTS

This collaboration also forms part of PhD research by PMN. Utah State University provided grants for sample analysis; University of Ghana supported field work for PMN whilst the International Student Exchange Programme provided travel grants. This work is dedicated to the memory of Kodjopa Attoh who passed on when this manuscript was being prepared. His untiring efforts in understanding the Dahomeyides of southeastern Ghana for the past thirty years or so provided the motivation for this work. Reviews by M. M. Ghazal and an anonymous journal reviewer greatly improved the manuscript and are very much appreciated.

REFERENCES

1.   Affaton, P. Kröner, A. & Seddoh, K.F., 2000. Pan-African granulite formation in the Kabye massif of northern Togo (West Africa): Pb-Pb zircon ages. International journal of Earth Science, 88, 778-790.

2.   Affaton, P., Rahaman, M.A., Trompette, R., & Sougy, J., 1991. The Dahomeyide orogen: Tectonothermal evolution and relationship with the Volta basin. In: The West African orogens and circum-Atlantic correlatives, Dallmeyer, R.D, Lecorche, J.P. (Eds.), 95-111 Springer, New York.

3.   Agbossoumonde , Y., Guillot , S. & Ménot , R. P. 2004. Pan-African subduction collision event evidenced by high-P corona in metanorites from Agou massif (southern Togo). Precambrian Research, 135, 1–25.

4.   Agbossoumonde, Y., Ménot, R.-P. & Guillot, S. 2001. Metamorphic evolution of Neoproterozoic eclogite from south Togo (West Africa). Journal of African Earth Sciences, 33, 227–244.

5.   Agbossoumonde, Y., Ménot, Pacquette J.L., Guillot, S, Yessoufou S., & Perrache C.,2007. Petrological and geochronological constraints on the origin of Palimé-Amlamé granitoids (South Togo, West Africa): A segment of the West African craton Palaeproterozoic margin reactivated during Pan-African collision. Gondwana Research, 12, 4750-488.

6.   Agyei, E.K., van Landewijk, J.E.J.M., Armstrong, R.L., Harakal, J.E., & Scott, K.L., 1987. Rb– Sr and K–Ar geochronometry of south-eastern

Ghana. Journal of African Earth Sciences, 6, 153–161.

7. Attoh, K., 1998. High-pressure granulite facies metamorphism in the Pan-African Dahomeyide Orogen, West Africa. Journal of Geology, 106, 236–246.

8. Attoh, K., & Morgan, J., 2004. Geochemistry of high-pressure granulites from the PanAfrican Dahomeyide orogen, West Africa: constraints on the origin and composition of lower crust. Journal of African Earth Sciences, 39, 201-208.

9. Attoh, K., & Nude, P.M. 2008. Tectonic significance of carbonatite and ultrahigh-pressure rocks in the Pan-African Dahomeyide suture zone, southeastern Ghana. In: The boundaries of the West African craton, Ennih, N., Liégeois, J. P (eds.), Geological Society of London Special. Publication, 297, 217-231.

10. Attoh, K., & Smith, M. D. 2005. Nd and Hf isotopic compositions of Pan-African highpressure mafic granulites. EOS Transactions, American Geophysical Union 86 (18) Joint Assembly Supplement V13B-02.

11. Attoh, K., Corfu, F., & Nude, P.M., 2007. U–Pb zircon age of deformed carbonatite and alkaline rocks in the Pan-African Dahomeyide suture zone, West Africa. Precambrian Research, 155, 251–260.

12. Attoh, K., Dallmeyer, R.D., & Affaton, P., 1997. Chronology of nappe assembly in the PanAfrican Dahomeyide orogen, West Africa: evidence from 40Ar/39Ar mineral ages. Precambrian Research, 82, 135–171.

13. Attoh, K., Hawkins, D., Bowring, S., & Allen, B., 1991. U-Pb zircon ages of gneisses from the Panafrican Dahomeyide Orogen, West Africa. EOS Transactions, American Geophysical Union, 72, 229.

14. Bell, K., & Tilton, G.R., 2001. Nd, Pb and Sr isotopic compositions of east African Carbonatites: evidence for mantle mixing and plume inhomogeneity. Journal of Petrology, 42, 1927–1945.

15. Berger, J., Caby, R., Liégeois, J-P., Mercier, C. J-C., & Demaiffe, D., 2011. Deep inside a neoproterozoic intra-oceanic arc: growth, differentiation and exhumation of the Amalaoulaou complex (Gourma, Mali). Contributions to Mineralogy and Petrology. DOI: 10.1007/s00410-011-0624-5.

16. Bernard-Grifiths, J., Peucat, J. J., & Menot, R. P., 1991. Isotopic (Rb-Sr, U-Pb, and Sm-Nd) and trace element geochemistry of eclogites from the Pan-African belt: a case study of REE fractionation during high grade metamorphism. Lithos, 27, 43-57.

17. Caby, R., 1987. The Pan-African belt of West Africa from the Sahara to the Gulf of Guinea. In: Anatomy of Mountain Ranges, Schaer, J.P.,

Rodgers, J. (Eds.), 129-170. Princeton University Press,.

18. Castaing, C., Triboulet, C., Feybesse, J-L., & Chevrement, P., 1993. Tectonometamorphic evolution of Ghana, Togo, and Benin in the light of the Pan-African/Brasiliano orogeny. Tectonophysics, 218, 323-342.

19. Chakhmouradian, A. R., Mumin, A. H., Deméy, A., & Elliott, B., 2007. Postorogenic carbonatites at Eden Lake, Trans-Hudson Orogen (northern Manitoba, Canada): geological setting, mineralogy and geochemistry Lithos. doi:10.1016/j.lithos.2007.11.004.

20. Cordani, U.G., D'Agrella-Filho, M.S., Brito-Neves, B. B., & Trindale, I.F., 2003. Tearing up Rodinia: the Neoproterozoic paleogeography of South American cratonic fragments. Terra Nova, 15, 350-359.

21. Dawson, J. B., Penkerton H., Norton G. E., & Pyle D. M., 1990. Physicochemical properties of alkali carbonatite lavas: Data from the 1988 eruption of Oldoinyo Lengai, Tanzania, Geology, 18, 260-263.

22. Dupuy, C., Liotard, J. M., & Dostal, J. 1992. Zr/Hf fractionation in intraplate basaltic rocks: carbonate metsaomatism in the mantle source. Geochimica et Cosmochimica Acta 56, 2417-2423.

23. Hammouda, T., Laporte, D., 2000. Ultrafast mantle impregnation by carbonatite melts, Geology 28, 283–285.

24. Hirdes, W. & Davis, D. W. 2002. U–Pb zircon and rutile metamorphic ages of the Dahomeyan garnet– hornblende gneiss in southeastern Ghana, West Africa. Journal of African Earth Sciences, 35, 445–449.

25. Hoffman, P.F., 1991. Did the breakout of Laurentia turn Gondwana inside-out? Science, 252, 1409–1412.

26. Holm, F.R., 1974. Petrology of alkalic gneiss in the Dahomeyan of Ghana. Geological Society of America Bulletin, 85, 1441–1448.

27. Hornig-Kjarsgaard, I., 1998. Rare earth elements in sovitic carbonatites and their mineral phases. Journal of Petrology, 39, 2105–2121.

28. Hunter, R.H., MacKenzie, D., 1989. The equilibrium geometry of carbonate melts in rocks of mantle composition. Earth and Planetary Science Letters, 92, 347–356.

29. Jochum, K. P., Seufert, H. M., Spettel, B., & Palme, H., 1986. The solar system abundances of Nb, Ta, and Y, and the relative abundances of refractory lithohpile elements in differentiated planetary bodies. Geochemica et Cosmochemica Acta 50, 1173–1183.

30. Kesse, G. O., 1985. The mineral and rock resources of Ghana. A.A. Balkema Publishers,ISBN 9061915899, Rotherdam.

31. McDonough, W. F., Sun, S., Ringwood, A. E., Jagoutz, E., & Hofmann,

A. W. 1991, K, Rb, ans Cs in the earth and moon and the evolution of the earth's mantle, Geochimica et Cosmochimca Acta, Ross Taylor Symposium volume.

32. Nelson, D. R., Chivas, A R., Chapell, B. W., & McCulloch, M T., 1988. Geochemical and isotopic systematics in carbonatites and implications for the evolution of ocean island sources. Geochemica et Cosmochemica Acta, 52, 1–17.

33. Nude P.M. ,Shervais J., Attoh K., Vetter S. K., & Barton C., 2009 Petrology and geochemistry of nepheline syenite and related carbonate-rich rocks in the Pan- African Dahomeyide orogen, southeastern Ghana, West Africa. Journal of African Earth Sciences, 55, 147-157.

34. Nude, P. M., Corfu, F. & Attoh, K. 2006. U–Pb zircon ages of deformed carbonatite and alkaline rocks in t he Pan-African Dahomeyide suture zone, West Africa. EOS Transactions, American Geophysical Union, 87, Fall Meeting Supplement, V31B- 0585.

35. Potter, L. E., 1996. Chemical variation along strike in feldspathoidal rocks of the eastern alkali belt, trans-pecos magamatic province, Texas and New Mexico. In: Alkaline rocks: petrology and mineralogy, Mitchel, R. H., Eby, G. N., & Martin, R. F. (Eds)., Canadian mineralogist, vol. 34 241-263.

36. Potts, P. J., Tindle, A. G., & Webb, P C., 1992. Geochemical Reference Material Composition. CRC Press, Boca Raton, FL. 313pp.

37. Rousseau, R. M., 1989. Concepts of influence coefficients in XRF analysis and calibration. In: Ahmedali, S.T. (Ed.), X-Ray Fluorescence Analysis in the Geological Sciences: Advances in Methodology. Geological Association of Canada GAC-MAC Short Course, vol. 7, pp. 141–220.

38. Rudnick, R. L. 1992. Restites, Eu anomalies, and the lower continental crust, Geochimica et Cosmochimica Acta, 56, 963-970

39. Rudnick, R. L., McDonough, W. F., & Chappell, B. W. 1993. Carbonatite metasomatism in the northern Tanzanian mantle: petrographic and geochemical characteristics. Earth and planetaryscience letters, 114, 463-475.

40. Sylvain J. P., Aregba A., Collart J., & Godonou K. S.. 1986. Notice explicative de la carte géologique du Togo 1//500,000. Direction Generale des Mines de la Géollogie et du Bureau National de Recherches Minières. Memoire 6.

41. Thompson, R. N., Smith, P. M., Gibson, S. A., Mattey, D. P., & Dickin, A. P., 2002. Ankerite carbonatite from Swartbooisdrif, Namibia: the first evidence for magmatic ferrocarbonatite. Contributions to Mineralogy

and Petrology, 143, 377–395.

42. Tohver, E., D'Agrella-Filho, M. S. & Trindale, R. I. F., 2006. Paleomagnetic record of Africa and South America for 1200-500 Ma interval, and evaluation of Rodinia and Gondwana Assemblies. Precambrian Research, 147, 193-222.

43. Wallace, M. E., & Green, D. H., 1988. An experimental determination of primary carbonatite magma composition. Nature, 335, 343–346.

44. Weyer, S., Munker, C., & Mezger, K., 2003. Nb/Ta, Zr/Hf and REE in the depleted mantle: implications for the differentiation history of the crust-mantle system. Earth and Planetary Science Letters, 205, 209–324.

45. Winter, J. D., 2001: An Introduction to Igneous and metamorphic petrology, Prentice Hall Inc, 697pp.

46. Woolley, A. R.., & Kemp, D. R. C., 1989. Carbonatites: nomenclature, average chemical compositions, and element distribution, In: Carbonatites—Genesis and Evolution, Bell, K. (Ed.), Unwin Hyman, London.

Chapter 7

ROUGHNESS RESEARCH OF CENTER PROFILE CURVE ON ROCK FRACTURE SURFACE BASED ON STATISTICAL METHOD

Xuezai Pan[1, 3], Zhigang Feng[2, 3], Guoxing Dai[3], and Hongguang Liu[4]

[1]School of Mathematics, Nanjing Normal University, Taizhou College, Taizhou, China

[2]State Key Laboratory of Coal Resources and Safe Mining, China University of Mining and Technology, Beijing, China

[3]Faculty of Science, Jiangsu University, Zhenjiang, China

[4]Faculty of Civil Engineering and Mechanics, Jiangsu University, Zhenjiang, China

ABSTRACT

In order to research roughness of rock fracture surfaces whether to depend on scale effect, Brazil discs were fractured under tensile and compression stresses in Brazil split test with MTS (Mechanics Test Systems) and a laser profilometer was used to scan rock fracture surfaces and coordinates datum of central profile were acquired. A figure of the central profile was plotted through the coordinates datum. A certain line segment length is regarded as a step length, which is called scale and the scale length is taken to connect pairs of closer peak points on the profile curve. The directional distribution of every scale's normal vector is analyzed by statistics and normal hypothesis test. Finally, some statistics of sample degrees datum are compared with other ones and reach a conclusion that roughness of center profile curve depends on scale effect. The distribution of degrees more and more approximates normal distribution along with increase of scale.

INTRODUCTION

Deformity and fracture of rock are involved in process of moving in earth crust, for example, earthquake, slide downhill, mud-stone flow and so on. In addition, rock fracture usually happens in rock project, for instance, explosion of rock, the project of tunnel, mining engineering etc. Studying rock fracture surfaces through morphology has been recognized by professionals since twenty-one century, because morphology of rock fracture surfaces implicates abundant information of rock fracture mechanics. Experts have discovered that rock fracture surface has characterization of roughness, irregularity and complexity. They attempt to depict the relation between complex morphology of rock fracture surfaces and roughness by all kinds of ways. For example, In order to assess the current state of rock masses and to predict the stability of jointed rock structures, the roughness of rock fracture surfaces has been studied to a higher level. Based on systematic experiments, Barton and Choubey in 1977 proposed a conceptual model to quantify the roughness of rock fracture surface [1]. They classified the roughness into ten categories and the Joint Roughness Coefficients (JRC) ranged from 0 to 20 [2-4]. Some investigators use fractal geometry and multifractal which have been developed since 1970's to describe rock fracture mechanism and have tried to establish the relationship between the fractal dimension and various mechanical parameters of rock fracture surfaces [5-8]. Some experts have indicated that the structural anisotropy of fracture surfaces in rocks greatly influences the mechanical behavior of rock joints under loading [9-11]. The following statement will study roughness of center profile curve on rock fracture surfaces from statistical view. Finally, three prospects will be put forward in the end of the paper.

EXPERIMENTAL METHOD AND ANALYSIS

Experimental Method

Firstly, a sort of special granites which were taken from Gansu province north mountain in China were used to experimental material, because the compactness of the rock material is relatively homogeneous. The granite material was made of the cylinder-shaped sample with rock drilling machine, and then the cylinder-shaped sample was cut into three Brazil discs samples with cutting off machine and buffing machine. The diameter and the height of the discs are equal to 112 mm and 28 mm respectively. Secondly, the rock discs were fractured under tensile and compression stresses in Brazil split test with MTS (Mechanics Test Systems). Loading speed was perminute 0.01 mm. When loading strength approximatively reaches 48 kN, the discs were fractured along vertical direction (refer with: Figure 1). Finally, according to rock mechanics principle, in indirect tensile stresses process of rock, the rock stress of the edge of disc is relatively centralized, so the edge of the disc was easily broken and a little stone chips fell. Whereas inner stress of the rock disc is relatively balanced [12-13], the inner of fracturerock has no stone chips fallen. So, 11 mm was removed from two ends of the rectangular fracture surfaces respectively. The length of center part of fracture surface is equal to 90 mm (refer with: Figure 2). The length of 90 mm is supposed to x axis direction. The center part of fracture surface was scanned by high-accuracy rock laser profilometer along x axis according to the way that interval of x axis is equal to 0.1 mm to acquired three dimension coordinates (x, y, z) of lattice.

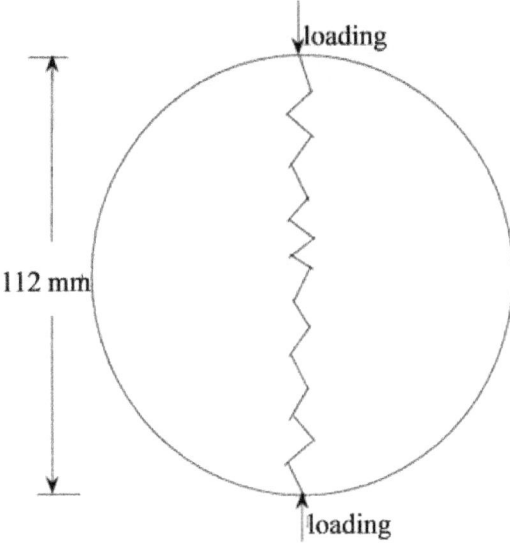

Figure 1: Indirect tensile diagrammatic sketch.

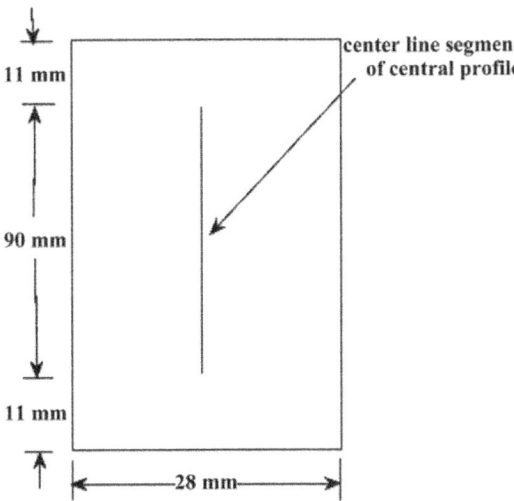

Figure 2: Center profile acquired datum.

Total 901 rope line segments were scanned through the above method, because the length of center part of fracture surface is 90 mm. Length of every line segment scanned is 28 mm, since the width of center part of rock fracture surface is equal to 28 mm (refer with: Figure 2).

Center Profile Curve Analysis on Rock Fracture Surface

The coordinates datum of the center segment of central profile curve on rock rectangle fracture surface (refer with: Figure 2) was extracted by computer procedure, and then the approximative two dimension curve figure of the profile was acquired by linear interpolating method (refer with: Figure 3). A unit length of step length is supposed to 0.1 mm. A step length was taken to connect pairs of points on the center profile [14-16]. Normal vectors' directional distribution of corresponding step length was considered and every angle of each normal vector departing from straight up vector is measured (refer with: Figure 4).The motivation for this approach is straightforward: for a perfectly straight profile the normal vectors will all be parallel and hence display zero dispersion, whereas the dispersion will increase as the profile departs from straight (i.e. becomes rougher). Suppose angle of straight up vector is zero degree. The degree of the angle in the direction skewing to left is negative, whereas that skewing to right is positive. Thus, these degree datum of angles can be obtained with computer procedure.

STATISTICAL METHOD

From statistical knowledge [17-20], mean and median of sample (refer with: Equation (1)) reflect concentrated tendency of sample datum. Range (refer with: Equation (2)), variance and sample standard deviation (refer with: Equation (3)) indicate the extent that sample datum depart from sample mean.

Figure 3: Linear interpolating figure of center profile.

Figure 4: Orientation of normal vector to a step length connecting two points on a profile curve.

Skewness (refer with: Equation (4)) and kurtosis (refer with: Equation (5)) are such statistic describing the shape of sample datum. Skewness reflects dispersive symmetrical characterization of sample datum. Finally, kurtosis indicates the situation that sample datum deviate normal distribution.

$$\overline{X} = \frac{1}{n}\sum_{i=1}^{n} X_i \tag{1}$$

$$R = \max\left(X_i\right) - \min\left(X_i\right) \tag{2}$$

$$s = \sqrt{\frac{1}{n-1}\sum_{i=1}^{n}\left(X_i - \overline{X}\right)^2} \tag{3}$$

$$g_1 = \frac{1}{s^3}\sum_{i=1}^{n}\left(X_i - \overline{X}\right)^3 \tag{4}$$

$$g_2 = \frac{1}{s^4}\sum_{i=1}^{n}\left(X_i - \overline{X}\right)^4 \tag{5}$$

where X_i denotes samples.

When $g_1 > 0$, the form is called right deviation, which illustrates the right datum of mean are more dispersive than those of the left datum; As $g_1 < 0$, the result is named left deviation, which illustrates the situation is opposite to that of right deviation. As g_1 approach zero, which is called impartiality, So, the distribution is regarded as symmetry. On the other hand, kurtosis of normal distribution is equal to 3. As $g_2 > 3$, there are a lot of datum departing from mean, whose shape of distributive curve is flatter than that of normal distribution accordingly; On the contrary, when $g_2 < 3$, the case is inverse to that of $g_2 > 3$. So, statistical method can be used to characterize roughness of profile curve subjected to fracture surfaces. The following discussion is concrete operation.

Suppose the step length is equal to 0.4 mm, 0.3 mm, 0.2 mm and 0.1 mm respectively, then the datum of angle variation are acquired by computer procedure corresponding to various step length. Furthermore, under the same scale, sample mean, median, range, variance, standard deviation, coefficient of skewness and kurtosis are computed respectively and frequency histogram [21-25] is drawn with computer program, which describes the distribution of orientation of normal vectors from the center profile curve. Frequency histogram under a step length is compared with that of other ones, which can show distributional differences each other. On the other hand, hypothesis test method is used to test whether the distribution of normal vectors obeys normal distribution or not and distribution function plots were drawn with computer program. For the degree datum input into computer, Jarque-Bera test is used to test these degree datum whether to obey normal distribution. Significance level α is supposed to 0.05. P is a probability value accepting original hypothesis. JBSTAT is test statistics value. CV is a threshold which can judge whether to refuse original hypothesis and H is test result. If H = 0, the distribution of the degree datum can be considered normal distribution; If H = 1, the distribution of the degree datum doesn't obey normal distribution. If $P < \alpha$, original hypothesis that the datum belong to normal distribution can be denied; If JBSTAT > CV, normal distributional original hypothesis can be negated. Every statistic in following tables is a mean value of corresponding statistic of three center profile curves datum under the same step length (i.e. the same scale), because there are three Brazil discs samples. The differences among the same statistic are compared within four tables under different scales.

1)   If the step length is equal to 0.4 mm, the following Tables 1 and 2 indicate a statistical result.

In Table 1, unit of mean, median, range and standard deviation is degree, whereas other statistics have no unit, because they are only coefficients (below affinity).

Variables in Table 2 have no unit (below affinity). From coefficient of skewness −0.0013 (≈ 0), dispersive extent of datum with left side and right one deviating mean is almost comparative. Distribution of angles' degrees approximatively summits to normal distribution from kurtosis coefficient 3.1511 (≈ 3) and its frequency histogram is referred with Figure 5. From hypothesis test view, where H = 0, $P < \alpha$ and JBSTAT < CV, the normal distributional original hypothesis can be accepted. From normal probability plot shown in Figure 6, the absolute major points gather on the red straight line, which illustrates the normal distributional suppose can be accepted.

In Figure 5, i denotes positive integer and $1 \leq i \leq 12$, because frequency histogram consists of twelve columns.

2) If the step length is equal to 0.3 mm, the statistical datum result is shown in the Tables 3 and 4.

The extent of departing from sample mean increases; The skewness coefficient increases and is more than 0, which illustrates the right datum of mean is more dispersive than that of the left, but the dispersive extent is faint; Kurtosis coefficient is equal to 2.8296 (≈ 3), which illustrates distribution of angle datum approximate normal distribution. The frequency histogram reflects that the distribution of angle datum close to normal distribution (shown in Figure 7). From normal hypothesis test view, $H = 0$, $P = 0.6492$ and JBSTAT $<$ CV indicate normal distributional hypothesis can be accepted with 64.92% probability. From normal probability plot shown in Figure 8, the absolute major points gather on the red straight line, which illustrates the normal distributional suppose can be accepted.

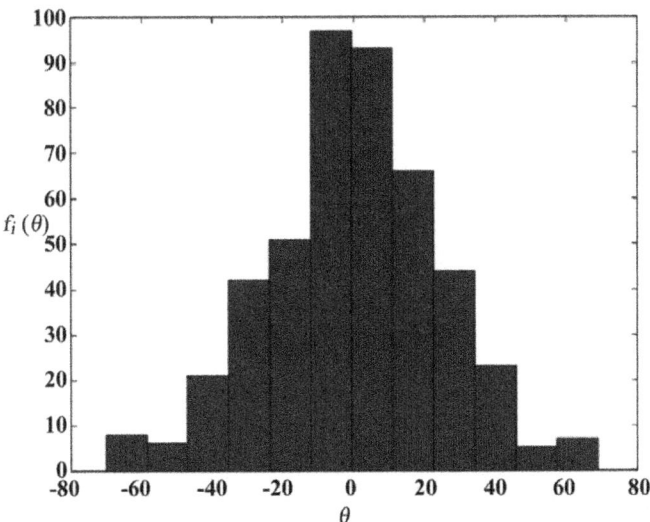

Figure 5: The histogram with the step length 0.4 mm (θ: degree; $f_i(\theta)$: frequency).

Table 1: Statistics of sample datum with the step length 0.4 mm

Mean	Median	Range	Variance	Standard deviation	Skewness	Kurtosis
−0.1333	0.0000	148.0381	603.3565	24.2938	−0.0013	3.1811

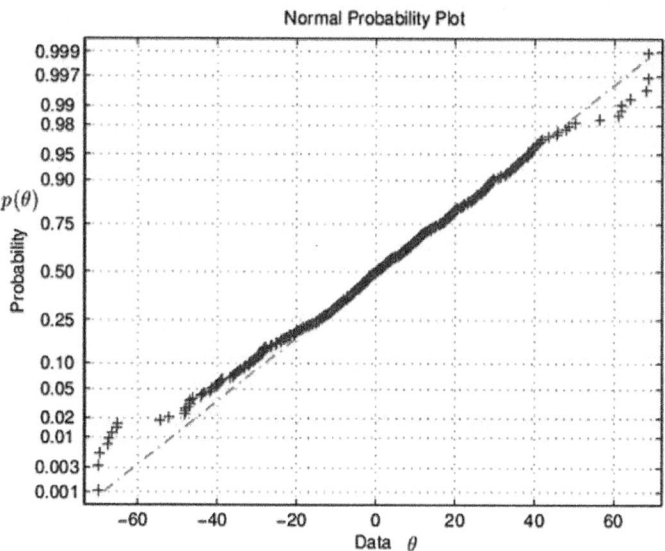

Figure 6: Normal probability plot with the step length 0.4 mm (θ: degree; $p(\theta)$: probability).

Table 2: Hypothesis test value with the step length 0.4 mm

H	P	JBSTAT	CV
0	0.2314	2.9269	5.9915

3)   If the step length is equal to 0.2 mm, the statistical result is shown in the next Tables 5 and 6.

The sample mean and median increase; Variance and standard deviation increase furthermore; Range hardly change; Coefficient of skewness reduces, however the decrement is very little; coefficient of kurtosis decreases furthermore and reaches 1.9127, which illustrates the distribution of angle datum continues to deviate from normal distribution.

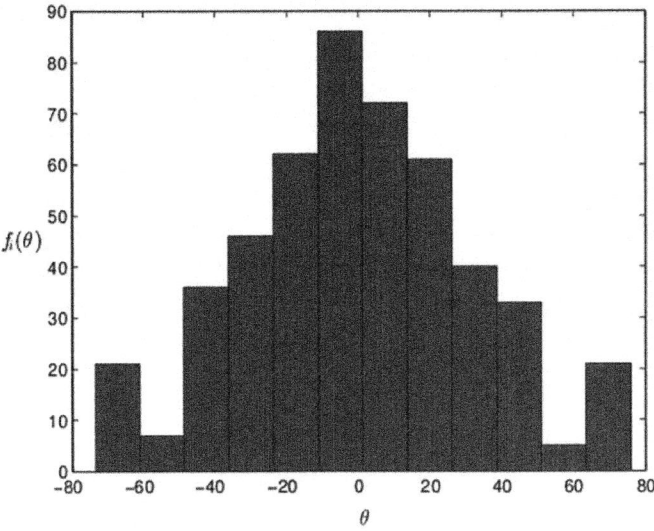

Figure 7: The histogram with the step length 0.3 mm (θ: degree; $f_i(\theta)$: frequency).

Frequency histogram is shown in Figure 9. From normal hypothesis test view, H = 1, the value of P and JBSTAT > CV indicate normal distributional hypothesis can be negated. Normal probability plot is referred with Figure 10.

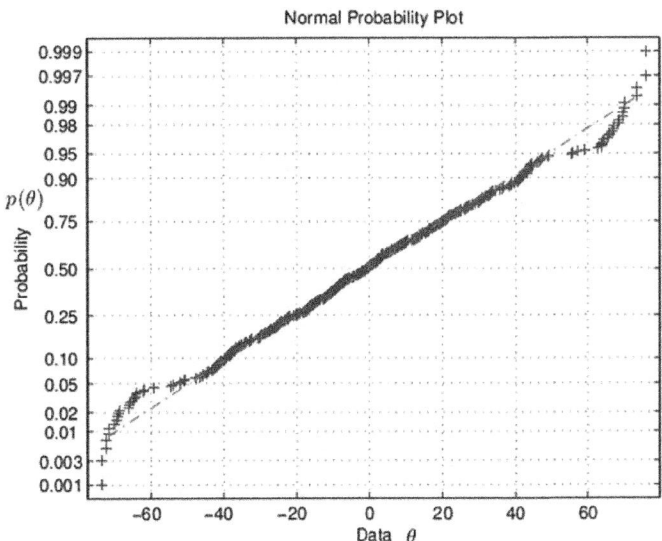

Figure 8: Normal probability plot with the step length 0.3 mm (θ: degree; $p(\theta)$: probability).

4)   If the step length is equal to 0.1 mm, the following Tables 7 and 8 indicate according statistical result.

The value of sample mean and median has a weak variation; Range still change rarely; Variance and standard deviation increase greatly; Coefficient of skewness continues to decrease, but it still fluctuates near 0, which still describes that two sides' datum of mean have the same dispersion characterization; Coefficient of kurtosis descends further and attains 1.6600, which indicates that the distribution of normal vector deviates from normal distribution; Frequency histogram is shown in Figure 11. From normal hypothesis test view, H = 1, the value of P = 0 and JBSTAT $\gg$ CV indicate normal distributional hypothesis can be negated completely.

Table 3: Statistics of sample datum with the step length 0.3 mm

Mean	Median	Range	Variance	Standard deviation	Skewness	Kurtosis
-0.3113	-0.8263	149.2868	660.3217	31.4666	0.0488	2.8526

Table 4: Hypothesis test value with the step length 0.3 mm

H	P	JBSTAT	CV
0	0.6762	0.8070	5.9915

Table 5: Statistics of sample datum with the step length 0.2 mm

Mean	Median	Range	Variance	Standard deviation	Skewness	Kurtosis
0.3166	0.0000	149.2881	1288.3798	36.8281	0.0081	1.6133

Table 6: Hypothesis test value with the step length 0.2 mm

H	P	JBSTAT	CV
1	2.5850e-011	48.7574	5.9915

Table 7: Statistics of sample datum with the step length 0.1 mm

Mean	Median	Range	Variance	Standard deviation	Skewness	Kurtosis
0.1318	0.2865	149.5898	1783.2761	42.2289	0.0072	1.6600

Table 8: Hypothesis test value with the step length 0.1 mm

H	P	JBSTAT	CV
1	0	112.9056	5.9915

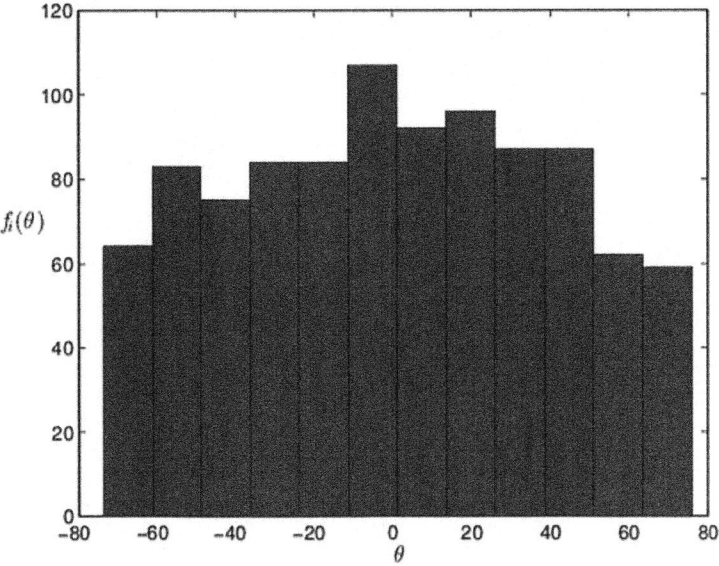

Figure 9: The histogram with the step length 0.2 mm (θ: degree; $f_i(\theta)$: frequency).

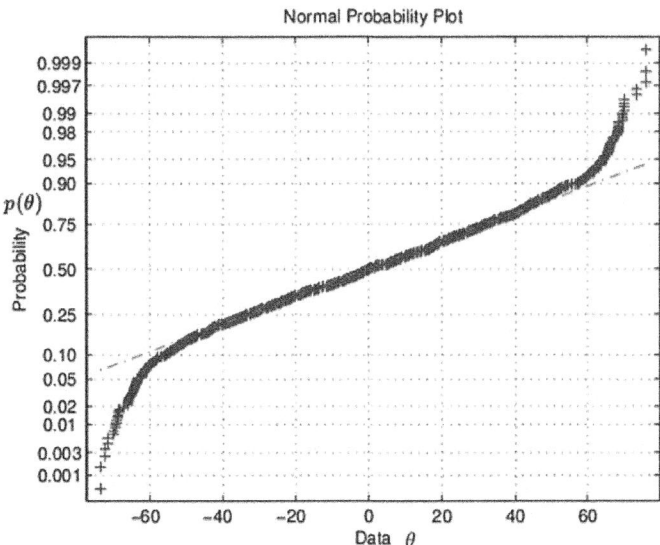

Figure 10: Normal probability plot with the step length 0.2 mm. (θ: degree; $p(\theta)$: probability).

Normal probability plot is referred with Figure 12.The absolute major points deviates from the red straight line.

In conclusion, the directional distribution of normal vectors is associated with the step length, which depends on scale effect.

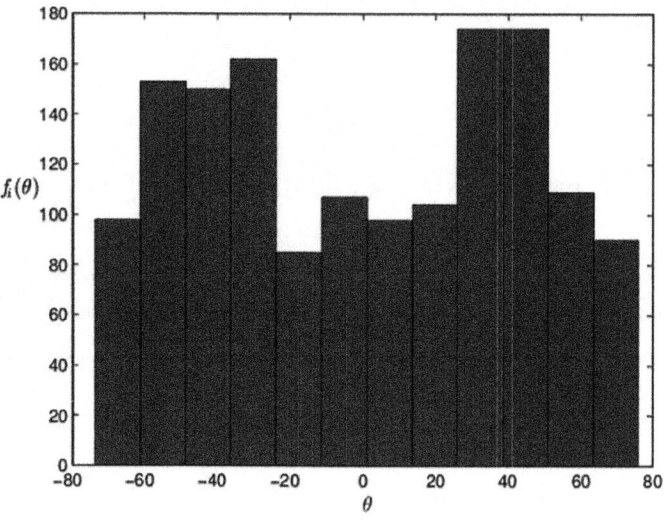

Figure 11: The histogram with the step length 0.1 mm (θ: degree; $f_i(\theta)$: frequency).

The smaller the kurtosis is, the smaller the scale is, which illustrates the distribution of angle datum deviates from normal distribution continuously. The case that range of sample datum is close to 149 is disco vered, which indicates that variable range of angle degrees is definite. And the dispersive extent between the left datum of sample mean and those of right is comparative, because skewness coefficient is almost equal to 0. In summary, the roughness of center profile curve could be supposed to standard state, when the distribution of angle datum belongs to standard normal distribution. Furthermore, the rougher center profile curve is, the smaller the step length is.

CONCLUSION AND PROSPECT

In generally, mean, median, range and skewness coefficient almost have no scale effect, however, variance and standard deviation will increase with decreasing of scale. Kurtosis coefficient will descent along with diminishing of scale. Accordingly, from normal hypothesis test view, H, P, JBSTAT have scale effect too. As a whole, H leaps from 0 to 1 along with decreasing of scale. P will decrease while scale drops. JBSTAT will raise as scale falls. That is to say, the rougher center profile curve is, the smaller the step length is. The ultimate purpose of researching morphology of rock fracture surfaces is that the information in process of rock fracture is acquired through methods of mathematical analysis. Successively, components of rock structure and limitation are discovered and further mechanics of rock fracture is posted. However, structure of rock and properties of mechanics exhibit nonlinear characterization. Rock fracture surface possesses fairly irregular and stochastic properties, hence the topic is developed so slowly that the results acquired through researching haven't been applied in forecasting and instructing practise of project. Therefore, the following three aspects of research will be evolved.

Firstly, advantages of approaches applied by experts now continue to be developed and perfected, and insufficiencies will be overcome extensively, which attempt to build the relationship between rock mechanics and topograph of rock fracture surfaces. Secondly, new approaches of studying rock fracture surfaces through morphology will be sought, which try to discover the mechanics of rock fracture. Finally, existing experimental results will be transformed into theoretical basis of guiding project practice possibly.

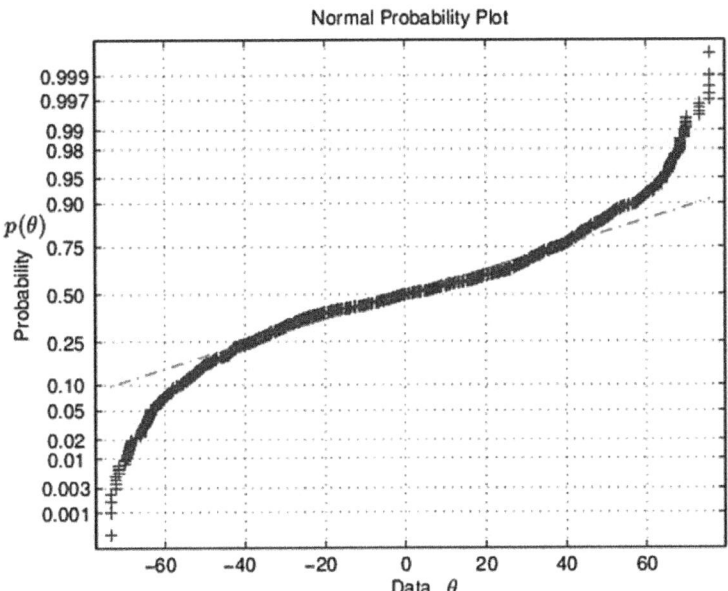

Figure 12: Normal probability plot with the step length 0.1 mm (θ: degree; $p(\theta)$: probability).

ACKNOWLEDGEMENTS

1.   Supported by the National Natural Science Foundation of China 51079064.

2.   Supported by State Key Laboratory of Coal Resources and Safe Mining, China University of Mining and Technology SKLCRSM10KFA02.

3.   Supported by Nanjing Normal University Taizhou College Q201234.

REFERENCES

1.   N. Barton and V. Choubey, "The Shear Strength of Rock Jounts in Theory and Practice," Rock Mechanics, Vol. 1977, Vol. 10, No. 2, pp. 1-45. doi:10.1007/BF01261801

2.   N. Barton, "Review of a New Shear Strength Criterion for Rock Joints," Engineering Geology, Vol. 7, No. 4, 1973, pp. 287-332. doi:10.1016/0013-7952(73)90013-6

3.   R. Tse and D. M. Cruden, "Estimating Joint Roughness Coefficients," International Journal of Rock Mechanics and Mining Sciences & Geomechanics Abstracts, Vol. 16, No. 5, 1979, pp. 303-307. doi:10.1016/0148-9062(79)90241-9

4.   N. H. Maerz, J. A. Franklin and C. P. Bennett, "Joint Roughness Measurement Using Shadow Profilometry," International Journal of Rock Mechanics and Mining Sciences & Geomechanics Abstracts, Vol. 27, No. 5, 1990, pp. 329-343. doi:10.1016/0148-9062(90)92708-M

5.   K. Falconer and W. G. Yang, "Fractal Geometry Mathematical Foundations and Applications," 2nd Edition, Posts & Telecom Press, Beijing, 2007.

6.   Y. H. Zhang, H. W. Zhou and H. P. Xie, "Improved Cubic Covering Method for Fractal Dimensions of a Fracture Surface of Rock," Chinese Journal of Rock Mechanics and Engineering, Vol. 24, No. 17, 2005, pp. 3192-3196.

7.   H. P. Xie, H. Q. Sun, Y. Ju, et al., "Study on Generation of Rock Surfaces by Using Fractal Interpolation," International Journal of Solid and Structure, Vol. 38, No. 32-33, 2001, pp. 5765-5787. doi:10.1016/S0020-7683(00)00390-5

8.   H. P. Xic, J.-A. Wang and M. A. Kwa Nicwski, "Multifractal Characterization of Rock Fracture Surfaces," International Journal of Rock Mechanics Mining Sciences, Vol. 36, No. 1, 1999, pp. 19-27. doi:10.1016/S0148-9062(98)00172-7

9.   S. C. Bandis, A. C. Lumsden and N. R. Barton, "Fundamentals of Rock Joint Deformation," International Journal of Rock Mechanics and Mining Sciences & Geomechanics Abstracts, Vol. 20, No. 6, 1983, pp. 249-268. doi:10.1016/0148-9062(83)90595-8

10.   P. H. S. W. Kulatilake, G. Shou, T. H. Huang and R. M. Morgan, "New Peak Shear Strength Criteria for Anisotropic Rock Joints," International

Journal of Rock Mechanics and Mining Sciences & Geomechanics Abstracts, Vol. 32, No. 7, 1995, pp. 673-697.doi:10.1016/0148-9062(95)00022-9

11. H. W. Zhou and H. Xie, "Anisotropic Characterization of Rock Fracture Surfaces Subjected to Profile Analysis," Physics Letters A, Vol. 325, No. 5-6, 2004, pp. 355-362.doi:10.1016/j.physleta.2004.04.006

12. W. Gao, "Mechanics of Rock," Peking University Press, Beijing, 2010.

13. Z. L. Fu, F. K. Xiao, Y. X. Liu and S. J. Chen, "Experiment Course on Rock Mechanics," Chemistry Engineering Press, Beijing, 2010.

14. V. Rasouli and J. P. Harrison, "Assessment of Rock Fracture Surface Roughness Using Riemannian Statistics of Linear Profiles," International Journal of Rock Mechanics Mining Sciences, Vol. 47, No. 6, 2010, pp. 940-948. doi:10.1016/j.ijrmms.2010.05.013

15. V. Rasouli and J. P. Harrison, "Scale Effect, Anisotropy and Directionality of Discontinuity Surface Roughness," Proceedings of the EUROCK Symposium, Aachen, 27-31 March 2000, pp. 751-756.

16. T. H. Wu and E. M. Ali, "Statistical Representation of Joint Roughness," International Journal of Rock Mechanics and Mining Science & Geomechanics Abstracts, Vol. 15, No. 5, 1978, pp. 259-262. doi:10.1016/0148-9062(78)90958-0

17. Z. S. Wei, "Probability and Statistics Tutorial," Higher Education Press, Beijing, 1983.

18. X. R. Chen, "Higher Statistics," China University of Science and Technology, Hefei, 1999

19. G. X. Liu, Z. F. He and J. L. Yang, "Probability and Statistics," Gansu Education, Lanzhou, 2002.

20. X. S. Liu, "Probability and Statistics," Sichuan University Press, Chengdu, 2009.

21. L. B. Wu and B. N. Li, "Mathematical Experiment and Modeling," Defense Industry Press, Changsha, 2007.

22. J. Zhao and Q. Dan, "Mathematical Modeling and Mathematical Experiment," Higher Education Press, Beijing, 2000.

23. H. G. Zhang, "Practical Tutorial of MATLAB/SIMULINK," Posts & Telecom Press, Beijing, 2009.

24. D. X. Zhang and L. S. Zhao, "Teaching of Language Procedure Design of MATLAB," China Railway Press, Beijing, 2010.

25. P. Wu, "Technology and Application of Effective Procedure of MATLAB: Analysis of 25 Cases," Beihang University Press, Beijing, 2010.

Chapter 8

FAILURE PROBABILITY MODEL CONSIDERING THE EFFECT OF INTERMEDIATE PRINCIPAL STRESS ON ROCK STRENGTH

Yonglai Zheng and Shuxin Deng

College of Civil Engineering, Tongji University, Shanghai 200092, China

ABSTRACT

A failure probability model is developed to describe the effect of the intermediate principal stress on rock strength. Each shear plane in rock samples is considered as a micro-unit. The strengths of these micro-units are assumed to match Weibull distribution. The macro strength of rock sample is a synthetic consideration of all directions' probabilities. New model reproduces the typical phenomenon of intermediate principal stress effect that occurs in some true triaxial experiments. Based on the new model, a strength criterion is proposed and it can be regarded as a modified Mohr-Coulomb criterion with a uniformity coefficient. New strength criterion can quantitatively reflect the intermediate principal stress effect on rock strength and matches previously published experimental results better than common strength criteria.

INTRODUCTION

The stress state in three-dimensional space is defined by three mutually perpendicular stress components $(\sigma_1, \sigma_2, \sigma_3)$ and it is an important subject in rock mechanics to study the rock failure behavior under complex stress conditions. The traditional considerations suggest that shear failure will occur when the shear stress along some plane in the sample is too large. The extreme values of shear stress in a material are related only to the largest and smallest principal stress (commonly denoted as σ_1 and σ_3). Based on such considerations, the influence of intermediate principal stress (σ_2) on the experimental results is not taken into account.

However three-dimensional unequal stress states are very common in engineering practice and it is very important to predict the rock strength with the effect of intermediate principal stress. Many researchers [1–6] have conducted a large number of true triaxial tests on different rock types such as Dunham dolomite, Solnhofen limestone, and granite, to investigate the behavior under triaxial stress conditions. They found that the strength first increases and then reduces with the increase of σ_2. At the same time, researchers have developed many numerical models or used commercial software to study the failure processes of rocks under polyaxial stress conditions [7–11]. In these numerical tests, it was clear that the intermediate principal stress has an effect on rock strength and the mechanism of such an effect is discussed. The phenomenon of intermediate principal stress effect seems to be related to the heterogeneity of materials [10, 11]. Without material heterogeneity, local tensile stress cannot be generated in an overall compressive stress environment so that there is no crack initiation and propagation [7]. The intermediate principal stress confines the rock in such a way that fractures can be easier to develop in the direction parallel to σ_1 and σ_2. Fjær and Ruistuen's numerical results [11] indicate that effect of is related to stress symmetry, rather than the stress level. When was close to σ_3 or σ_1, there were more possible directions for the failure planes with higher stress symmetry. The directions for the failure plane for which the theoretical failure criterion is first fulfilled may not coincide with the directions preferred by the rock's heterogeneity. The rock will be weaker where the stress state is more symmetric and there are several equivalent directions of the failure plane to choose. These considerations are supported by Chang and Haimson's observations [1]. They found that stress induced micro cracks in amphibolites were randomly oriented when , σ_3 while micro cracks became more aligned with the direction of σ_2 when $\sigma_2 > \sigma_3$. Similar phenomenon was also observed by Cai [7] and Pan et al. [10].

According to Fjær and Ruistuen's considerations [11], σ_2 affects the failure probabilities in different directions under three-dimensional unequal stress. Which direction the failure plane eventually takes is determined by the heterogeneities of the rock. The heterogeneity is always incorporated by assuming the micro-units' properties complying with a certain distribution and here Weibull distribution [12] is selected. Weibull distribution is introduced to explain the statistical size effect and later justified theoretically on the basis of some reasonable hypotheses about the statistical distribution and the role of microscopic flaws or micro cracks [13]. Unlike previous Weibull analysis [14–17] with volume consideration in this paper each potential shear failure plane is regarded as a micro-unit and failure probabilities of all directions are calculated. Thus, a new failure probability model considering the failure probabilities in different directions is proposed. By combining the new model with Mohr-Coulomb criterion, a new strength criterion is developed, which can quantitatively describe the effect of intermediate principal stress on rock strength.

MODELLING

Failure Probability Model with Volume Considerations

Assume that under the nominal stress R the density function of micro flaws per unit volume is (R). The amount of micro flaws in materials sample with a volume of V could be expressed as

$$N(R) = V \int_0^R n(x)\,dx.$$

(1)

The strength of micro-units complies with Weibull distribution [12] and the probability density function is

$$f(R) = \frac{m}{R_0} \left\langle \frac{R - R_u}{R_0} \right\rangle^{m-1} \exp\left(-\left\langle \frac{R - R_u}{R_0} \right\rangle^m\right),$$

(2)

where $\langle \cdot \rangle$ is Macaulay bracket; when $x \le 0$, $\langle x \rangle = 0$; when $x > 0$, $\langle x \rangle = x$; R_u is stress threshold; R_0 is scalar parameter, $R_0 > 0$; m is shape parameter and it can be considered as the uniformity coefficient, $m > 1$.

Stress threshold R_u is often set to zero. In this case, (2) has only two parameters (m and R_0). The failure probability of materials sample with one flaw can be obtained:

$$F(R) = \int_0^R f(x)\,dx = 1 - \exp\left(-\left\langle \frac{R}{R_0} \right\rangle^m\right).$$

(3)

According to weakest link theory, the failure probability of materials sample with a volume of V can be represented:

$$F_V(R) = 1 - [1 - F(R)]^{N(R)}$$

(4)

By inserting (1) and (3) into (4), we obtain

$$F_V(R) = 1 - \exp\left[-\left\langle \frac{R}{R_0} \right\rangle^m V \int_0^R n(x)\,dx\right]$$

$$= 1 - \exp\left[-\frac{V}{V_0} \left\langle \frac{R}{R_0} \right\rangle^m\right],$$

(5)

where $V_0 = 1/\int_0^R n(x)dx$ is considered as a reference volume. From (5), the average strength can be calculated:

$$\overline{R} = \int_0^\infty sf(s)\,ds = \int_0^\infty s\,dF(s) = \int_0^\infty \left[1 - F_V(s)\right]ds.$$

(6)

By inserting (5) into (6) and using variable substitution, (6) can be represented:

$$\overline{R} = R_0 \left(\frac{V_0}{V}\right)^{1/m} \Gamma\left(1 + \frac{1}{m}\right),$$

(7)

where $\Gamma(\cdot)$ is gamma function, $\Gamma(x) = \int_0^\infty t^{x-1}\exp(-t)dt.$

The variance s^2 and variation coefficient ω can be obtained from (5) and (7):

$$s^2 = \int_{R_u}^\infty \left(x - \overline{R}\right)^2 f(x)\,dx = \int_{R_u}^\infty \left(x - \overline{R}\right)^2 dF_V(x)$$

$$= \overline{R}^2 \left[\frac{\Gamma(1 + 2/m)}{\Gamma^2(1 + 1/m)} - 1\right],$$

(8)

$$\omega = \frac{s}{\overline{R}} = \sqrt{\frac{\Gamma(1 + 2/m)}{\Gamma^2(1 + 1/m)} - 1}.$$

(9)

Considering the nonuniform force field, strength of each point in xyz coordinates can be expressed as (x, y, z). Here R is the maximum strength and (x, y, z) is a dimensionless coordinate function for each point. Thus (5) becomes

$$F_V(R)$$

$$= 1 - \exp\left\{-\frac{1}{V_0}\int_{Rf(x,y,z)} \left[\frac{Rf(x,y,z)}{R_v}\right]^m dV\right\}.$$

(10)

Accordingly, (6) becomes

$$\overline{R_*} = \int_0^\infty \exp\left[-\frac{g(R)}{V_0}\right]dR,$$

(11)

$$g(R) = \int_{Rf(x,y,z)} \left[\frac{Rf(x,y,z)}{R_0}\right]^m dV.$$

(12)

Comparing the average strength $\overline{R}$ in uniform force field and in nonuniform force field, from (6) and (11) we obtain

$$\overline{R_*} = \overline{R}\left(\frac{V}{V_*}\right)^{1/m},$$

$$\tag{13}$$

$$V_* = \int_V [f(x, y, z)]^m \, dV.$$

$$\tag{14}$$

Failure Probability Model with Direction Considerations.

Shear failure is a basic failure mode of rocks under triaxial compression. Following Fjær and Ruistuen's considerations [11], σ_2 affects the failure probabilities in different directions under three-dimensional unequal stress. In order to describe the intermediate principal stress effect, direction considerations instead of volume considerations are made to express failure probabilities in different directions. The shear planes in rock samples are considered as potential failure planes. In order to calculate the probability of each direction, each potential shear failure plane is regarded as a micro-unit. The effect of the intermediate principal stress can quantitatively be estimated by calculating the failure probabilities for all the shear planes and combining these into the total probability for failure.

Figure 1 shows a normal vector (ON) of a shear plane defined by the spherical coordinates α and φ, and the normal and shear stresses are given as

$$\sigma = \sigma_1 \cos^2\alpha + \sigma_2 \sin^2\alpha \cos^2\varphi + \sigma_3 \sin^2\alpha \sin^2\varphi,$$

$$\tag{15}$$

$$\tau$$

$$= \sqrt{\sigma_1^2 \cos^2\alpha + \sigma_2^2 \sin^2\alpha \cos^2\varphi + \sigma_3^2 \sin^2\alpha \sin^2\varphi - \sigma^2}.$$

$$\tag{16}$$

The ranges of α and φ are $0 \le \alpha \le \pi/2$ and $0 \le \varphi \le \pi/2$. For each shear plane, it is assumed that shear failure will occur when the shear stress along some surface in the sample is too large. The maximum strength is

$$R_{\max} = \frac{\sigma_1 - \sigma_3}{2}.$$

$$\tag{17}$$

The nominal stress for each shear plane can be expressed as $R_{\max}(\alpha, \varphi)$ with a direction function $f(\alpha, \varphi)$. Based on the assumption, failure occurs when

$$R_{\max} f(\alpha, \varphi) = \tau.$$

$$\tag{18}$$

According to (16), (17), and (18), (α, φ) can be expressed as

$$f(\varphi, \alpha) = 2\sqrt{\sin^2\alpha \left[\cos^2\varphi (1-b)^2 + \sin^2\varphi\right] - \sin^4\alpha \left[\cos^2\varphi (1-b) + \sin^2\varphi\right]^2},$$

$$\tag{19}$$

where b is intermediate stress ratio, $b = (\sigma_2 - \sigma_3)/(\sigma_1 - \sigma_3)$.

For the direction consideration instead of volume consideration, the (x, y, z) in (10) could be replaced by (α, φ). Considering the special condition $b=0$ $(\sigma_3 = \sigma_2 < \sigma_1$, the conventional triaxial test), $f(\alpha, \varphi)$ becomes

$$f(\alpha, \varphi) = \sin 2\alpha. \tag{20}$$

By inserting (20) into (14), the reference volume (denoted by V_0 for $b=0$) can be obtained as

$$V_0 = \frac{\pi}{2} \int_0^{\pi/2} \sin^m (2\alpha)\, d\alpha \tag{21}$$

when $b>0$, and according to (13), we obtain

$$\overline{R_*} = \overline{R_0} \left(\frac{V_0}{V_*} \right)^{1/m}, \tag{22}$$

where $\overline{R_0}$ is the mean strength when $b=0$, namely, the strength in conventional triaxial test. Vb can be calculated by

$$V_b = \int_0^{\pi/2} \int_0^{\pi/2} f^m (\alpha, \varphi)\, d\alpha\, d\varphi. \tag{23}$$

In volume considerations, materials heterogeneity represents the different properties between different points. In direction considerations, materials heterogeneity represents the different properties between different shear planes. Equations (22) and (13) have similar form. However, meanings of "volume" are different. $V*$ and V in (13) are integrals of the dimensionless coordinate function $f(x, y, z)$ for volume considerations while $V*$ and V_0 in (22) are integrals of the direction function $f(\alpha, \varphi)$ for direction considerations. Also, $f(x, y, z)$ in (10) is replaced by $f(\alpha, \varphi)$ and the macro strength of materials sample is a synthetic consideration of all directions' probabilitiesThe variance $s2$ and variation coefficient ω can be obtained from (8) and (9). Figure 2 shows the relationship curve between ω and m according to (9). From (19), (22), and (23) the relationship between R_b/R_0 and b can be obtained which is shown in Figure 3. The results shown in Figure 3 indicate that the intermediate principal stress significantly affects rock strength. The failure probability model in this paper is observed to reproduce the typical phenomenon of intermediate principal stress effect that occurs in some true triaxial experiments. The effect of intermediate principal stress is controlled by m. As m increases, the materials become more homogeneous and the effects of intermediate principal stress become less prominent and the variation coefficients of the results are lower (see Figure 2). When $m \to \infty$, the materials are absolutely homogeneous, there is no intermediate principal stress effect, and the curve of $R_b/R_0 \sim b$ is a horizontal line. The model developed in this paper becomes Mohr-Coulomb criterion, which does not include any effect of intermediate principal stress.

Strength Criterion considering Intermediate Principal Stress Effect.

For the shear failure of rock materials, computed strength R can be expressed by the shear stress. When $b=0$, R_0 can be calculated by Mohr-Coulomb criterion:

$$R_0 = \frac{\sigma_{10} - \sigma_3}{2} = \frac{\sigma_3 \sin\varphi + c\cos\varphi}{1 - \sin\varphi},$$

(24)

where φ is inner friction angel, c is cohesion, and $\sigma 10$ is axial stress when $b=0$ ($\sigma_2 = \sigma_3$).

From (24), (22) becomes

$$\frac{\sigma_1 - \sigma_3}{2} = \frac{\sigma_3 \sin\varphi + c\cos\varphi}{1 - \sin\varphi} \left(\frac{V_0}{V_b}\right)^{1/m}.$$

(25)

Thus, a new strength criterion considering intermediate principal stress effect and heterogeneity of materials is obtained:

$$\tau_{13} = f_1(\sigma_3) \cdot f_2(\mu_\sigma),$$

(26)

Where

$$\tau_{13} = \frac{\sigma_1 - \sigma_3}{2},$$

$$f_1(\sigma_3) = \frac{\sigma_3 \sin\varphi + c\cos\varphi}{1 - \sin\varphi},$$

$$f_2(\mu_\sigma) = \left[\frac{2\pi \int_0^{\pi/2} \sin^m(2\alpha)\, d\alpha}{\int_0^{2\pi}\int_0^{\pi/2} f_3^m(\varphi,\alpha,\mu_\sigma)\, d\alpha\, d\varphi}\right]^{1/m},$$

$$f_3(\varphi,\alpha,\mu_\sigma) = 2\sqrt{\sin^2\alpha\left[\frac{(1-\mu_\sigma)^2}{4}\cos^2\varphi + \sin^2\varphi\right] - \sin^4\alpha\left[\frac{1-\mu_\sigma}{2}\cos^2\varphi + \sin^2\varphi\right]^2},$$

$$\mu_\sigma = \frac{2\sigma_2 - \sigma_1 - \sigma_3}{\sigma_1 - \sigma_3}.$$

(27)

μ_σ is Lode parameter, which is used to distinguish between the different shear stress states in three dimensions (3D), ranging from axisymmetric tension to biaxial tension with axisymmetric compression and passing through inplane shear. The relationship between μ_σ and b is $b = (1 + \mu_\sigma)/2$. When $m\to\infty$, $f2(\mu_\sigma)=1$, strength is independent of μ_σ or b and (26) is equivalent to Mohr-Coulomb criterion. Thus, (26) could be regarded as a modified Mohr-Coulomb criterion that can reflect the effect of intermediate principal stress.

MODEL VALIDATION AND VERIFICATION

Comparison with Experimental Results

Comparisons between the failure probability model as described in this paper and results of the experimental results [5] are plotted in Figure 4. Fjær and Ruistuen [11] have developed a numerical method to calculate the intermediate principal stress effect on rock strength. Their results are also plotted in Figure 4 for comparison. The strength parameters can be obtained by fitting the strength envelopes. Three strength parameters are as follows: $c = 31$ MPa, $\varphi = 27.37\circ$, and $m = 2.5$. From Figure 4, both model predictions in this paper and Fjær and Ruistuen's numerical results [11] are observed to be in good agreement with the results from the experimental results [5]. Failure probability model in this paper matches the experimental results [5] better than Fjær and Ruistuen's numerical results [11], especially when σ_2 is low.

Comparison with Common Strength Criteria

Common strength criteria are Mohr-Coulomb criterion, DruckerPrager criterion, 3D Griffith criterion, and twin shear stress criterion. Mathematical forms of these strength criteria are as follows.

Mohr-Coulomb criterion is as follows:

$$\frac{\sigma_1 - \sigma_3}{2} = \frac{\sigma_1 + \sigma_3}{2} \sin \varphi + c \cos \varphi.$$

(28)

Drucker-Prager criterion [18] is as follows:

$$\sqrt{J_2} - \alpha I_1 - k = 0,$$

(29)

where α and k are material constants, $I_1 = \sigma_1 + \sigma_2 + \sigma_3$ is the first stress invariant, and $J_2 = [(\sigma_1 - \sigma_2)^2 + (\sigma_2 - \sigma_3)^2 + (\sigma_3 - \sigma_1)^2]/6$ is the second deviatoric stress invariant.

3D Griffith criterion [19] is as follows:

$$(\sigma_1 - \sigma_2)^2 + (\sigma_2 - \sigma_3)^2 + (\sigma_3 - \sigma_1)^2 = 24T_0 (\sigma_1 + \sigma_2 + \sigma_3),$$

(30)

where T_0 is a material constant

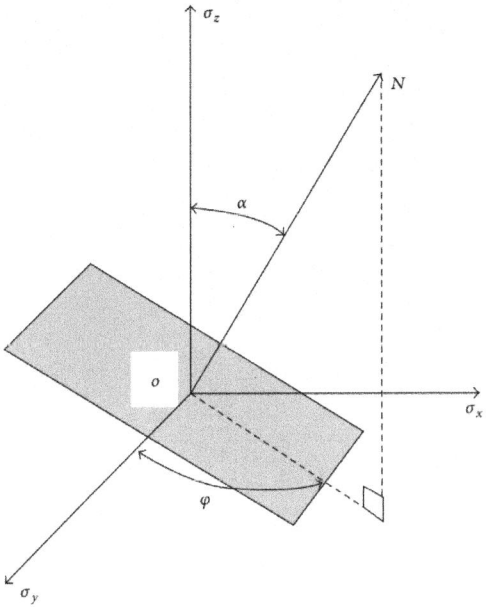

Figure 1: Normal vector of a shear plane.

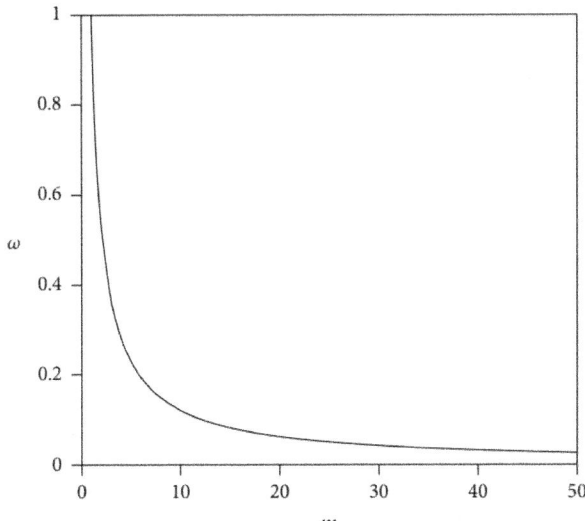

Figure 2: Relationship between ω and m.

Twin shear stress criterion [20] is as follows:

$$\sigma_1 - \frac{1}{2}(\sigma_2 + \sigma_3) = \sigma_t, \quad \text{when}$$

$$\sigma_2 \le \frac{1}{2}(\sigma_1 + \sigma_3),$$

$$\frac{1}{2}(\sigma_1 + \sigma_2) - \sigma_3 = \sigma_t, \quad \text{when}$$

$$\sigma_2 \ge \frac{1}{2}(\sigma_1 + \sigma_3),$$

(31)

where σ_t is a material constant.

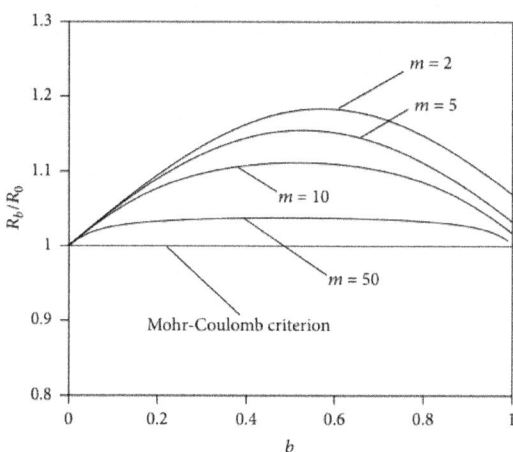

Figure 3: Relationship between R_b/R_0 and b for various m.

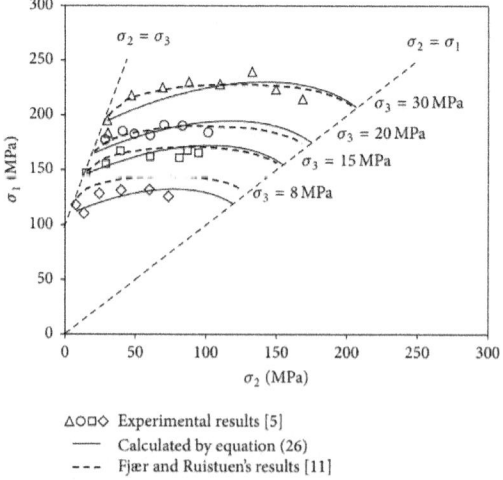

Figure 4: Relationships between σ_1 and σ_2 for constant σ_3.

A comparison of the criteria listed above and the strength criterion developed in this paper is shown in Figure 5 for σ_3 = 15 MPa. The material constants for each criterion have been chosen so that all predict the same strength when $\sigma_2 = \sigma_3$ = 15 MPa. The material constants are as follows: c = 31 MPa, φ = 27.37°, m = 2.5, α = 0.209, k = 38.51 MPa, T_0 = 8.05 MPa, and σ_t = 137.58 MPa. Figure 5 shows that Mohr-Coulomb criterion does not depend on the intermediate principal stress, contrary to the experimental observations. Drucker-Prager criterion and 3D Griffith criterion overestimate the effect of σ_2. Both twin shear stress criterion and the strength criterion developed in this paper can well describe the intermediate principal stress effect on rock strength. However, twin shear stress criterion cannot reflect the observation [19] that the strength of a sample under extension loading ($\sigma_2 = \sigma1$) is greater than that

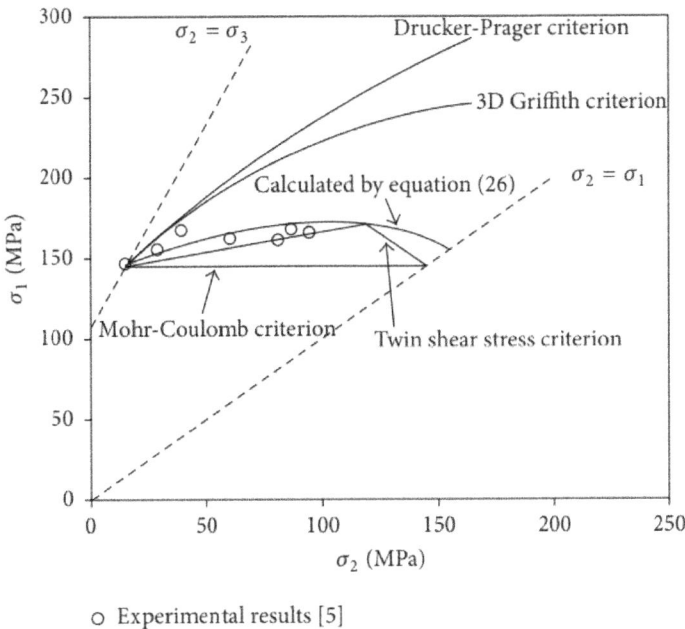

o Experimental results [5]

Figure 5: Comparison of common criteria and the strength criterion developed in this paper.

under compression loading ($\sigma_2 = \sigma_3$). On the other hand, the failure surface for twin shear stress criterion is a set of planes, forming sharp corners at the intersections. As such, the failure surface is not differentiable at the corners, which sometimes causes problems in numerical calculations involving the criterion. New strength criterion developed in this paper could overcome these problems.

CONCLUSIONS

This paper has developed a new failure probability model to predict the effect of the intermediate principal stress on rock strength. Each shear plane in rock samples is considered as a micro-unit. The strengths of these micro-units are assumed to match Weibull distribution. Direction considerations instead of the volume considerations are made to express failure probabilities in different directions. The macro strength of rock sample is a synthetic consideration of all directions' probabilities. New failure probability model reproduces the typical phenomenon of intermediate principal stress effect that occurs in some true triaxial experiments. Material heterogeneity plays a major role for intermediate principal stress effect. The intermediate principal stress effect becomes more prominent for more heterogeneous rocks. When materials are absolutely homogeneous, there is no intermediate principal stress effect and the failure model becomes Mohr-Coulomb criterion.

Based on the new model with direction considerations, computed strength is expressed by the shear stress and the computed strength when $\sigma_2 = \sigma_3$ is calculated by Mohr-Coulomb criterion. In this way a strength criterion is developed to quantitatively describe the effect of intermediate principal stress on rock strength. When uniformity coefficient m is infinitely large, the new criterion is equivalent to Mohr-Coulomb criterion. Therefore, the proposed strength criterion can be regarded as a modified Mohr-Coulomb criterion that can reflect the effect of intermediate principal stress.

By comparing with the experimental results and common strength criteria, it is found that strength criterion developed in this paper can well describe the intermediate principal stress effect on rock strength. New strength criterion matches the experimental results better than common strength criteria. In addition, new strength can avoid the nondifferentiable problem in numerical calculations involving the criterion.

REFERENCES

1. C. Chang and B. Haimson, "True triaxial strength and deformability of the German Continental Deep Drilling Program (KTB) deep hole amphibolite," Journal of Geophysical Research: Solid Earth, vol. 105, no. 8, pp. 18999–19013, 2000.

2. B. Haimson, "True triaxial stresses and the brittle fracture of rock," in Rock Damage and Fluid Transport, Part I, Pageoph Topical Volumes, pp. 1101–1130, Birkhäuser, Basel, Switzerland, 2006.

3. B. C. Haimson and C. Chang, "True triaxial strength of the KTB amphibolite under borehole wall conditions and its use to estimate the

maximum horizontal in situ stress," Journal of Geophysical Research: Solid Earth, vol. 107, no. 10, pp. 15.1–15.14, 2002.

4. K. Mogi, "Fracture and flow of rocks under high triaxial compression," Journal of Geophysical Research, vol. 76, no. 5, pp. 1255–1269, 1971.

5. M. Takahashi and H. Koide, "Effect of the intermediate principal stress on strength and deformation behavior of sedimentary rocks at the depth shallower than 2000 m," in Proceedings of the ISRM International Symposium, ISRM-IS-1989-003, International Society for Rock Mechanics, Pau, France, August 1989, https://www.onepetro.org/conference-paper/ISRM-IS-1989-003.

6. M. You, "True-triaxial strength criteria for rock," International Journal of Rock Mechanics and Mining Sciences, vol. 46, no. 1, pp. 115–127, 2009.

7. M. Cai, "Influence of intermediate principal stress on rock fracturing and strength near excavation boundaries—insight from numerical modeling," International Journal of Rock Mechanics and Mining Sciences, vol. 45, no. 5, pp. 763–772, 2008.

8. L. Shi and X.-C. Li, "Analysis of end friction effect in true triaxial test," Rock and Soil Mechanics, vol. 30, no. 4, pp. 1159–1164, 2009.

9. C. Tang, Rock Failure Mechanisms, CRC Press, 2010, http://cds.cern.ch/record/1614355.

10. P.-Z. Pan, X.-T. Feng, and J. A. Hudson, "The influence of the intermediate principal stress on rock failure behaviour: a numerical study," Engineering Geology, vol. 124, no. 1, pp. 109–118, 2012.

11. E. Fjær and H. Ruistuen, "Impact of the intermediate principal stress on the strength of heterogeneous rock," Journal of Geophysical Research: Solid Earth, vol. 107, no. 2, pp. ECV 3.1–ECV 3.10, 2002.

12. W. Weibull, "A statistical theory of the strength of materials," Journal of Applied Mechanics, vol. 18, pp. 293–297, 1951.

13. A. M. Freudenthal, "Statistical approach to brittle fracture," in Fracture, H. Liebowitz, Ed., vol. 2, pp. 591–619, 1968.

14. Z. P. Bažant, "Size effect on structural strength: a review," Archive of Applied Mechanics, vol. 69, no. 9-10, pp. 703–725, 1999.

15. M. R. Wisnom, "Relationship between strength variability and size effect in unidirectional carbon fibre/epoxy," Composites, vol. 22, no. 1, pp. 47–52, 1991.

16. Z. P. Bazant, Y. Xi, and S. G. Reid, "Statistical size effect in quasi-brittle structures. I. Is Weibull theory applicable?" Journal of Engineering

Mechanics, vol. 117, no. 11, pp. 2609–2622, 1991.

17. M. E. Saleh, J. L. Beuth, and M. P. de Boer, "Validated prediction of the strength size effect in polycrystalline silicon using the three-parameter weibull function," Journal of the American Ceramic Society, vol. 97, no. 12, pp. 3982–3990, 2014.

18. D. C. Drucker, Some Implications of Work Hardening and Ideal Plasticity, Division of Applied Mathematics, Brown University, 1949.

19. S. A. F. Murrell, "A criterion for brittle fracture of rocks and concrete under triaxial stress and the effect of pore pressure on the criterion," in Proceedings of the 5th Rock Mechanics Symposium, pp. 563–577, 1963.

20. M.-H. Yu, "Twin shear stress yield criterion," International Journal of Mechanical Sciences, vol. 25, no. 1, pp. 71–74, 1983.

Chapter 9

ROCK MASS BLASTABILITY CLASSIFICATION USING FUZZY PATTERN RECOGNITION AND THE COMBINATION WEIGHT METHOD

Shuangshuang Xiao, Kemin Li, Xiaohua Ding, and Tong Liu

State Key Laboratory of Coal Resources and Safe Mining, School of Mines, China University of Mining

ABSTRACT

Rock mass blastability classification provides a theoretical basis for rock mass blasting design, which is used to select blasting explosives, to estimate the unit explosive consumption, and to determine blasting design parameters. The primary factors that affect rock mass blastability were analyzed by selecting five indexes for rock mass blastability classification, that is, the rock Protodyakonov coefficient, rock tensile strength, rock density, rock wave impedance, and integrity coefficient of rock mass, and by identifying standards for the rock mass blastability classification and a method for testing the blasting classification indexes. The index weights were calculated using the combination weight method, which is based on game theory. A model for rock mass blastability classification was developed in combination with a fuzzy pattern recognition method. This classification method was applied to a Heidaigou open-pit coal mine, where mudstone, fine sandstone, medium sandstone, and coarse sandstone were determined to have a blastability degree of II, which corresponds to a blastability characterization of "easy," and the unit explosive consumption of mudstone, fine sandstone, medium sandstone, and coarse sandstone was determined to be 0.44, 0.42, 0.40, and $0.36 \, \text{kg/m}^3$, respectively. These results were used to develop a loose blasting design that was effective for loose blasting.

INTRODUCTION

Rock mass blastability is a measure of the resistance of a rock mass to blasting and crushing. The physical and mechanical properties and structural characteristics of rocks synchronize to various extents and in different ways to impede blasting and crushing under blasting loading [1]. Thus, the rock mass blastability is also a comprehensive indicator of several inherent properties of a rock mass under dynamic loading. The rock mass blastability reflects the degree of difficulty in rock blasting [2]. An understanding of rock mass blastability and systematic rock mass blastability classifications form the theoretical basis of blasting optimization design [3]. The rock mass blastability can be used to select suitable explosives, estimate the explosive unit consumption, and determine reasonable blasting parameters, which can reduce blasting costs and improve labor productivity by ensuring predictable blasting characteristics.

Foreign and Chinese scholars have conducted numerous research studies on rock mass blastability classification using different methods from various perspectives and have developed a variety of indexes and methods for rock mass blastability classification [4]. There are currently two methods for rock mass blastability classification. In the first method, the analysis and calculation involve one or more parameters, and a numerical value, such as a blastability index or the crushing energy, is chosen as a measure of the blastability of the rock mass [5]. In the second method, various parameters are chosen to describe the rock mass, and the rock mass blastability classification is performed using statistical mathematics, fuzzy mathematics, or other mathematical methods. The characteristics and internal mechanisms that affect rock mass blastability can be identified more accurately using several indexes to systematically evaluate the rock mass blastability. Therefore, many classification schemes and evaluation algorithms have been applied to rock mass blastability classification, including neural networks [6–8], projection pursuit [9], genetic algorithms [10], fuzzy set theory [11–13], cluster analysis [14], and attribute recognition [15]. Each algorithm has its advantages and disadvantages. For example, a neural network has considerable fault-tolerance ability and a rapid evaluation speed but requires a representative learning sample. In addition, the learning parameters and number of hidden layers are difficult to identify, and the number of hidden layers affects the convergence rate, the convergence properties of the network, and its applicability to nonlinear problems. The index weights do not need to be identified when the rock mass blastability is classified using the projection pursuit algorithm, thereby ensuring that the classification is objective. However, when optimizing the projection direction, this scheme can easily converge to a local optimum, which results in early maturing or early convergence, among other problems. Genetic algorithms can be used to accurately classify the respective categories but has additional parameter requirements, such as gene variables and genetic generations. The challenge encountered in using cluster analysis and attribute recognition is to determine reasonable index weights.

There are three essential requirements for developing a rock mass blastability classification model. First, the most representative characteristic must be chosen as the classification index, and classification standards must be developed. Second, each index should be assigned a reasonable weight. Finally, a suitable evaluating algorithm should be chosen. The rock blasting mechanism and the factors affecting rock mass blastability for the aforementioned research scenario were used to identify the classification indexes and classification standards for rock mass blastability. The blastability of a rock mass was described using the following values: "easy," "moderate," "difficult," and other fuzzy values, depending on practical production requirements. The indices of two rock samples typically have similar values but are characterized by different rating categories by observation. This result is not reasonable. Thus, rock blastability can be characterized using transitional values that lie in between different levels; that is, the values are fuzzy. There is no distinct boundary between different levels. The same rock mass could be assigned to different classifications by different people or based on different situations. Thus, it is more suitable to use fuzzy mathematics to classify rock mass blastability. Rock mass classification can then be based on this developed rock mass rating and the rock characteristics; that is, rock mass blastability classification is a pattern recognition problem. Therefore, fuzzy pattern recognition was used to develop a rock mass blastability classification model. However, the weights of the indexes are not considered in pattern recognition, which prevents the application of this method to cases with unequal index weights. The combination weight method was used to identify the index weights to reduce the effects of subjective factors and avoid irrelevant factors.

INDEXES AND STANDARDS FOR THE CLASSIFICATION OF ROCK MASS BLASTABILITY

Selection of Classification Index

Explosive blasting can fracture a rock mass in two ways. First, the cohesive force between rock granules can be overcome, thereby rupturing the internal rock structure and producing a new fracture surface. Second, primary and secondary fractures can be exacerbated via further expansion. Therefore, the primary influential factors of rock mass blastability are the physical and mechanical properties of the rock and the structural characteristics of the rock mass [16]. Typical indexes for classifying rock mass blastability include the rock density, rock wave impedance, rock tensile strength, integrity coefficient of the rock mass, and the mean crack interval of the rock mass. These indexes reflect different aspects of rock mass blastability. However, to simplify rock mass blastability classification and enable its practical application, all of the indexes are not used. The characteristics of a rock mass must be considered when choosing indexes for rock mass blastability classification. Minor representative indexes can comprehensively reflect different aspects of the rock mass blastability. There should be little or no correlation between the indexes. The chosen indexes should be easy to obtain using various methods, such as experiments and field measurements. The aforementioned considerations were used to select the following final indexes for when considering rock mass blastability.

Protodyakonov Coefficient and Tensile Strength of Rock

The shock wave and detonation gas produced by explosive blasting can typically rupture a rock mass through pulling and pressing. Therefore, the Protodyakonov coefficient and tensile strength of the rock are important parameters in rock mass blastability. During blasting, the rock is subject to temporary impact loading, for which the rock dynamic loading strength is clearly higher than the rock static loading strength. Therefore, the rock mass blastability can be accurately measured by indexes for the dynamic loading strength that are affected by the triaxial effect of the rock. However, the dynamic loading strength of rock is difficult to measure and exhibits a strong linear correlation with the uniaxial static loading compression strength and tensile strength [17]. Thus, the static loading strength is chosen as one of the indexes for rock mass blastability classification. The Protodyakonov coefficient of the rock, which is determined from the uniaxial compressive strength of the rock (1), is an objective measure of rock fastness that is widely applied in China. Therefore, the Protodyakonov coefficient and compressive strength of the rock are chosen as indexes for rock mass blastability classification:

$$f = \frac{\sigma_p}{K}.$$

(1)

In the equation above, f is the Protodyakonov coefficient of the rock; σ_p is the rock's uniaxial compressive strength in MPa; and K is a constant equal to 10 MPa.

Rock Density

The energy produced from rock blasting is transferred into kinetic energy in the rock block, which can result in the displacement or thrusting of the rock block. A higher rock density causes more of the energy produced in rock blasting to be consumed by the displacement and thrusting of the rock. Therefore, the amount of energy consumed is indicative of the difficulty of the rock blast; that is, the rock mass blastability decreases with increasing rock

Table 1: Classification standards of rock mass blastability

Blastability class	σ_i (MPa)	f	P (t/m³)	K_v	Z (10⁶ kg/m³ × m/s)	Characterization of blastability
I	≤1.5	≤2.5	≤2.0	≤0.15	≤3	Very easy
II	1.5–3	2.5–6	2.0–2.4	0.15–0.35	3–6	Easy
III	3–6	6–10	2.4–2.75	0.35–0.55	6–9	Moderate
IV	6–12	10–18	2.75–3.0	0.55–0.75	9–12	Difficult
V	≥12	≥18	≥3.0	≥0.75	≥12	Very difficult

density. Therefore, the rock density is generally used as an index for rock mass blastability classification.

Rock Wave Impedance.

The dynamic Poisson's ratio, dynamic elastic modulus, bulk modulus, and Lame param- ́ eter for rock can be derived from the P- and S-wave velocities of the rock. All of the physical property indexes of rocks such as the mineral composition, porosity, water-bearing, and weathering degree are captured in the P-wave velocity of the rock. The P-wave velocity of the rock can be easily measured. The rock wave impedance can be obtained by multiplying the P-wave velocity of the rock by the rock density (2). The impedance is a measure of the force of the disturbance required to produce a unit speed of a moving rock particle during the transmission of a stress wave in the rock and is a measure of the resistance of the rock to momentum transfer. Therefore, the rock wave impedance is chosen as one of the indexes for the rock mass blastability classification:

$$z = \rho V_{pr}.$$

In the equation above, z denotes the rock wave impedance (10^6 kg/m³ × m/s), ρ denotes the rock density (kg/m), and Vpr denotes the P-wave velocity of the rock (m/s).

Integrity Coefficient of Rock Mass.

The geological properties of a rock mass, such as the integrity, fissure, and degree of development of a joint fissure, are captured in the Pwave velocity of the rock mass. A fast wave propagation velocity in a rock mass typically corresponds to mild rock densification, hardness, integrity, and weathering. In contrast, a slow wave propagation velocity corresponds to severe rock porosity, weakness, fragmentation, structural development, and weathering.

The integrity coefficient of a rock mass is given by the square of the ratio of the P-wave velocity of a rock mass to the P-wave velocity of the rock (3), which reflects the extent of fracturing for a geological discontinuity, such as a joint fissure. A rock mass with a small integrity coefficient is susceptible to a large amount of rock mass crushing, and the rock mass can be easily blasted. Therefore, the integrity coefficient of the rock mass is chosen as one of the indexes for the rock mass blastability classification:

$$K_V = \left(\frac{v_{pm}}{v_{pr}} \right)^2 .$$

(3)

In the equation above, K_V denotes the integrity coefficient of the rock mass, V_{pm} denotes the P-wave velocity of the rock

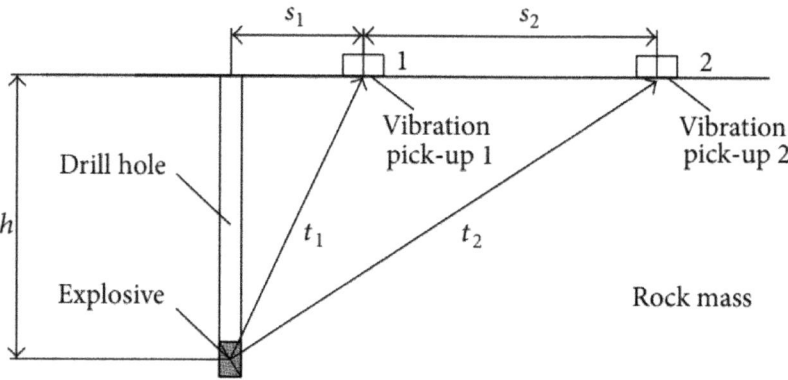

Figure 1: Schematic for testing the P-wave velocity of the rock mass.

mass (m/s), and V_{pr} denotes the P-wave velocity of the rock (m/s).

In conclusion, five indexes were chosen for the rock mass blastability classification: the Protodyakonov coefficient and tensile strength of the rock, rock density, rock wave impedance, and the integrity coefficient of the rock mass. Among these indexes, the Protodyakonov coefficient and tensile strength of the rock are mechanical property indexes of the rock, the rock density and rock wave impedance are physical property indexes of the rock; and the integrity coefficient of the rock mass is a measure of the geological properties of the rock mass. These five indexes primarily reflect the relevant physical and mechanical properties and characteristics of the geological structure of a rock mass and blasting and can be easily obtained by field measurements and experiments.

Determination of the Standards for Rock Mass Blastability Classification.

The value selection and designation of classification standards play an important role in the development of models for rock mass blastability classification. In the literature, the standards for classification indexes are determined using five ranks for rock mass blastability: very easy, easy, moderate, difficult, and very difficult. These classification standards for rock mass blastability are shown in Table 1 [18].

Measurements and Results for the Indexes

P-Wave Velocity of the Rock Mass.

Figure 1 illustrates a hole that is drilled by a geological drilling rig to measure a rock mass. A small quantity of explosive is placed at the bottom of the hole, and two vibration pick-up instruments are placed around the hole. These two instruments should be

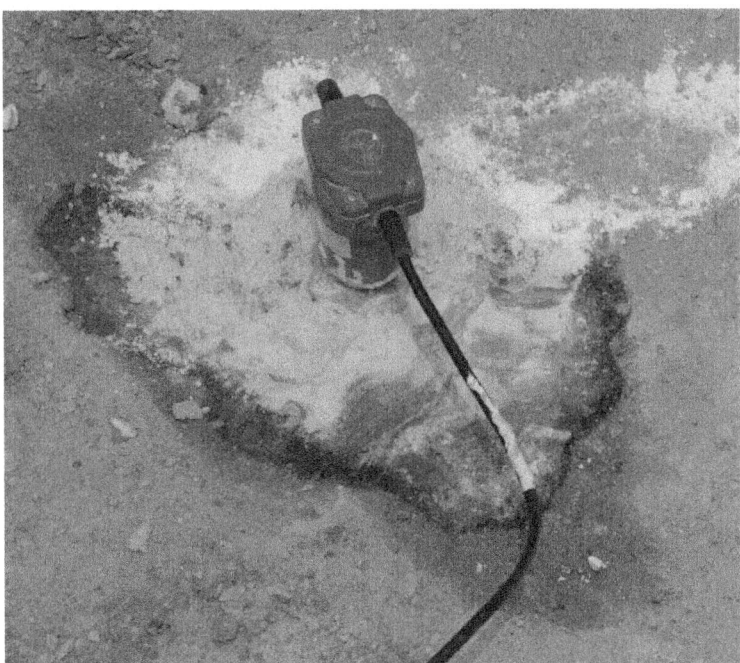

Figure 2: Vibration pick-up.

placed as far apart as possible and in a straight line with the hole. Detonating the explosive in the hole produces a surface wave and volume wave in the rock mass, where the volume wave consists of P-wave and S-wave components. The velocity of the P-wave is higher than that of the S-wave, and, thus, the P-wave reaches the measuring point before the S-wave. When the P-wave reaches measuring point 1, it triggers the "vibration pick-up 1," which records the time $t1$ at which the P-wave reaches point 1. The "vibration pick-up 2" records the time t_2 at which the P-wave reaches measuring point 2. The P-wave velocity of the rock mass is calculated using the difference between the times that the P-wave reaches measuring points 1 and 2 and the distances between the two measuring points and the center of the explosive cartridge. Equation (4) shows the formula used to calculate the P-wave velocity of the rock mass for the geometry shown in Figure 1 [19]:

$$V_{pm} = \frac{\sqrt{h^2 + (s_1 + s_2)^2} - \sqrt{h^2 + s_1{}^2}}{t_2 - t_1}.$$

(4)

In the equation above, h denotes the depth of the hole (in m), 1 denotes the distance between measuring point 1 and the hole (in m), s_2 denotes the distance between the two points (in m), t_1 denotes the time at which the P-wave reaches point 1 (in s), and t_2 denotes the time at which the P-wave reaches point 2 (in s).

Following the aforementioned principles, field tests were performed at different locations in the Heidaigou open-cast coal mine, where the drilling hole diameter was 127 mm and the hole depth was between 5 and 17 m. A vibration velocity transducer (model CD-21, manufactured by Beijing Instrument Industry Group Co., Ltd., in Beijing, China) was used as the vibration pick-up instrument, and a blasting vibration instrument (model EXP3850, Chengdu VIDTS Dynamic Instrument Co., Ltd., in Chengdu, China) was used as the blasting vibration recording instrument, as shown in Figures 2 and 3, respectively.

The bench consisted primarily of medium sandstone, and a hole was drilled in the bench at an altitude of 1,185 m, as shown in Figure 4. The depth of the hole was 16.3 m, and the distances s_1 and s_2 were 7.1 and 16.7 m, respectively

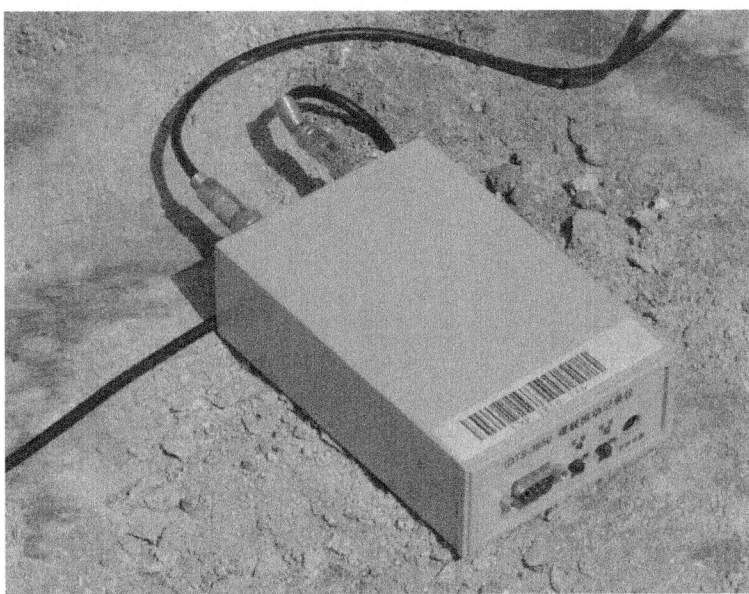

Figure 3: Blasting vibration recorder.

Figure 4: Photograph of the rock mass.

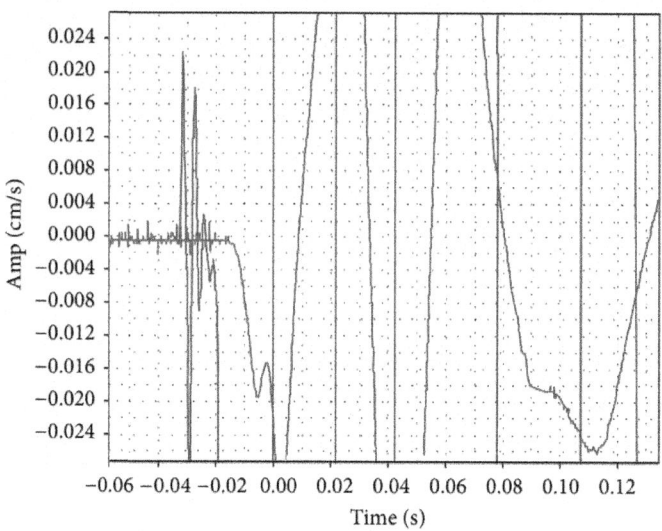

Figure 5: Test pattern of the P-wave velocity of the rock mass.

The two times that were obtained from the test, $t_1 = -23.65$ ms and $t_2 = -13.75$ ms, are shown in Figure 5. The Pwave velocity of the medium sandstone mass was calculated to be vpm = 1,118 m/s using (4).

Similarly, the measurements indicated that the wave velocity in the mudstone mass was 1,273 m/s, the wave velocity in the sandstone mass was 944 m/s, and the wave

Figure 6: Rock samples.

Table 2: Test results for the rock mass blastability indexes

Rock mass	σ_t (MPa)	σ_p (MPa)	ρ (t/m^3)	σ_{pm} (m/s)	v_{pr} (m/s)
Mudstone	2.19	27.6	2.68	127 3	374 4
Fine sandstone	2.72	43.6	2.41	957	322 6
Medium sandstone	2.26	27.3	2.36	1118	272 4
Coarse sandstone	1.50	24.3	2.23	796	236 0

velocity in the coarse sandstone rock mass was 796 m/s in the Heidaigou open-cast coal mine.

P-Wave Velocity of the Rock Mass.

Different types of rocks were collected from the Heidaigou open-cast coal mine and processed into standard samples with specifications of ϕ50 mm × 100 mm, as shown in Figure 6. A TICO ultrasonic detector was used to determine the P-wave velocity of the rock. The TICO detector had two sensors, a sender and a receiver. The distance l between the two sensors and the propagation time t of the test sound wave to travel between the two sensors were used to calculate the wave velocity Vpr = l/t. The wave velocities were as follows: 3,744 m/s for the mudstone, 3,226 m/s for the fine sandstone, 2,724 m/s for the medium sandstone, and 2,360 m/s for the coarse sandstone.

Rock Density, Tensile Strength, and Compressive Strength.

The rock density was determined by weighing the samples and calculating the sample volumes. The tensile strength and uniaxial compressive strength of the rock were measured using microcomputer control electron universal testing machines (modelWDW300, KeXin Testing Machine Co., Ltd., in Changchun, China). The specifications for the test samples for the uniaxial compressive strength measurements were ϕ50 mm × 100 mm. The Brazil splitting method was used to measure the tensile strength of the rock with specifications of ϕ50 mm × 25 mm.

The test results for the rock density, rock tensile strength, and rock compressive strength are provided in Table 2.

Table 3: Rock mass blastability classification index values

Rock mass	σ_t (MPa)	f	P (t/m^3)	K_v	Z (10^6 kg/m^3 × m/s)
Mudstone	2.19	2.76	2.68	0.12	10.03
Fine sandstone	2.72	4.36	2.41	0.09	7.77
Medium sandstone	2.26	2.73	2.36	0.17	6.43
Coarse sandstone	1.50	2.43	2.23	0.11	5.26

Table 4: Judgment matrix P

Variables	σ_t	f	p	K_v	z
σ_t	1	1/5	1/3	1/4	1/2
f	5	1	3	2	4
P	3	1/3	1	1/2	2
K_v	4	1/2	2	1	3
z	2	1/4	1/2	1/3	1

The classification index values of the rock masses were calculated and are shown in Table 3.

DETERMINATION OF THE INDEX WEIGHTS USING THE COMBINATION WEIGHT METHOD

To circumvent the deficiencies of the subjective weight method and the objective weight method, the index weight was systematically determined using the analysis hierarchy process (AHP) tso calculate the subjective weight, the entropy method (EM) was used to calculate the objective weight, and the combination weight method (CWM), which is based on game theory (GT), was used to determine the index weights for the rock mass blastability classification.

Determination of Subjective Weight Using the Analysis Hierarchy Process.

AHP is a systematic, hierarchical analysis method that incorporates both qualitative and quantitative analyses; thus, AHP is simple and practical and avoids the uncertainties and errors that can arise when determining the index weights.The methodology used to calculate the weights using the AHP is detailed in Saaty's papers [20, 21]. A pairwise comparison of the Protodyakonov coefficient, tensile strength, density, wave impedance, and integrity coefficient of the rock mass was performed to construct a judgment matrix, as shown in Table 4.

The calculation produced the eigenvector W = (0.06, 0.42, 0.16, 0.26, 0.10)^T, maximum eigenvalue λmax = 5.068, and the consistency index CI = 0.017. For $n = 5$ elements, the random index was found to be RI = 1.12, which was used to calculate the consistency ratio CR = 0.017/1.12 = 0.015 < 0.1. Therefore, the judgment matrix satisfied the consistency check. The consistency ratio weight vector was determined using AHP to be W_1 = (0.06, 0.42, 0.16, 0.26, 0.10).

Determination of the Objective Weight Using the Entropy Method.

Information theory states that the information entropy is a measure of the degree of disorder in a system. Lower information entropy indicates a lower degree of disorder and a higher utility value of the information; the opposite holds true for higher information entropy [22]. An indicator that does not produce different effects for different levels of rock mass blastability is not useful for rock mass blastability classification. Thus, if an indicator produces a smaller difference for different levels of rock mass blastability, the effect of the indicator on the rock mass blastability classification is smaller, and the corresponding information entropy is higher; the opposite holds true for an indicator that produces a larger difference for different levels of rock mass blastability. That is, the difference degree of the indicator for the rock mass blastability classification is inversely proportional to the information entropy. Therefore, the index weight was determined from the difference degree of the rock mass blastability classification indicator using the EM. The calculation procedure is detailed in the papers of PeiYue et al. [23] and Zou et al. [24].

Using the sample data for mudstone, fine sandstone, medium sandstone, and coarse sandstone that are given in Table 3, the entropy of each indicator was calculated to be H = (0.984 8, 0.979 6, 0.998 4, 0.979 9, 0.979 5), which yielded the consistency ratio weight vector W_2 = (0.20, 0.26, 0.02, 0.26, 0.26).

Combination Weight from Game Theory

The weights obtained from different weight methods may not be consistent with each other. GT can be used to minimize the sum of the differences between the final determined weight and the weight determined from each method [25]. The procedure for determining the GT-based combination weight is given below [26]. Assume that L methods are used to weight the indexes of the rock mass blastability classification, where the L weight vectors are given by $W_k = (\omega_{k1}, \omega_{k2}, ..., _n)$ (where $1 \leq k \leq L$, and n indicates the number of indexes of the rock mass blastability classification). A combination weight vector can be constructed from a random linear combination of the L weight vectors as follows:

$$\mathbf{W}_c = \sum_{k=1}^{L} \alpha_k \cdot \mathbf{W}_k^{\mathrm{T}}.$$

(5)

In the equation above, W_c denotes the combination weight vector and α_k denotes the coefficient of the linear combination, where $\alpha_k > 0$.

The most satisfactory combination weight vector is obtained by minimizing the deviation between the combination weight vector W_c and each weight vector W_k by optimizing the coefficient α_k of the L linear combinations. The game model is given as

$$\min \left\| \sum_{k=1}^{L} \alpha_k \cdot W_k^T - W_l^T \right\|_2 \quad (l = 1, 2, \ldots, L).$$

(6)

The differential attribute of the matrix is used to calculate the optimum first derivative condition in (6), which can be written in the following matrix form:

$$\begin{bmatrix} W_1 \cdot W_1^T & W_1 \cdot W_2^T & \cdots & W_1 \cdot W_L^T \\ W_2 \cdot W_1^T & W_2 \cdot W_2^T & \cdots & W_2 \cdot W_L^T \\ \vdots & \vdots & \ddots & \vdots \\ W_L \cdot W_1^T & W_L \cdot W_2^T & \cdots & W_L \cdot W_L^T \end{bmatrix} \begin{bmatrix} \alpha_1 \\ \alpha_2 \\ \vdots \\ \alpha_L \end{bmatrix}$$

$$= \begin{bmatrix} W_1 \cdot W_1^T \\ W_2 \cdot W_2^T \\ \vdots \\ W_L \cdot W_L^T \end{bmatrix}.$$

(7)

The matrix is used to calculate the normalization processing of (α_1, α_2,...,), as shown in (8), to obtain the optimal linear combining coefficient α_k^* $(1 \le k \le L)$ as follows:

$$\alpha_k^* = \frac{\alpha_k}{\sum_{k=1}^{L} \alpha_k}.$$

(8)

Therefore, the optimal combination weight vector is

$$W^* = \sum_{k=1}^{L} \alpha_k^* \cdot W_k.$$

(9)

In the equation above, W^* denotes the optimal combination weight vector, $W^* = (\omega_j^*)_{1 \times n}$, where ω_j^* is the optimal combination weight vector for the jth index, with $\sum_{i=1}^{n} \omega_j^* = 1$

The calculated subjective and objective weights were used to determine $\alpha_1^* = 0.691\ 2$ and $\alpha_2^* = 0.308\ 8$ using (7) and (8), and $\mathbf{W}^* = (0.10, 0.37, 0.12, 0.26, 0.15)$ was calculated using (9). The index weights that were determined using the AHP, EM, and CWM are shown in Figure 7.

Figure 7 illustrates that the index weights that were determined using the combination weight method lay between those calculated using the AHP and those calculated using EM. The CWM balanced and coordinated the impacts that the subjective and objective methods exerted on the weights, thus overcoming the one-sidedness of each method; the finalized CWM produced more realistic results than the individual methods.

FUZZY PATTERN RECOGNITION MODEL FOR ROCK MASS BLASTABILITY

Model Principles

We assume that the rock mass blastability can be divided into m levels for n indexes, where each blastability has its own grading standards (the value range of each index) and that the m grading standards of the rock mass blastability can serve as the m fuzzy subsets $\widetilde{A}_1, \widetilde{A}_2, \ldots, \widetilde{A}_m$ that constitute a standard sample database $\{\widetilde{A}_1, \widetilde{A}_2, \ldots, \widetilde{A}_m\}$. The value of the jth grading standard of the rock mass

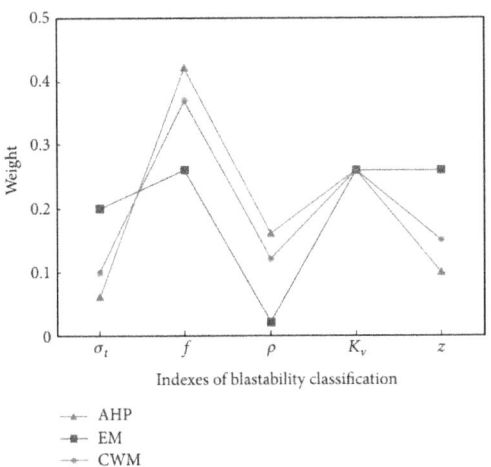

Figure 7: Comparison of weights.

blastability classification is denoted by x_j ($j = 1, 2, \ldots, n$), and the universe $U = \{u \mid u = (x_1, x_2, \ldots, x_n)\}$, namely, u, denotes the set of n indexes for a certain rock mass sample.

The membership functions $\mu_i(x_j)$ and $\mu_i(u)$ are constructed for a particular fuzzy subset $\tilde{A}i$. For these functions, i ($i = 1, 2, \ldots, m$) denotes the number of the rock mass blastability classification for a standard sample, j ($j = 1, 2, \ldots, n$) denotes the index number, and x_j indicates the value of the jth index for a specific rock mass sample. (x_j) denotes the membership degree of the jth index of a specific rock mass sample that relatively belongs to the blastability degree of the ith rock mass, and $\mu_i(u)$ is the membership degree of a specific rock mass sample that relatively belongs to the blastability degree of the ith rock mass.

If $u_0 \in U$ and $k \in \{1, 2, \ldots, m\}$ exist such that $\mu_k(u_0) = \max\{\mu_1(u_0), \mu_{2}(u_0), \ldots, u_m(x0)\}$, u_0 is considered to belong to $\tilde{A}_k$, which means that the blastability degree of rock mass sample u_0 belongs to degree k.

Development of the Membership Function

To establish a fuzzy relation between the classification indexes and the standard samples, the membership functions must first be developed between each index and each standard sample [27]. The fuzziest principle and clearest principle must be obeyed while formulating the membership function. That is, the membership degree is 0.5 at the endpoint of the interval for the fuzziest state, and the membership degree is 1 at the midpoint of the interval for the clearest state. Moreover, the sum of the membership degrees at any point is 1 [28].

Commonly used membership functions include triangular, trapezoidal, normal distribution and mountain-shaped membership functions [29]. Mikkili and Panda concluded that there are no considerable differences among the membership degrees that correspond to different membership

Table 5: Value range of the index

Index number	Degree 1	Degree 2	$\cdots$	Degree $m - 1$	Degree m
j	$\leq a_{1j}$	$a_{1j} \sim a_{2j}$	$\cdots$	$a_{(m-2)j} \sim a_{(m-1)j}$	$\geq a_{(m-1)j}$

functions, and the analysis results are consistent [30]. Therefore, a trapezoidal membership function was used to formulate the fuzzy assessment matrix. Based on the principle used to identify the membership function, the trapezoidal membership function degenerated into a triangular membership function.

Table 1 illustrates that the values of the classification indexes that were chosen in this study increased with the degree of rock mass blastability, namely, the incremental index. The assumed value range of the jth index for the ith degree is provided in Table 5. Let us consider the trapezoidal membership function as an example, where the membership functions for each index of each degree are given as follows:

$$
\mu_1(x_j) = \begin{cases}
1 & x_j < b_{1j} \\[2mm]
\dfrac{x_j + b_{1j} - 2a_{1j}}{2(b_{1j} - a_{1j})} & b_{1j} \le x_j < a_{1j} \\[2mm]
\dfrac{b_{2j} - x_j}{2(b_{2j} - a_{1j})} & a_{1j} \le x_j < b_{2j} \\[2mm]
0 & x_j \ge b_{2j},
\end{cases}
$$

$$
\mu_i(x_j)
$$

$$
= \begin{cases}
0 & x_j < b_{(i-1)j},\; x_j \ge b_{(i+1)j} \\[2mm]
\dfrac{x_j - b_{(i-1)j}}{2(a_{(i-1)j} - b_{(i-1)j})} & b_{(i-1)j} \le x_j < a_{(i-1)j} \\[2mm]
\dfrac{x_j + b_{ij} - 2a_{(i-1)j}}{2(b_{ij} - a_{(i-1)j})} & a_{(i-1)j} \le x_j < b_{ij} \\[2mm]
\dfrac{b_{ij} - 2a_{ij} - x_j}{2(a_{ij} - b_{ij})} & b_{ij} \le x_j < a_{ij} \\[2mm]
\dfrac{b_{(i+1)j} - x_j}{2(b_{(i+1)j} - a_{ij})} & a_{ij} \le x_j < b_{(i+1)j},
\end{cases}
$$

$$
\mu_m(x_j) = \begin{cases}
0 & x_j < b_{(m-1)j} \\[2mm]
\dfrac{x_j - b_{(m-1)j}}{b_{mj} - b_{(m-1)j}} & b_{(m-1)j} \le x_j < b_{mj} \\[2mm]
1 & x_j \ge b_{mj}.
\end{cases}
$$

(10)

In the equations above, x_j denotes the value of the jth index for a given rock sample, (x_j) denotes the membership degree of index x, that relatively belongs to the ith blastability degree, and $b_{mj} = 2a_{(m-1)j} - b_{(m-1)j}$, where $b_{1j} = \max\{2a_{1j} - b_{2j}, a_{1j}/2\}$, and $b_{ij} = (a_{(i-1)j} + a_{ij})/2$ $(2 \leq i \leq m - 1)$.

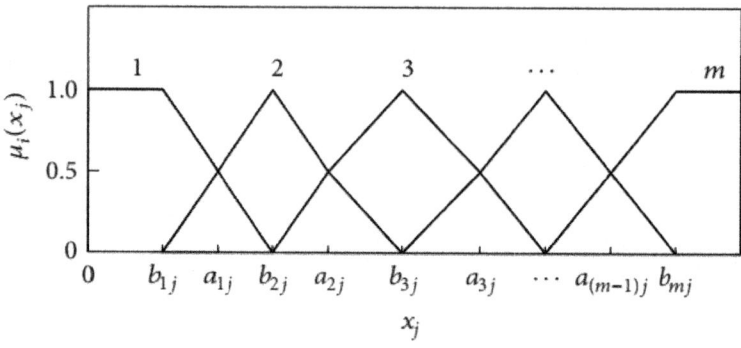

Figure 8: Pictorial representation of trapezoidal membership functions.

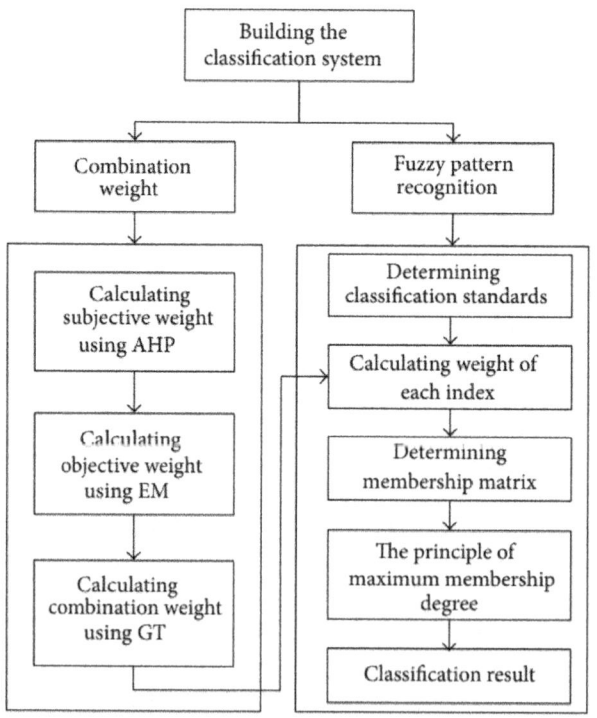

Figure 9: Rock mass blastability classification procedure.

Equations(10) correspond to the membership function of the incremental index, which is shown in Figure 8. The membership function of the descending index can be determined using the same method and is thus not presented here. 4.3. Identification of the Rock Mass Blastability Degree. The procedure for the rock mass blastability classification is summarized in Figure 9. The membership degree of every blastability degree can be calculated using (11) together with the classification index and the classification standard provided in section two, the combination weight provided in section three, and the membership function provided in section four for any type of rock as long as the values of the n indexes are known, and the blastability degree of the rock can be determined using the principle of the maximum membership degree. Consider

$$\mu_i(u) = \frac{1}{n}\sum_{j=1}^{n}\omega_j^* \mu_i(x_j) \quad 1 \le i \le m.$$

(11)

Table 6: Membership degree for each index for mudstone

Blastability class	σ_t	f	ρ	K_v	z
I	0.04	0.43	0	0.73	0
II	0.96	0.57	0	0.27	0
III	0	0	0.70	0	0.16
IV	0	0	0.30	0	0.84
V	0	0	0	0	0

Table 7: Membership degree of each rock mass

Rock mass	Blastability class				
	I	II	III	IV	V
Mudstone	0.35	0.38	0.11	0.16	0.00
Fine sandstone	0.24	0.51	0.24	0.01	0.00
Medium sandstone	0.27	0.59	0.14	0.00	0.00
Coarse sandstone	0.44	0.51	0.05	0.00	0.00

In the equation above, ω_j^* denotes the satisfactory combination weight of the jth index.

PROJECT APPLICATION

Rock Mass Blastability Classification

The identified classification index provided in Table 1 was used to calculate the specific membership function using (10). Let us consider the rock mass blastability classification of mudstone from the Heidaigou open-cast coal mine as an example. The membership degree between each index and each rock mass blastability classification was calculated by substituting the relevant indexes of the rock mass of mudstone into the membership function in Table 3 and is shown in Table 6. Equation (11) was used to calculate the following values: $\mu_1(u) = 0.35$, $\mu_2(u) = 0.38$, $\mu_3(u) = 0.11$, $\mu_4(u) = 0.16$, and $\mu_5(u) = 0$. The principle of the maximum membership degree was used to determine that the largest rock mass blastability degree of mudstone was $\mu_2(u)$, which corresponds to a degree of 2, the "easy" blasted rock mass. The membership degree between each index for the rock masses of fine sandstone, medium sandstone, and coarse sandstone and the rock mass blastability classification was similarly confirmed, and the membership degree between each rock mass and the rock mass blastability classification was calculated, as shown in Table 7 and Figure 10. The principle of the maximum membership degree indicated that the rock masses of fine sandstone, medium sandstone, and coarse sandstone could all be categorized as easy blasted rock masses.

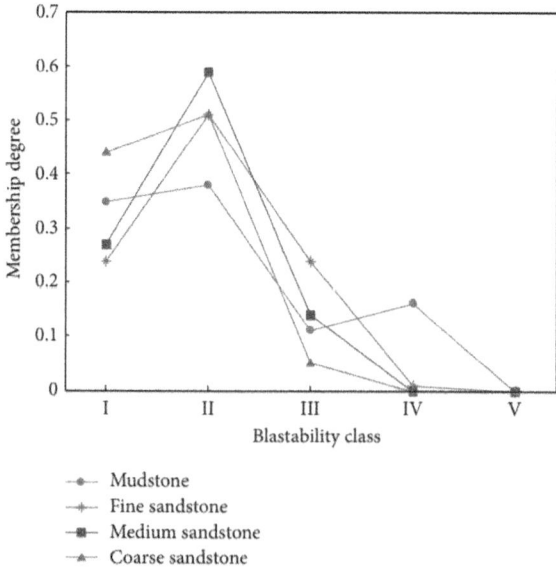

Figure 10: Membership degree of each rock mass.

Figure 10 illustrates that although each rock mass had a maximum membership degree for a blastability of II, which belonged to the easy blasted rock mass, the mudstone and coarse sandstone rock mass also had a large membership degree for a blastability of I, indicating that the degree for mudstone and coarse sandstone was between the very easy blasted and easy blasted rock masses. Although the fine sandstone and coarse sandstone had equal membership degrees for a blastability of II, the coarse sandstone had a

Table 8: Unit explosive consumption of rock mass of each classification

Blastability class	Unit explosive consumption (kg/m^3)
I	≤0.35
II	0.35–0.45
III	0.45–0.65
IV	0.65–0.90
V	≥0.90

larger membership degree for a blastability of I and a smaller membership degree for a blastability of III, illustrating that coarse sandstone was more explosive than fine sandstone despite both sandstones being categorized as easy blasted rock mass.

Unit Explosive Consumption

The relevant literature and material and blasting experience from open-cast mines was used to determine the unit explosive consumption (ANFO) for the loose blasting of a rock mass for each classification, as shown in Table 8.

Let q_2, q_3, and $q4$ denote the midpoints of the interval of the unit explosive consumption for degrees II, III, and IV in Table 8, respectively, where $q_1 = 0.35 - (q_2 - 0.35)$ and $q_5 = 0.90 + (0.90 - q_4)$ form the vector $Q = (q_i)1×5 = (0.30, 0.40, 0.55, 0.78, 1.02)$.

The membership matrix $U = (\mu_i(u))1×5$ between the rock mass and the rock mass blastability degree was calculated. The unit explosive consumption for the loose blasting of the rock mass was then calculated using

$$q = \begin{cases} \leq 0.3 & \mu_1(u) = 1 \\ \geq 1.02 & \mu_5(u) = 1 \\ Q \cdot U^T & \mu_1(u) \neq 1, \mu_5(u) \neq 1. \end{cases}$$

$$(12)$$

Consider the mudstone rock mass in the Heidaigou open-cast coal mine as an example, where the unit explosive consumption for loose blasting was calculated to be q_{mud} = (0.30, 0.40, 0.55, 0.78, 1.02) · (0.35, 0.38, 0.11, 0.16, 0) = 0.44 kg/m3 . Similarly, the unit explosive consumption values for fine sandstone, medium sandstone, and coarse sandstone were determined to be qf = 0.42 kg/m³ , qmed = 0.40 kg/m³ , and q_c = 0.36 kg/m³ , respectively

When there are several types of rock masses in a blasting area, the unit explosive consumption can be approximated using

$$q_a = \frac{\sum_{i=1}^{n} q_{mi} V_{mi}}{\sum_{i=1}^{n} V_{mi}}.$$

(13)

In the equation above, q_a denotes the average unit explosive consumption (in kg/m³) for loose blasting in the blasting area, q_{mi} denotes the unit explosive consumption (in kg/m³) for loose blasting of the rock mass i, V_{mi} denotes the volume (in m³) of the rock mass i, and n denotes the number of types of rock masses in the blasting area. When there are several rock masses that are approximately level or when there is a gentle incline from the top to the bottom of the blasting area, V_{mi} can be replaced by h_{mi} in (13); thus, h_{mi} denotes the thickness of the rock mass i (m).

CONCLUSIONS AND FURTHER RESEARCH

(1)There are no clear boundaries between different classifications for the blastability of a rock mass; that is, there is fuzziness in the classification problem. Thus, a fuzzy pattern recognition method was used to develop a model for rock mass blastability classification. The classification results from the model were obtained as a vector, which provided more information than a point value. Thus, a simple algorithm was used to make full use of the information, and the classification procedure was easy to understand.(2)The relevant indexes of each rock mass that was tested in the Heidaigou open-cast coal mine were used with the developed rock mass blastability classification model to demonstrate that mudstone, fine sandstone, medium sandstone, and coarse sandstone in the Heidaigou open-cast coal mine could all be considered easy blasted rock mass. The unit explosive consumption of mudstone, fine sandstone, medium sandstone, and coarse sandstone was determined to be 0.44, 0.42, 0.40, and 0.36 kg/m³, respectively. Blasting experiments in the Heidaigou open-cast coal mine demonstrated the accuracy of the aforementioned conclusions. These conclusions were used to develop a loose blasting design, which yielded good results.(3)The standards for the blastability classification of the rock masses were determined from the literature. These standards can be applied to actual situations but are not based on sufficiently large amounts of sample data; therefore, the scientific basis of these standards has not been confirmed. In addition, a sufficiently large amount of sample data was not used with the objective weight method. Thus, this insufficient sample size likely affected the weights that were determined by the objective weight method to some extent. Therefore, the next step is to collect a large amount of sample data of rock masses in typical areas for use in a more accurate calculation of the weight of each classification index and a more accurate calculation and analysis of the standards for the blastability classification of rock masses.

ACKNOWLEDGMENT

Financial support for this work, provided by the National High Technology Research and Development Program of China (no. 2012AA062002), is gratefully acknowledged.

REFERENCES

1. J. Zhou and X. Li, "Integrating unascertained measurement and information entropy theory to assess blastability of rock mass," Journal of Central South University, vol. 19, no. 7, pp. 1953–1960, 2012.

2. Q. Y. Ma and Z. H. Zhang, "Classification of frozen soil blastability by using perception neural network,"Journal of Coal Science & Engineering, vol. 8, no. 1, pp. 54–58, 2002.

3. J.-L. Shang, J.-H. Hu, R.-S. Mo, X.-W. Luo, and K.-P. Zhou, "Predication model of game theory-matter-element extension for blastability classification and its application," Journal of Mining & Safety Engineering, vol. 30, no. 1, pp. 86–92, 2013.

4. J.-P. Latham and P. Lu, "Development of an assessment system for the blastability of rock masses,"International Journal of Rock Mechanics and Mining Sciences, vol. 36, no. 1, pp. 41–55, 1999.

5. S. R. Chen, S. Q. Xie, Y. L. Li, Y. X. Yu, and P. Z. Wu, "Experimental study of physical properties and drillability and blastability of special ore-bodies," Journal of Central South University of Technology, vol. 35, no. 4, pp. 667–669, 2004.

6. Y. D. Cai and L. S. Yao, "Artificial neural network approach of determining the grade of the blasting classification of rocks," Blasting, vol. 10, no. 2, pp. 50–52, 1993.

7. X. T. Feng, "A study on neural network on rock blastability," Explosion and Shock Waves, vol. 14, no. 4, pp. 298–306, 1994.

8. X. T. Feng, "A neural network approach to comprehensive classification of rock stability, blastability and drillability," International Journal of Surface Mining, Reclamation and Environment, vol. 9, no. 2, pp. 57–62, 1995.

9. C. Fang, X.-G. Zhang, and Z.-H. Dai, "Projection pursuit regression method of rock blastability classification based on artificial fish-swarm algorithm," Blasting, vol. 26, no. 3, pp. 14–17, 2009.

10. Y. D. Cai, "Application of genetic programming in determining the blasting classification of rocks,"Explosion and Shock Waves, vol. 15, no. 4, pp. 329–334, 1995.

11. P. P. Huang, "Rock blastability classification with fuzzy synthesis," Quarterly of the Changsha Institute of Mining Research, vol. 9, no. 4, pp. 63–72, 1989.

12. Y. Azimi, M. Osanloo, M. Aakbarpour-Shirazi, and A. Aghajani Bazzazi, "Prediction of the blastability designation of rock masses using fuzzy sets," International Journal of Rock Mechanics and Mining Sciences, vol. 47, no. 7, pp. 1126–1140, 2010.

13. A. Aydin, "Fuzzy set approaches to classification of rock masses," Engineering Geology, vol. 74, no. 3-4, pp. 227–245, 2004.

14. S. J. Qu, S. L. Mao, W. S. Lu, M. Y. Xin, Y. J. Gong, and X. Y. Jin, "A method for rock mass blastability classifcation based on weighted clustering analysis," Journal of University of Science and Technology Beijing, vol. 28, no. 4, pp. 324–329, 2006.

15. J.-G. Xue, J. Zhou, X.-Z. Shi, H.-Y. Wang, and H.-Y. Hu, "Assessment of classification for rock mass blastability based on entropy coefficient of attribute recognition model," Journal of Central South University (Science and Technology), vol. 41, no. 1, pp. 251–256, 2010.

16. Y. L. Yu, D. S. Wang, and S. J. Qu, "Zoning of the blasting compliance of rocks in shuichang open pit,"Chinese Journal of Rock Mechanics and Engineering, vol. 9, no. 3, pp. 195–201, 1990.

17. S.-J. Qu, M.-Y. Xin, S.-L. Mao et al., "Correlation analyses of blastability indexes for rock mass," Chinese Journal of Rock Mechanics and Engineering, vol. 24, no. 3, pp. 468–473, 2005. ·

18. P. Lu, The characterisation and analysis of in-situ and blasted block size distribution and the blastability of rock masses [Ph.D. thesis], University of London, London, UK, 1997.

19. K. M. Li, Z. H. Guo, and Y. Zhang, Research and Application of Casting Blast Technologies in Opencast Mine, China Coal Industry Publishing House, Beijing, China, 2011.

20. T. L. Saaty, "How to make a decision: the analytic hierarchy process," European Journal of Operational Research, vol. 48, no. 1, pp. 9–26, 1990.

21. T. L. Saaty, "Decision making with the analytic hierarchy process," International Journal of Services Sciences, vol. 1, no. 1, pp. 83–98, 2008.

22. J. Wu, J. Sun, L. Liang, and Y. Zha, "Determination of weights for ultimate cross efficiency using Shannon entropy," Expert Systems with Applications, vol. 38, no. 5, pp. 5162–5165, 2011.

23. L. Pei-Yue, Q. Hui, and W. Jian-Hua, "Application of set pair analysis method based on entropy weight in groundwater quality assessment—a case study in dongsheng city, northwest China," E-Journal of Chemistry, vol. 8, no. 2, pp. 851–858, 2011.

24. Z. H. Zou, Y. Yun, and J. N. Sun, "Entropy method for determination of weight of evaluating indicators in fuzzy synthetic evaluation for water quality assessment," Journal of Environmental Sciences, vol. 18, no. 5, pp. 1020–1023, 2006.

25. W. Zhang, J.-P. Chen, Q. Wang et al., "Susceptibility analysis of large-scale debris flows based on combination weighting and extension methods," Natural Hazards, vol. 66, no. 2, pp. 1073–1100, 2013. ·

26. D. C. Chi, T. Ma, and S. Li, "Application of extension assessment method based on game theory to evaluate the running condition of irrigation areas," Transactions of the CSAE, vol. 24, no. 8, pp. 36–39, 2008.

27. H. Karimnia and H. Bagloo, "Optimum mining method selection using fuzzy analytical hierarchy process–Qapiliq salt mine, Iran," International Journal of Mining Science and Technology, vol. 25, no. 2, pp. 225–230, 2015.

28. N. A. Bahri, F. M. A. Ebrahimi, and S. G. Reza, "A fuzzy logic model to predict the out-of-seam dilution in longwall mining," International Journal of Mining Science and Technology, vol. 25, no. 1, pp. 91–98, 2015.

29. Y. H. Su, M. C. He, and X. M. Sun, "Equivalent characteristic of membership function type in rock mass fuzzy classification," Journal of University of Science and Technology Beijing, vol. 29, no. 7, pp. 670–675, 2007.

30. S. Mikkili and A. K. Panda, "Simulation and real-time implementation of shunt active filter id-iq control strategy for mitigation of harmonics with different fuzzy membership functions," IET Power Electronics, vol. 5, no. 9, pp. 1856–1872, 2012.

Chapter 10

EFFECTS OF ROCK MASS CONDITIONS AND BLASTING STANDARD ON FRAGMENTATION SIZE AT LIMESTONE QUARRIES

Takashi Sasaoka[1*], Yoshiaki Takahashi[1], Wahyudi Sugeng[1], Akihiro Hamanaka[2], Hideki Shimada[1], Kikuo Matsui[1], Shiro Kubota[3]

[1]Department of Earth Resources Engineering, Faculty of Engineering, Kyushu University, Fukuoka, Japan

[2]Center of Environmental Science and Disaster Mitigation for Advanced Research, Muroran Institute Technology, Muroran, Japan

[3]National Institute of Advanced Industrial Science and Technology, Ibaraki, Japan

ABSTRACT

The size distribution of fragmented rocks depends on not only the blasting standard but also the mechanical properties, joint system and crack density of rock mass. As, especially, the cracks in the rock mass are heavily developed at the limestone quarries in Japan, the joints and/or cracks in the rock mass have big impacts on the blasting effects such as the size of fragmented rocks. Therefore, if the joint system and/or crack density in the rock mass can be known and evaluated in quantity, the blasting operation can be done more effectively, efficiency and safety. However, the guideline for designing the appropriate blasting standard based on the rock mass condition such as mechanical properties, joint system and/or distribution of cracks, discontinuities, from the scientific point of view, has not been developed yet. Therefore, a series of blasting tests had been conducted in different mines and faces, geological conditions and blasting standards in order to know the impacts of each factor on the blasting effects. This paper summarizes the results of blasting tests and describes the impacts of rock mass conditions and blasting standard on the size of fragmented rocks.

INTRODUCTION

Rock blasting is the rock excavation technique most widely adopted in the various fields of the mining and construction industries because of its economical and efficient aspects [1] .

In surface mines and quarries, the main objective is to exact the largest possible quantity at minimum cost. The material may include ore, coal, aggregates for construction and also the waste rock required to remove the above useful material. The blasting operations must be carried out to provide quantity and quality requirements of production in such a way that overall profits of mining or quarrying operation are maximized. In-situ rock is reduced in size by blasting and crushing into the required size or with additional grinding, into a fine powder suitable for mineral processing. Large blocks needing secondary breakage or an excess of fines can result from poorly designed blasts or due to adverse geological conditions. A well designed blast should produce shapes and sizes that can be accommodated by the available loading and hauling equipment and crushing plant with little or no need for secondary breakage [2] while optimizing the fragmentation is also important for safety and ease of loading.

Blasting designs in mines are still optimized over months or years by trial and error [3] Hence, a series of blasting tests were conducted in different mines, locations, and rock mass conditions by using a borehole camera and software for fragmentation analysis. Then, the effects of rock mass conditions such as fracture, discontinuity, mechanical properties of rocks and the blasting standard on the size of fragmented rocks were discussed in order to develop the guidelines for designing optimal blasting standard based on the rock mass condition.

Field Experiment

Overview of Field Experiments

In order to know the effects of rock mass conditions and blasting standard on the blasting effects, a series of field experiments were conducted in four different limestone quarries in Japan. Mines A, B, C are located at the northern part of Kyushu Island, Japan. Mine D is located at the southern part of Kyushu Island, Japan.In the testing sites, clay strata, joint systems/many cracks were observed in the faces at Mine A, Mine B and Mine C. On the other hand, quite a few clay strata was observed at Mine D. Table 1 shows the general blasting standards of each mine.

Figure 1 shows the procedures of field tests. At first, the blast holes and observation holes drilled in vertical and parallel to the face were observed by using borehole camera in order to know the rock mass condition especially joint system and cracks. Then, after blasting, the photograph of fragmented rocks was taken for the rock fragmentation analysis. Finally, the observation holes were observed again in order to compare the rock mass conditions before and after blasting. Besides, the rock samples were obtained at every face after blasting. Table 2 shows the results of rock property tests.

Borehole Camera

Figure 2 shows the borehole camera system we have developed and used so far [3] . This system consists of probe, control unit and digital video. The diameter and the length of the probe are 80 mm and 1.0 m, respectively. The motorized mirror, for the single micro camera, allows one to view to the side 360 degrees of the wall of borehole.

Table 1: General blasting standard of each mine

	Mine A Face 1	Mine A Face 2	Mine A Face 3	Mine A Face 4	Mine B Face 1	Mine B Face 2	Mine C Face 3	Mine D Face 1-3
Bench height (m)	10	15	13	15	15	15	10	10
Bench angle (deg.)	70	70	70	70	70	70	70	70
Hole spacing (m)	4.0	3.8	4.0	4.0	5.5	6.0	5.0	2.0
Burden (m)	4.1	3.6	4.2	4.0	6.0	6.0	5.0	2.5
Hole diameter (mm)	110	90	110	90	140	140	130	75
Powder factor (g/t)	91.7	77.5	92.2	52.8	112	102	66.7	158

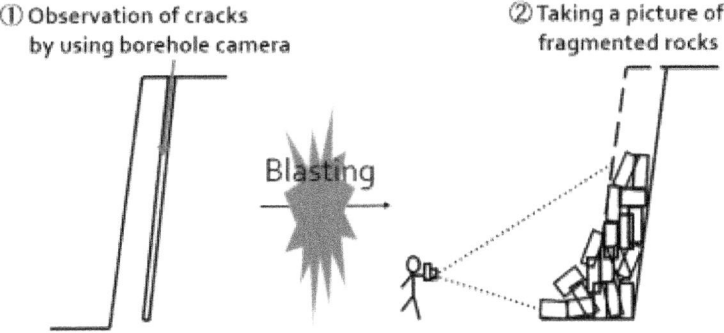

Figure 1: Illustration of test procedures.

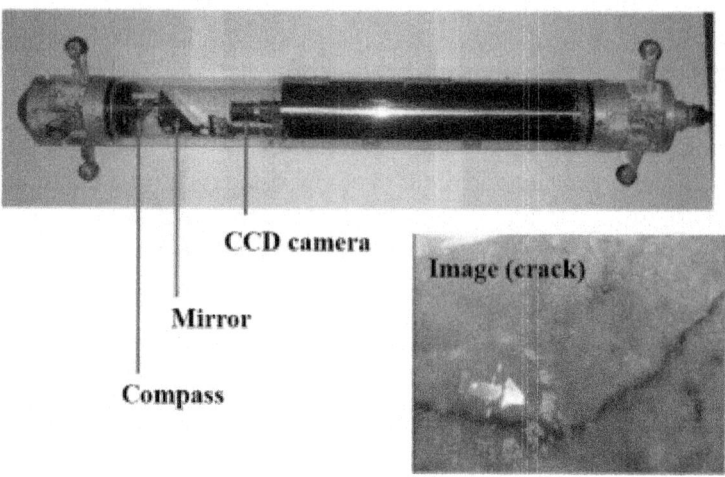

Figure 2: Borehole camera system and image.

Table 2: Mechanical properties of rocks in each mine

	Mine A Face 1	Mine A Face 2	Mine A Face 3	Mine A Face 4	Mine B Face 1	Mine B Face 2	Mine C Face 1	Mine D Face 1	Mine D Face 2	Mine D Face 3
Dencity (g/cm³)	2.69	2.70	2.69	2.69	2.72	2.68	2.65	2.46	2.51	2.60
Uniaxial compressive strength (MPa)	94.1	91.6	72.8	108	57.0	44.4	65.4	107	117	134
Brazilian tensile strength (MPa)	6.58	7.41	3.36	6.64	3.00	5.90	6.28	7.10	6.34	11.3
RIHN (-)	37.6	39.0	5.4	38.8	3.8	3.9	56.8	41.7	36.8	62.5

The image comes from the probe is displayed on the screen of video camera and recorded on digital video tape. The features of this system are cheap, simple and easy to carry and use it.Figure 2 also shows the image of crack and describes how to calculate the strike and dip of joints/discontinuities based on the image of borehole camera.

Fragmentation Analysis

Fragmentation assessment was achieved by the analysis of scaled photograph taken from the fragmented rocks. Paley recommended a procedure for taking photographs of fragmented rocks so as to minimize errors due to distortion [4] .

Two balls with diameter of 24 cm were used to provide scale in the photograph. The balls were placed in the same vertical line down the fragmented rocks, preferably with one ball near the top of the fragmented rocks and the other near the bottom. The balls should not be placed randomly in the fragmented rocks nor in a horizontal line across them. The camera was held such that the long axis of the photograph is vertical. The photograph was then taken with the camera as perpendicular to the surface of fragmented rocks as possible. By having two balls on the surface of fragmented rocks, allowance was made for variable scale within the photograph when the camera could not be positioned perpendicular to the surface of fragmented rocks.

The scaled fragmentation photographs were manually digitized from the original photograph on computer screen by software known as Split-Desktop developed by Split Engineering as illustrated in Figure 3 [5] . The outlines of visible rocks above a certain minimum resolution, 3 mm in diameter on the photograph, were traced by mouse. After the digital image is analyzed, the particle size distribution of fragmented rock was derived as shown in Figure 4. The representative particle size at 50% of the gain size accumulation curve; Xp_{50} was used in this research. Table 3 shows the representative particle size at 50% of the gain size accumulation curve at each face.

Results and Discussions

Figure 5 shows the relationship between Xp_{50} and average crack space. Here, an average crack space is an average space of nearby cracks and the crack density is a number of the crack per 1 m along the borehole. From this figure, they have a good correlation and it can be said that crack space and/or crack frequency has an obvious impact on the size of fragmented rock. Figures 6(a)-(d) show the relationships between Xp_{50} and uniaxial compressive strength, Brazilian tensile strength, RIHN, powder factor, respectively. Although it has been already pointed out that the size of fragmented rock depends on the blasting standard and mechanical properties of rock [6] , it can be seen from this figure that the data vary widely and no obvious correlation can be recognized. This may be because the crack condition of rock mass, mechanical properties of rocks, blasting standard such as powder factor are different at different test sites.

Figure 3: Digitized fragment outlines.

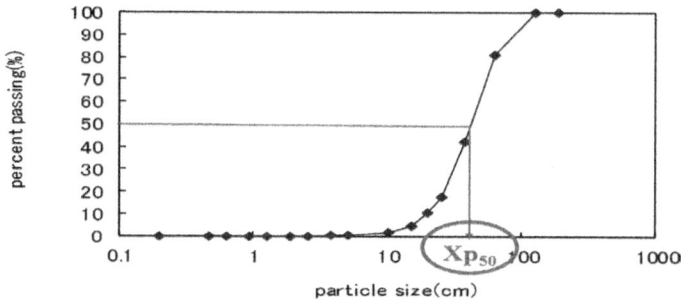

Figure 4: Particle size distribution (Mine B Face 2).

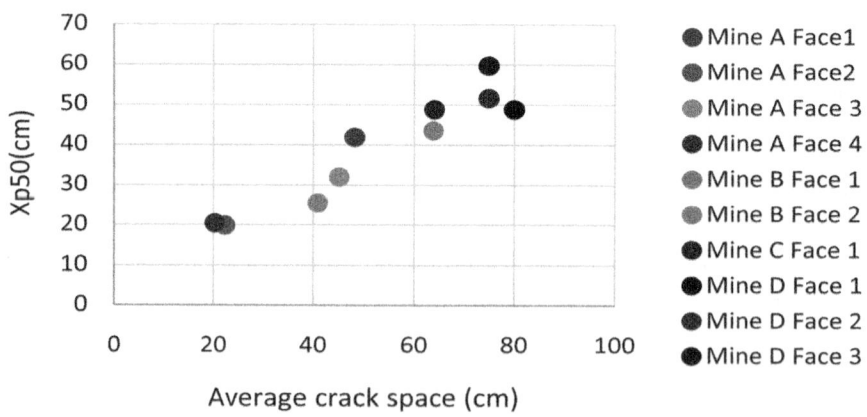

Figure 5: Relationship between representative particle size: Xp_{50} and average crack space.

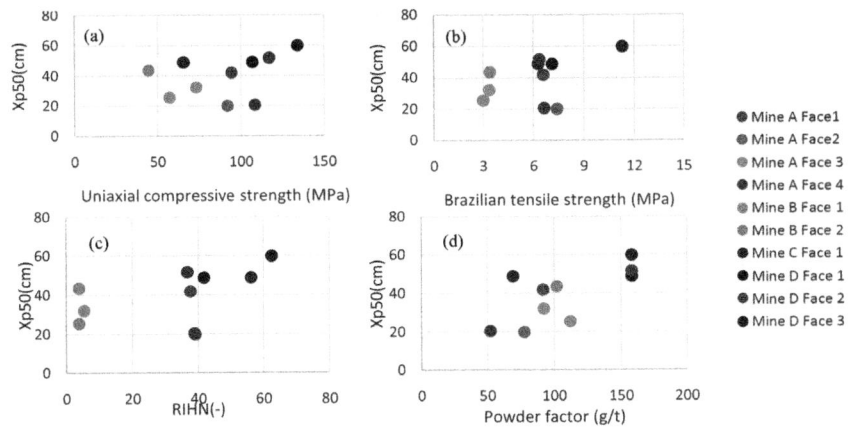

Figure 6: Relationships between representative particle size: Xp_{50} and (a) uniaxial compressive strength, (b) Brazilian tensile strength, (c) RIHN and (d) powder factor.

Table 3: Representative particle size at 50% of the gain size accumulation curve: Xp_{50} at each face

	Mine A Face 1	Mine A Face 2	Mine A Face 3	Mine A Face 4	Mine B Face 1	Mine B Face 2	Mine C Face 1	Mine D Face 1	Mine D Face 2	Mine D Face 3
Xp_{50} (cm)	41.8	19.8	32.0	20.3	25.4	43.5	48.7	51.5	59.6	73.5

Here, it is assumed that the crack condition of rock mass has an obvious impact on the size of fragmented rock. It attempts to eliminate this impact from the value of Xp_{50} as follows. The distance from the mouth of the hole to the first crack is X_1, that from the first crack and second one is X_2, that from the $(n - 1)^{th}$ and n^{th} ones is X_n. Here, it is assumed that a rock mass is fragmented according to the existing cracks/fractures, the number of rock particles produced by blasting is n and the shape of them is circular. Besides, it is also assumed that the density of rock mass is constant. In this case, the percent passing is represented as $(X_n/L) \times 100$. Where, L is the length of blast hole. Table 4 shows there presentative particle size at 50% of the grain size accumulation curve; Xb_{50} and the value of $(Xb_{50} - Xp_{50})$. Xb_{50} reflects the effect of existing cracks/fractures on the fragmentation size. As it were, the value of $(Xb_{50} - Xp_{50})$ should reflect the effect of mechanical properties of rock on fragmentation size. By using the value of $(Xb_{50} - Xp_{50})$ and considering the mechanical properties of rock and powder factor, the data was reanalyzed. Figures 7(a)-(d) show the relationships between the value of $(Xb_{50} - Xp_{50})$ and uniaxial compressive strength,/Brazilian tensile strength, /RIHN,/ powder factor, respectively. Compared with Figure 6, it can be recognized that the value of $(Xb_{50} - Xp_{50})$ is correlated strongly with the mechanical properties of rock and powder factor. The weaker the rock is and/or the larger the powder factor is, the larger the value of $(Xb_{50} - Xp_{50})$ is. Therefore, if the effect of crack condition of rock mass on the size of fragmented rock is eliminated, it can be seen clearly that the mechanical properties of rock and blasting standard also affect the size of fragmented rock.

From the above results, it can be concluded that made clear that the crack conditions of rock mass has an obvious impact on the size of fragmented rock. Moreover, the size of fragmented rock also depends on the mechanical properties of rock and blasting standard. These results fit with empirical knowledge. It can be concluded that the crack conditions of rock mass has to be grasped and evaluated in quantity in order to estimate the size of fragmented rock more accurately.

Design Methodology for Blasting Standard Case Study

At limestone quarry, it is preferred that the size of fragmented rocks is close to that of final products. Even though the research on control blasting for fragmentation size is conducted by many researchers, the solution has not been developed yet.

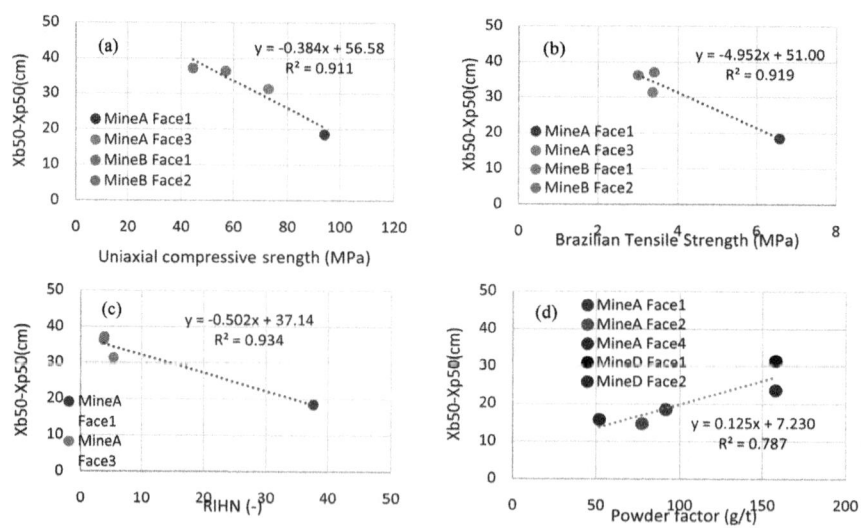

Figure 7: Relationships between representative particle size: Xp_{50} and (a) uniaxial compressive strength, (b) Brazilian tensile strength; (c) RIHN and (d) powder factor.

Table 4: Xb50 – Xp50, powder factor and mechanical characteristic values of each mine and face

	Mine A Face 1	Mine A Face 2	Mine A Face 3	MineA Face 4	Mine B Face 1	Mine B Face 2	Mine C Face 1	Mine D Face 1	Mine D Face 2	Mine D Face 3
Xb_{50} (cm)	60.1	34.5	63.2	36.0	61.7	137	64.1	80.0	75.0	120
$Xb_{50} - Xp_{50}$ (cm)	18.4	14.7	31.3	15.7	36.3	37.1	15.4	31.3	23.5	46.5
Uniaxial compressive strength (MPa)	94.1	91.6	72.8	108.4	57.0	44.4	65.4	107	117	134
Brazilian tensile strength (MPa)	6.58	7.41	3.36	6.64	3.00	3.40	6.28	7.10	6.34	11.3
RHN(-)	37.6	39.0	5.4	38.8	3.8	3.9	56.2	41.7	36.8	62.5
Powder factor (g/t)	91.7	77.5	92.2	52.1	112	102	69.2	158	158	158

Appropriate blasting standard and/or measures can still be determined place by place through trial and error. From the results obtained in the above, it can be said that different powder factor have an obvious impact on the size of fragmented rocks. Moreover, it was reported that the size of fragmented rocks can be controlled by changing the length of burden [7] . In order to develop the methodology of designing blasting standard, especially focused on the length of burden, additional field tests were conducted at Mine A. In this field experiment, the burden was changed from 3.7 m to 4.7 m which range seems to be not affected on the surrounding environment in terms of the blast-induced ground vibration [8] . The effect of burden on the size of fragmented rocks is discussed as follows.

Figure 8 shows the size distributions of fragmented rocks in different burdens. Here, the curve of "Ideal" is that obtained in Mine A, empirically. From this figure, it can be seen that the larger the burden is the larger the size of fragmented rocks is.

The relationship between the average fragmentation size of rocks and the energy of explosive can be represented as following equation (1) [9] .

$$X = A(K)^{-0.8} Q_e^{1/6} \left(\frac{115}{E}\right)^{19/30}$$

(1)

where, X; average size of fragmented rocks, A; a factor determined by the hardness of rock, K; powder factor, Q_e; the amount of explosive per hole, E; a parameter for indicating the strength of explosive. As a series of field tests in mine E was conducted at the same face and under the same amount of explosive charge (40 kg of ANFO), the value of Q_e and E can be considered as a constant. From Equation (1), it can be considered that the average size of fragmented rocks depends only on powder factor and the larger the burden is the larger the size of fragmented rocks is. This result obtained from Equation

(1) agrees with that of field tests. Table 5 shows the representative particle size at 80% of the gain size accumulation curve;

Xp_{80} (cm). The reason why the representative particle size at 80% of the gain size accumulation curve was used in this research is as follows. If less than 80% of total amount of fragmented rocks can pass the crushing facilities such as crusher and/or grizzly bear, there is an obstacle to operating the preparation plant in Mine A, and its size is about 60 cm. From Figure 4, it can be said that the appropriate burden for obtaining ideal size distribution of fragmented rocks in Mine A is from 4.2 m to 4.7 m. On the other hand, if the burden is larger than 4.2 m, the more fragmented rocks which size is over 60cm are produced and it seems to have an effect on the operation of the preparation plant.

Figure 9 shows the relationship between Xp_{80} and powder factor. From this figure, Equation (2) can be derived.

$$Xp_{80} = 490e^{-(-0.035t)}$$

(2)

where, t; the amount of explosive per volume of rock (kg/m³). From this equation, the appropriate amount of

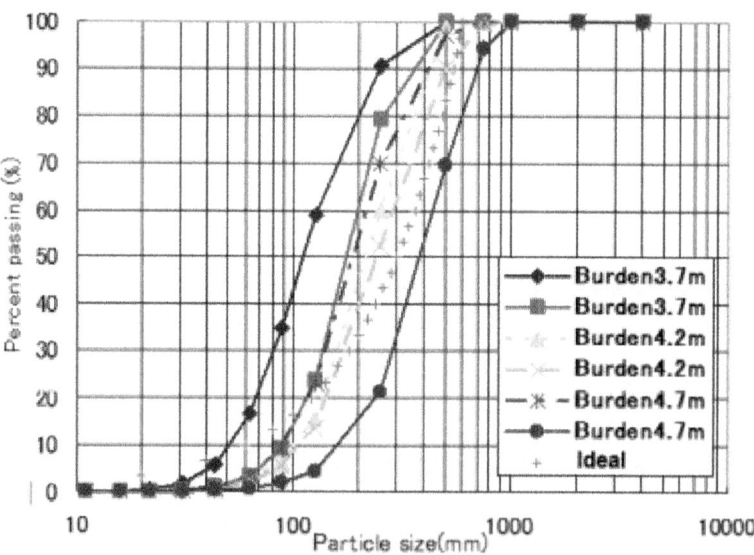

Figure 8. Grain size accumulation curves.

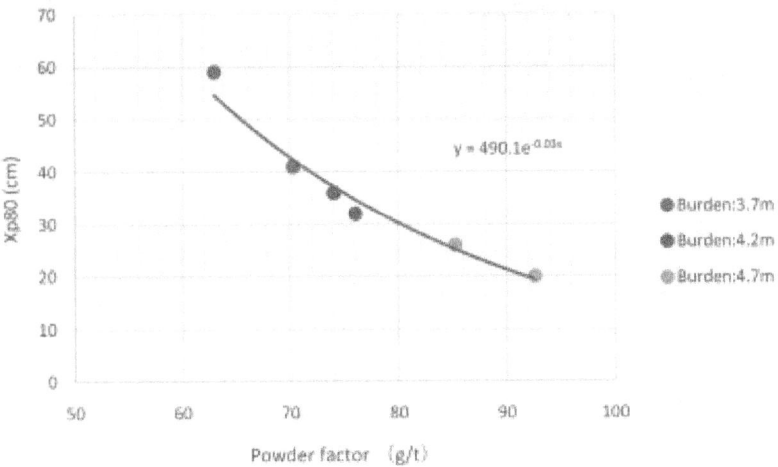

Figure 9. Relationship between Xp_{80} and powder factor at Mine A.

Table 5. Representative particle size at 80% of the grain size accumulation curve, Xp_{80}

Burden	3.7 m	4.2 m	4.7 m
Xp_{80}(cm)	20, 26	32, 36	40, 59

explosive can be determined in order to obtain desired size of fragmentation rocks and this equation is helpful to determine the appropriate blasting deign.

CONCLUSIONS

Based on the results of a series of field experiments, the following conclusions can be obtained:

1) In order to predict the size of fragmented rock more accurately, not only basting standard and mechanical properties of rocks, but also crack/fracture condition of rock mass have to be evaluated quantitatively.

2) If the variability of the length of burden is relatively small, as the effect of it on the blast-induced ground vibration is small, the length of burden can be determined based on the desirable size of fragmented rocks in terms of mine operating costs.

In order to verify these trends and develop the guidelines for designing blasting standard, more fields experiments and study have to be conducted.

ACKNOWLEDGEMENTS

The authors would like to express our thanks to staffs of Mines A, B, C and D. We also really appreciate cooperation of Dr. Yuji OGATA and Dr. Tei SABURI in Advanced Industrial Science and Technology.

REFERENCES

1. Bhandari, S. (1997) Engineering Rock Blasting Operations. A.A.Balkema, Rotterdam.

2. Hustrulid, W. (1999) Blasting Principles for Open Pit Mining, Volume 1-General Design Concepts. A.A.Balkema, Rotterdam.

3. Shimada, H., Matsui, K., Ichinose, M., Sasaoka, T. and Kubota, S. (2006) Study on Effect of Blast Vibration on Size of Fragmentation at Limestone Quarry. Journal of the Mining and Materials Processing Institute of Japan, 122, 11-19. (In Japanese)http://dx.doi.org/10.2473/shigentosozai.122.573

4. Paley, N.L. (1990) Image Based Fragmentation Assessment. MEngSc Thesis, The University of Queensland (JKMRC), Brisbane.

5. Sprit Engineering LLC (2003) Manual of Split-Desktop Software.http://www.spliteng.com/download/2102/

6. Matsui, K. and Shimada, H. (2004) Evaluation of Rock Mass Condition and Effect of Fracture of Rock Mass on Blasting Effect. Sekkaiseki, 330, 57-61. (In Japanese)

7. Inoue, H., Sasaoka, T., Shimada, H., Hamanaka, A., Matsui, K., Tanaka, H. and Inoue, M. (2014) Study on Control of Fragmentation Size by Changing Blasting Pattern at Open Pit Metal Mine. Proceedings of International Symposium on Earth Science and Technology 2014, Fukuoka, 4-5 December 2014, 214-217.

8. Sasaoka, T., Shimada, H., Hamanaka, A. and Matsui, K. (2011) Study on Blast Vibration and Size of Fragmentation at Limestone Quarry. Proceedings of 20th International Symposium on Mine Planning and Equipment Selection, Almaty, 12-14 October 2011, 714-730.

9. Persson, P.A., Homberg, R. and Lee, J. (1993) Rock Blasting and Explosives Engineering. CRC Press, Boca Raton.

Chapter 11

STUDY ON HYSTERETIC FRACTURE OF NATURALLY CRACKED SURROUNDING ROCK

Zhibin Zhong,[1] Ronggui Deng,[1] Fang Lin,[1] Ying Zhang,[1] Lei Lv,[2] Jing Yin,[1] andXiaomin Fu[3]

[1]School of Civil Engineering, Southwest Jiaotong University, Chengdu, Sichuan 610031, China

[2]Chengdu Southwest Jiaoda Yaosen Engineering Technology Co., Ltd., Chengdu, Sichuan 610031, China

[3]State Key Laboratory of Geohazard Prevention and Geoenvironment Protection, Chengdu University of Technology, Chengdu, Sichuan 610059, China

ABSTRACT

To determine the hysteretic fracture mechanism of the hard and naturally cracked surrounding rock mass, uniaxial and biaxial compression tests were performed on rhyolite specimens. In the biaxial compression test, displacements and strains around the U-shaped opening were monitored throughout to study the fracture pattern and distribution of stress. To compare with the experimental results, the finite difference code $^{FLAC^{3D}}$ was used to simulate a perfectly intact rock model with the same geometric and mechanical conditions in a continuum model. Stress-strain curves under uniaxial compression and the surrounding rock stresses of numerical results were compared with laboratory test. On one hand, laboratory test and numerical results all showed that tensile fracture regions were found at the crown and floor of the opening while shear regions are found at the sidewall. On the other hand, due to microcracks in the laboratory specimen, the laboratory test showed lower ultimate compressive strength. However, its vertical displacement of initial fracture was larger than those of numerical model which did not consider about the microcracks. They revealed the hysteretic fracture behaviour of underground opening in hard and cracked rock.

INTRODUCTION

Niba mountain tunnel, located on the expressway between Beijing and Kunming in Sichuan province, is 11 km long with a maximum overburden of 1660 m. Due to the complex geologic structure of the east Qinghai-Tibetan Plateau, the greatest challenge is the failure of cracked surrounding rock. During its excavation, rock burst often occurred behind the working face of the rock which is named hysteretic fracture.

Joints and cracks have great influence on the mechanical characteristics of rock. Extensive experimental studies have been conducted on rock or rock-like material specimens with preexisting regular artificial cracks. These studies have indicated that the fracture characteristics of preexisting cracks are closely related to the lithology, length, and distribution of cracks and the test loading path. The cracks lead to the coalescence and unstable failure of rock and rock-like materials [1–3]. Jia et al. [4] performed uniaxial compression test on brittle carbonate rock, and acoustic emission (AE) and CT imaging technology were used to investigate the interconnection of microcracks, forming macroscopic fracture failure. Eberhardt et al. [5] believed that cracks in rock could degenerate its strength and proposed detailed analysis of the crack initiation and propagation thresholds. Based on PMMA, molded gypsum and Hwangdeung granite containing preexisting cracks, Lee and Jeon [3] considered that different material and geometry of preexisting cracks held the different crack initiation, propagation, and coalescence.

Furthermore, some studies have focused on the fracture characteristics of jointed surrounding rock with openings. These studies showed that the cracks propagate to macrocracks, which lead to failure of the surrounding rock. Ewy and Cook [6] observed the growth of small, opening-mode, splitting cracks oriented parallel to the tangential stresses, starting very close to the hole wall and occurring deeper in the rock with increasing stress. Fakhimi et al. [7] conducted a biaxial compression test on sandstone with a circular opening. Two notches were developed at the lateral boundaries of the hole due to the stress concentration and the failure planes were inclined to the axial direction. Based on the ubiquitous-joint model, Wang and Huang [8, 9] examined the distribution and mechanical properties of joints in intact rock and analyzed the failure characteristics of regularly heterogeneous cracked surrounding rock in theory. Using the dislocation model and strain energy density factor theory, Zhou et al. [10] analyzed the nucleation, growth, interaction, and coalescence of cracks. Moreover, numerical simulation methods have been employed to simulate the fracture of surrounding rock. Zhou et al. [11] studied the effect of cracks and joints on fracture of surrounding rock for the underground caverns of Jinping I hydropower station. Sagong et al. [12] investigated the rock fracture and joint sliding behaviour of persistent jointed rock masses with an opening under biaxial compression. Zhang et al. [13] studied the deformation and failure characteristics surrounding the deep caverns by model test and theoretical analysis methods. Souley et al. [14], Hao and Azzam [15], and Jiang et al. [16] described the fracture process of rock around a tunnel using discrete element modelling (DEM). Meanwhile, Zhu et al. [17] and Jia and Tang [18] examined the microscopic damage and heterogeneity of rock using the finite element modelling (FEM) software RFPA. Progressive fracturing processes around circular, elliptical, and U-shaped underground excavations were simulated.

In summary, many experimental and theoretical studies have been conducted to reveal the fracture characteristics of cracked and jointed rock [1–5]. Then, fracture and failure modes of surrounding rock in homogeneous and preexisting regular jointed rock with circular openings were also investigated [6–12, 19]. Furthermore, numerical simulations have been performed to investigate the progressive fracturing process of different geometric opening and the effect of joint geometric parameters [14–18]. These studies have produced many substantial achievements and contributed to the study of the fracture mechanism of cracked underground surrounding rock. However, homogeneous surrounding was idealized model while cracks significantly affect the mechanical characteristics of rock. In addition, circular tunnel sections are rarely found in underground engineering projects. Above all, there was hardly any study forced on the difference between cracked and intact surrounding rock. Due to the numbers of hysteretic fractures observed during the excavation of Niba mountain deeply buried and long tunnel, it was necessary to reveal the mechanism of hysteretic fractures in cracked and hard surrounding rock.

In this paper, experimental and numerical simulation are conducted to evaluate the deformation and fracture modes of rock with an inverted U-shaped opening in natural original cracked and noncracked rhyolite, respectively. In Section 2, experimental processes which include uniaxial and biaxial compression tests are described in detail. In Section 3, the fracture modes of uniaxial and biaxial compression specimens are analyzed; meanwhile, surrounding rock stresses around the opening are illustrated. In Section 4, numerical models with intact rock are established to compare with the experimental results in fracture modes and variations of surrounding rock stresses. Finally, conclusions are drawn in Section 5. The stress distribution, failure process, fracture mechanism, and characteristics are examined by biaxial compression loading to simulate the hysteretic fracture after excavation. Numerical results show the difference between cracked and intact surrounding rock regarding the stress field and fracture behaviour.

EXPERIMENTAL PROCESS

Experimental Specimen Setup

The rhyolite used to manufacture the specimen was obtained from the Daxiangling Niba mountain tunnel on the expressway from Beijing to Kunming (G5) in Sichuan province. The tunnel is 10,017 m long with a maximum overburden of 1660 m. Due to the complex geological structure, the rhyolite in the tunnel region is highly cracked. Rock burst, hysteretic fracture occurred frequently during excavating of the tunnel [20]. In this section, firstly uniaxial compression test was conducted to investigate the basic mechanical characteristic. The specimen used to perform uniaxial compression test was dug from the biaxial specimen (see Figure 1). Then biaxial compression test, whose specimen contained an inverted U-shaped opening, was carried out to investigate the fracture modes and distribution laws of stresses around the opening.

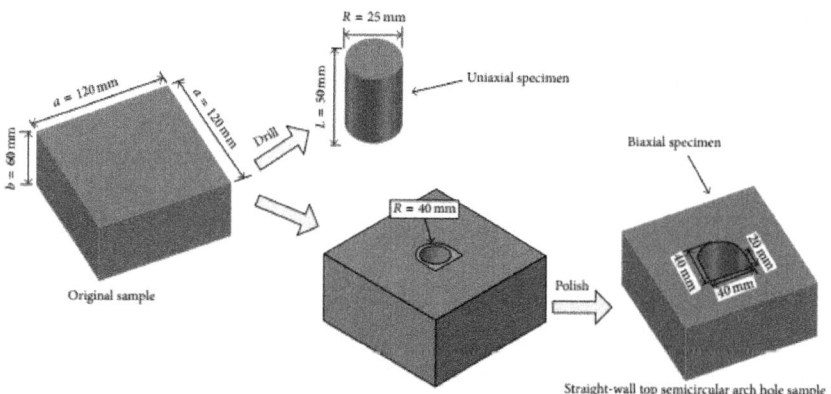

Figure 1: Manufacturing of specimens.

Based on the load capacity and dimensions of the loading apparatus, the dimensions of the biaxial specimen were 120 mm × 120 mm × 60 mm. As it was difficult to perfectly cut a horseshoe-shaped opening in a hard and cracked rock specimen and considered about the boundary effect, simplified 20 mm straight-wall 40 mm top semicircular arch inverted U-shaped opening was manufactured in the middle. Li et al. [21] and Zhu et al. [17] used this kind of model to successfully analyze the mechanics and failure behaviour of tunnels.

Loading and Monitoring Conditions

All the tests were performed in the State Key Laboratory of Geohazard Prevention and Geoenvironment Protection of Chengdu University of Technology. The uniaxial compression tests were performed on the MTS815 Rock Mechanics Test System (Figure 2(a)). The maximum loading capacity of axial servocontrolled system is 3000 kN. During the uniaxial tests, the axial force and deformation were measured by a loading cell and LVDTs, respectively. The biaxial compression test was performed on the self-designed Rock Biaxial Compressive Test System (RBCTS) (Figure 2(b)), which was designed in and controlled by a computer program. The applied load and boundary displacement of the specimen data were automatically recorded at a rate of 10 samples per second. Loading was applied through horizontal and vertical hydraulic lifting jacks with maximum loads of 300 kN and 600 kN, respectively, while the left and bottom restraint plates were fixed. Steel balls were installed at the left, upper, and bottom boundaries of the loading frame, and a restraint plate lubricated with butter with its surface facing the specimen was placed on the specimen to eliminate the boundary effects. The loading apparatus was constructed of high-duty steel, and reinforcement measures were implemented to satisfy the stiffness.

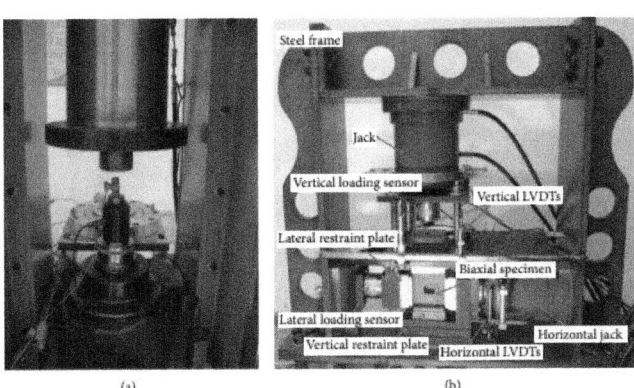

(a)   (b)

Figure 2: Experimental equipment and scheme: (a) MTS815, (b) RBCTS.

In addition, the loading rate of uniaxial compression was first controlled by axial force with a rate of 0.5 kN/min until the axial force arrived at 5 kN, and then the loading rate was transferred to axial deformation controlling with a loading rate of 0.1 mm/min to obtain the postpeak deformation. To the biaxial compression test, the lifting-loading mode in this test consisted of the vertical pressure (σ_v) and horizontal pressure (σ_h) increasing simultaneously by 0.5 kN/s until a force of 7.2 kN was reached, after which the horizontal pressure was kept constant (7.2 kN) while the vertical pressure continued to increase until significant specimen failure was achieved. Finally, the horizontal and vertical pressures were removed to unload the specimen.

To monitor the fractures in the specimen, the indirect method, acoustic emission (AE) monitoring techniques, is used. The SAEU2S AE monitoring system contains AE event monitoring sensor and loading monitoring sensor. They can collect and display AE event emissions and the corresponding vertical loading timely (Figure 3(a)). The technique has been widely used to study the failure mechanism for rocks, concrete, and other materials. For biaxial compression test, based on the GMTS theory [22], the changes of tangential stresses around the opening were important for studying the fracture of rock during the test. Therefore, strain gauges were used to measure the strains around and inside the opening. The distribution of the strain gauges was depicted in Figure 3(b). All the strain gauge, AE events, LVDT data, and loading records were transferred to and stored by computers. A digital camera was used to monitor and record the cracking process of the opening during the loading process.

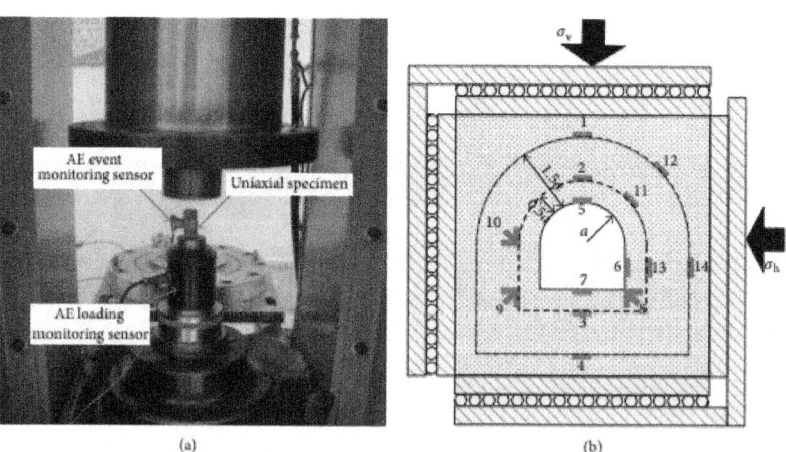

(a)   (b)

Figure 3: Experimental monitoring measures: (a) AE monitoring in uniaxial test, (b) strain gauges.

EXPERIMENTAL RESULTS AND DISCUSSION

Uniaxial Compression Test

Figure 4 showed the AE monitoring results during the test. It can be seen from Figure 4 that, in the elastic deformation stage, there was hardly any AE count measured, and the loading curve was relatively straight without remarkable fluctuations. Afterwards, AE counts suddenly increased and reached the highest when the axial stress was 31.1 MPa; it indicated that the microcracks in the specimen began to fracture; then $\sigma_i = 31.1$ MPa was the crack initiation stress. For the uniaxial compression strength $\sigma_{ucs} = 72.3$ MPa, σ_i MPa, was 43% of. σ_{ucs}. Once the crack initiation stress reached, the AE counts increased continuously until the specimen fractured to failure. Additionally, sharply increased AE counts occurred when microcracks propagated to macrofracture with the external loading. As the stress reached the peak strength, a large number of AE events occurred abruptly, which indicated that isolated microfracture formed macrofracture plane in the rock specimen. In the postpeak stage, the AE counts were still high and more frequent, showing that microfractures continued to propagate and formed ultimate failure of specimen. They indicated that there were numbers of original microcracks in the specimen; once they reached the initiation stress, the cracks fractured and propagated continually with a mount of AE counts released.

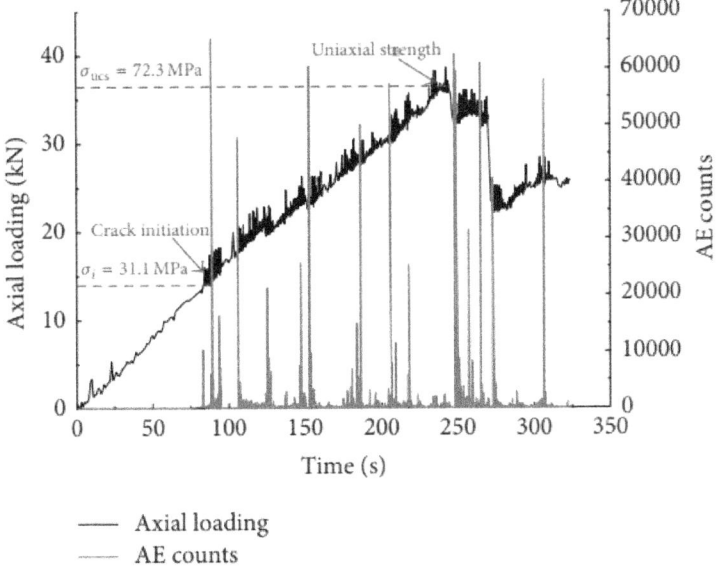

Figure 4: AE monitoring results of uniaxial compression test.

Eberhardt et al. [5] developed the moving point regression technique to determine the stress thresholds of rock samples. It made use of a "sliding window" to scan through axial strain and axial stress data sets and superimposes a straight line over a user-defined regression interval. Using the method, Figure 5 showed the axial stress-strain curve and axial stiffness-strain curve. Based on the method, the axial stiffness curve was horizontal while the corresponding axial stress curve was oblique. Thus the slope of stress-strain curve was the elastic modulus in elastic deformation stage. Point A should be the crack initiation which was 39.0 MPa. It was larger than that observed by AE results. Comparing Figure 5 with Figure 4, AE monitoring results can investigate the fracture process more accurately.

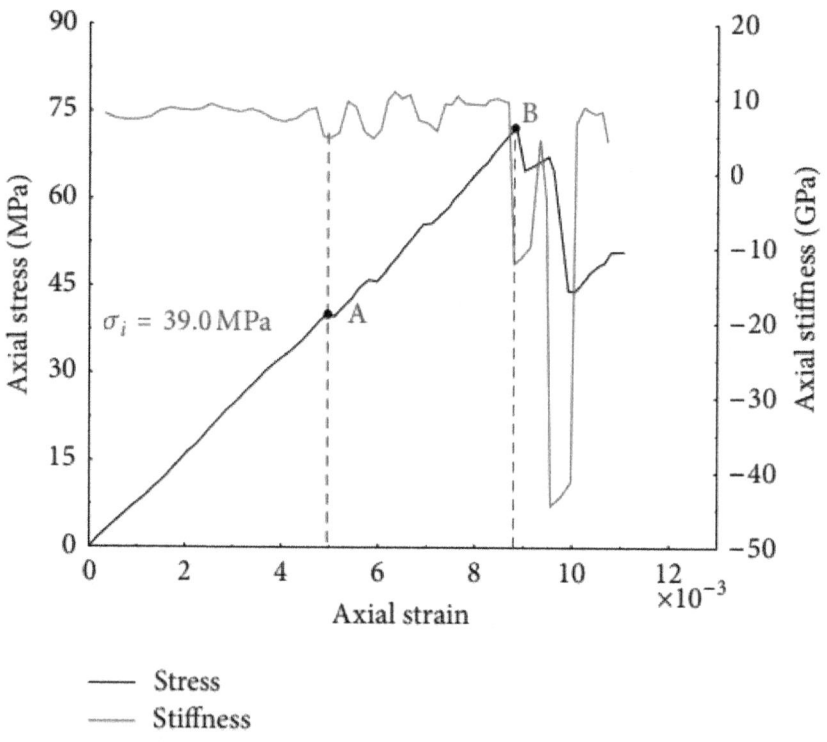

Figure 5: Experimental results of uniaxial compression test.

Figure 6 showed the ultimate failure mode of the specimen. There was a distinct fracture plane approximately along the vertical which controlled the final failure of the specimen. It indicated that the rhyolite was brittle and splitting failure under uniaxial compression test.

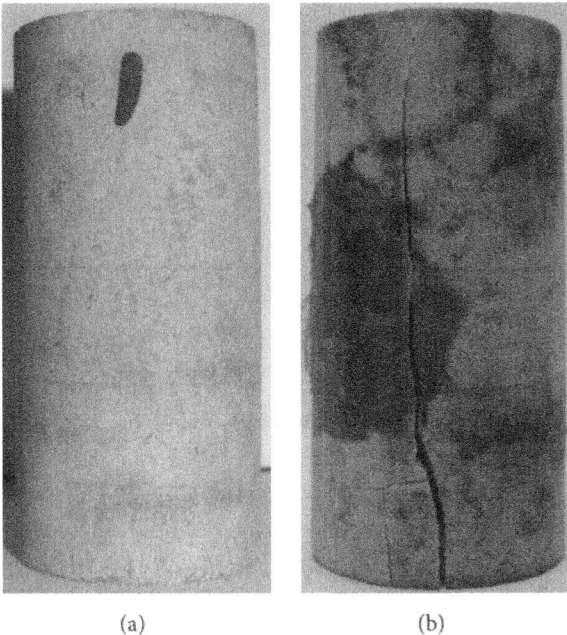

(a)   (b)

Figure 6: Ultimate failure mode of uniaxial compression test (a) before and (b) after the test.

Biaxial Compression Test

Fracture Characteristics and Processes of the Surrounding Rock

Figure 7 depicts the failure mode after the test. Three regions are distinguished according to the crack initiation sequence in the fractured area. With increasing vertical pressure, fractures were observed along regions ①, ②, and ③. Shear cracks initiate and propagate to macrofracture at arch spring. Further macrocrack planes shear off the sidewall and then slip into the opening. A large amount of rock detritus is observed in front of the shear opening, as shown on the left side of Figure 7. Meanwhile, a complete macrocrack plane is observed on the side of the specimen, as shown on the right side of Figure 7. It indicated that, during the test, the boundary effect is limited, and the cracks subperpendicular to the axial opening do not control the failure of specimen. The final failure of the specimen is controlled by the macrofractures in the loading plane.

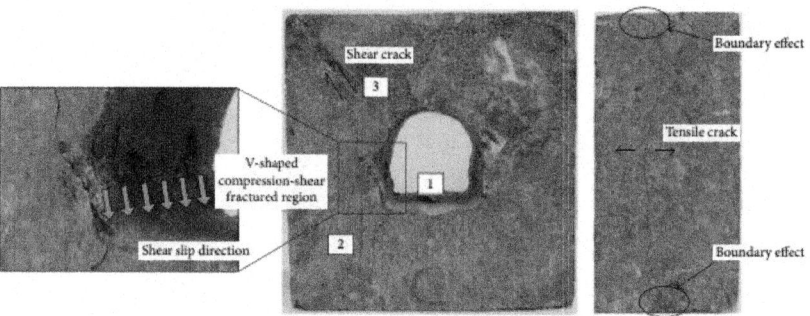

Figure 7: Final fracture mode of a biaxial specimen.

In this test, the "V-shaped" compression-shear fractured region is formed at the sidewall near the arch springing point, and macro- and microscopic cracks are well developed. The shear and tensile cracks propagate towards the surrounding rock at the crown and floor, and then a shear wedge is observed at the crown and shears through the crushing area of the "V-shaped" notch, which results in more cracks at the reentrant and leads to its final fracture.

Stress Distribution Characteristics

According to above uniaxial and previous conventional triaxial compression tests [20, 23], the elastic modulus E and Poisson's ratio υ of rhyolite are 8.7 GPa and 0.1, respectively. In biaxial compression tests, the normal strains in the tangential direction of the opening profile of the surrounding rock at crown and floor ε_{nt} were monitored, as shown in Figure 3(b). The subscript "nt" means "normal value in the tangential direction of the opening profile." Then the normal stresses in the tangential direction of the opening profile, defined as σ_{nt}, can be obtained based on Hooke's law, which is expressed as $\sigma_{nt} = E \cdot \varepsilon_{nt}$, and then their variations with vertical displacement, d_v, can be plotted.

As shown in Figure 8, at either the crown or the floor, cracking begins with increasing tensile stress in the inner surface of the opening at 0a from the opening profile. Next, σ_{nt} located at 0.5a and 1.5a from the opening profile orderly and gradually transform from compressive stress to tensile stress until the surrounding rock is tensile fractured. Meanwhile, strain energy is released to cause a sudden decrease in tensile stress. This indicates that the initiation, propagation, and coalescence of cracks can release the strain energy in the rock accumulated during the loading and then reduce the surrounding rock stresses. Table 1 summarises the vertical displacement values when the surrounding rock cracked at the crown and floor, where , d_{v0a}, $d_{v0.5a}$, and $d_{v1.5a}$, were the corresponding values of vertical displacements when the cracks were initiated at locations 0a, 0.5a, and 1.5a from the opening profile, respectively, and d_{vf} represented the final fracture vertical displacement.

Table 1: Fracture vertical displacements of surrounding rock

Location	d_{v0a}/mm	$d_{v0.5a}$/mm	$d_{v1.5a}$/mm
The crown	1.13	1.26	Unfractured
The floor	0.81	0.93	2.27

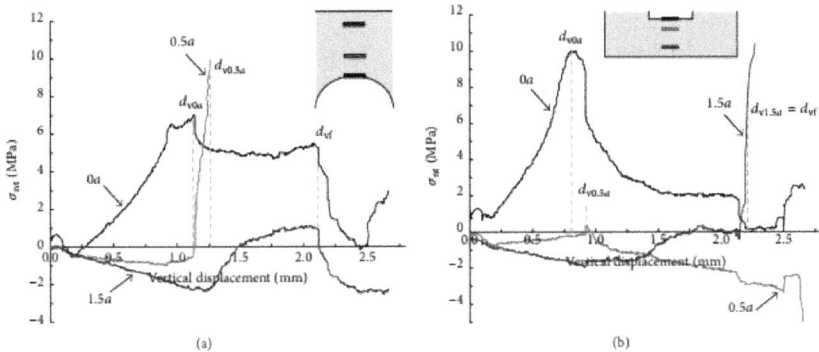

(a)   (b)

Figure 8: Variation of surrounding rock normal stresses in the tangential direction of the opening profile (a) at the crown and (b) at the floor.

In accordance with the variation of σ_{nt} at the crown and floor shown in Figure 8 and the corresponding vertical displacements, we can determine that the rock surrounding the floor fractures first and the cracks propagated to deeper locations markedly faster than those at the crown. Macrofractures occur once the cracks in the floor propagated to the boundary of the specimen, and the surrounding rock stresses were released. They indicated that the surrounding rock undergoes the stress adjustment processes of "bearing → initiation of cracks → stress release → bearing" during the test. Tensile and compressive stresses alternating occurred in the surrounding rock, with the addition of new initiation cracks, which facilitated the cracking of the surrounding rock.

NUMERICAL SIMULATION RESULT AND DISCUSSION

Joints, cracks, and other discontinuities in a rock mass play a key role in the responses of a tunnel. G. Barla and M. Barla [24] discussed and compared continuum and discontinuum modelling in tunnel engineering. The experimental results confirmed that the stress and fracture behaviour of naturally cracked surrounding rock are significantly affected by the yet-unknown distribution of cracks. Meanwhile, almost all natural rock contains different geometrical cracks. Therefore, it is difficult to prepare perfectly intact rock specimens to compare with cracked rock specimens under the same loading conditions in laboratory tests. On the contrary, numerical model can provide the perfectly intact rock specimen by continuum medium model [24]. To conduct the comparison, the finite difference code $^{FLAC^{3D}}$ was used to simulate a perfectly intact rock model with the same geometric and mechanical conditions in a continuum model.

Simulation of Uniaxial Compression Test

Using a numerical model, a strain-softening model was selected to simulate the intact rhyolite (Figure 9). This method has been extensively used to describe the nonlinear mechanical properties of rock around a tunnel [25]. The equivalent physical properties of the simulated material referenced to the experimental results [20,23, 26] are shown in Table 2. To simulate the quasibrittle characteristics of rhyolite, the residual cohesive strength, residual internal friction angle, and residual tensile strength were achieved with an equivalent plastic shear strain of 1.0×10^{-3}. Because the rock specimen was intact, cracks were not considered.

The uniaxial compression numerical model was the same size as the experimental model and was divided into a total of 20000 eight-noded hexahedron elements. The upper and bottom boundaries were both constrained, and then vertical compressive displacement was applied in the upper boundary at a rate of 1.0×10^{-8} m per step for the total steps of 60000. Then the axial stress-strain curve of the numerical result was compared with the test result which was mentioned in Section 3.1. As shown in Figure 10, the numerical result showed a good agreement with the laboratory test results. It indicated that the parameters used in the numerical model were acceptable. Therefore, they could be used to simulate the biaxial compression test.

Table 2: Physical properties of the numerical model material

	Elastic modulus/GPa	Poisson's ratio	Tensile strength/MPa	c/MPa	φ/°
Initial value	8.7	0.1	7.0	15.0	50
Residual value	8.7	0.1	0.01	13.0	40

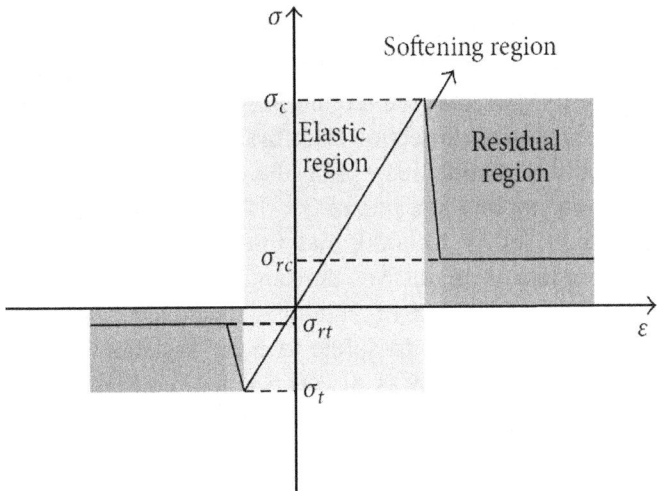

Figure 9: The strain-softening model.

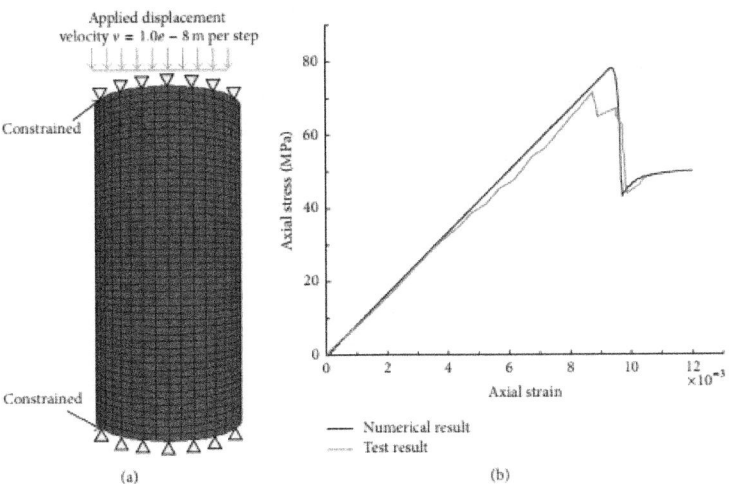

Figure 10: Comparison of the experimental and numerical results under uniaxial compression: (a) numerical model; (b) stress-strain curves.

Simulation of Biaxial Compression Test

Figure 11 showed the typical mesh pattern generated when simulating perfectly intact surrounding rock. In $^{\text{FLAC}^{3D}}$, concentrated loads were specified at given surface nodes, and imposed boundary displacements were prescribed in terms of nodal velocities [27]. In this numerical model, the frontage and backside of the numerical model were both free boundaries, while the displacements along the x,y,, , and z directions of the upper and bottom boundaries were both constrained. Otherwise, the left and right sides were only constrained by the displacement of the x direction with the application of 1.0 MPa of lateral pressure, which was maintained during the calculation. Vertical compressive displacement was applied at a rate of 2×10^{-8} m per step. For the simulation, the step size was set as 130,000, and the total vertical displacement was 2.6 mm. The numerical model was the same size as the experimental model and was divided into a total of 32,640 eight-noded hexahedron elements. When the model was solved, to define velocity variations and corresponding space intervals, the medium was discretized into constant tetrahedral strain-rate elements with vertices represented by the nodes of the mesh [27].

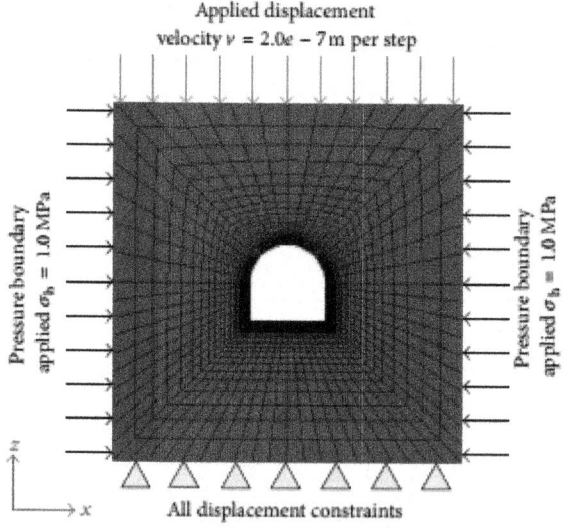

Figure 11: Grid and boundary conditions of the biaxial numerical model.

Figures 12 and 13 showed the principal stresses and plastic region of numerical simulation results. It can be seen that the crown and floor of the opening were subject to tension stress which controlled the tensile fracture of the locations. On the contrary, the region of sidewall was subject to principal compressive stress and led to shear fracture in this region. In addition, an oblique shear fracture plane was observed near the arch spring line. The fracture behaviour of numerical simulation results showed good agreement with the experimental results.

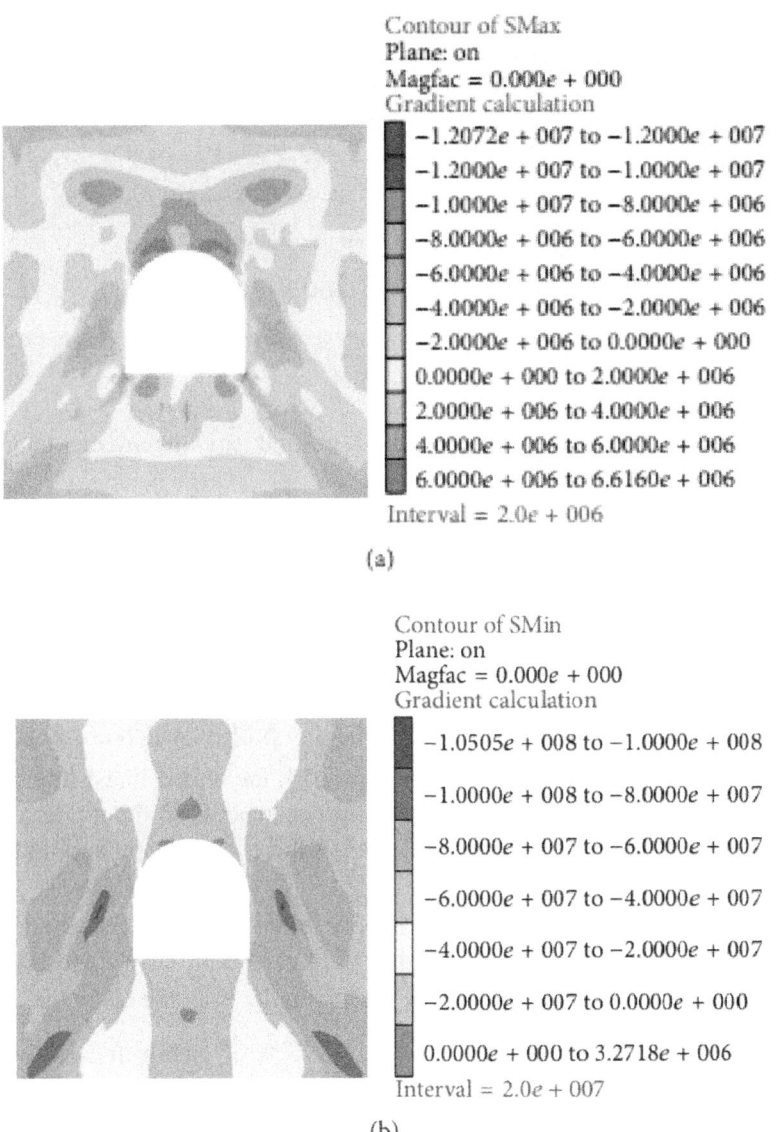

Figure 12: Numerical simulation results: (a) principal tensile stress, (b) principal compressive stress.

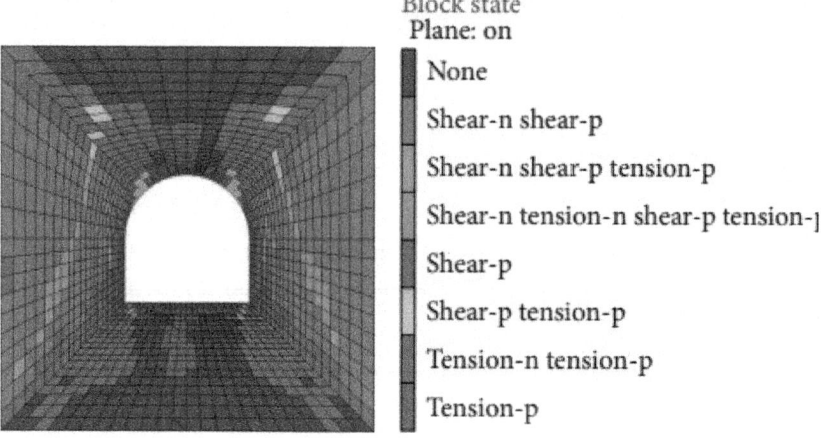

Figure 13: Plastic region of numerical result.

Figure 14 depicts the variations of normal stresses in the tangential direction of the opening σ_{nt}, profile monitored by the test and numerical simulation at the same locations in the specimen. Generally, the test and simulation results had the same variation tendencies, especially in the deep surrounding rock, such as 1.5a from the opening profile. However, their vertical displacements of fracture were different. Table 3 and Figure 15 summarise the corresponding vertical displacements when the surrounding rock was fractured at locations 0a, 0.5a, and 1.5a from the opening profile, which are denoted by d_{v0a}, $d_{v0.5a}$, and $d_{v1.5a}$, respectively. In Figure 15, the dotted lines stood for the unfractured values. To investigate the differences between the numerical and experimental results, the difference values of the experimental and numerical d_{v0a} and $d_{v0.5a}$ are denoted as Δd_{v0a} and $\Delta d_{v0.5a}$ (Δd_{v0a} = experimental d_{v0a} − numerical d_{v0a}, $\Delta d_{v0.5a}$ = experimental $d_{v0.5a}$ − numerical $d_{v0.5a}$ = experimental − numerical), respectively. It can be seen from Table 3 and Figure 15 that, in both crown and floor, the numerical model (intact rock) d_v was much smaller than the test value. In the floor, Δd_{v0a} was approximately equal to $\Delta d_{v0.5a}$, while it was different at the crown. Because the numerical model was continuum without cracks, this finding indicated that the original natural cracks in the surrounding rock (experimental specimen) can dissipate the strain energy and then postponed fracture in the initial fracture stage. However, once the cracks were initiated, they propagated to fracture faster than those in the intact specimen (numerical model), especially at the crown.

Table 3: Comparison of the fractured vertical displacements units: mm

Location	Result	d_{v0a}	Δd_{v0a}	$d_{v0.5a}$	$\Delta d_{v0.5a}$	$d_{v1.5a}$
The crown	Numerical	0.22	0.91	0.80	0.46	Unfractured
	Experimental	1.13		1.26		Unfractured
The floor	Numerical	0.18	0.63	0.34	0.69	Unfractured
	Experimental	0.81		0.93		2.27

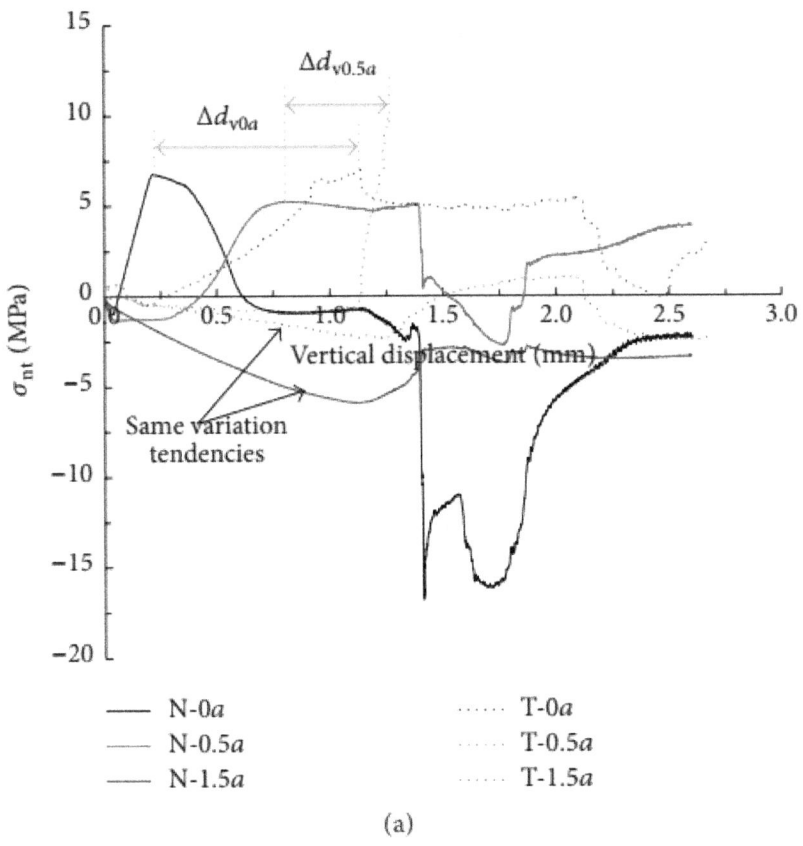

(a)

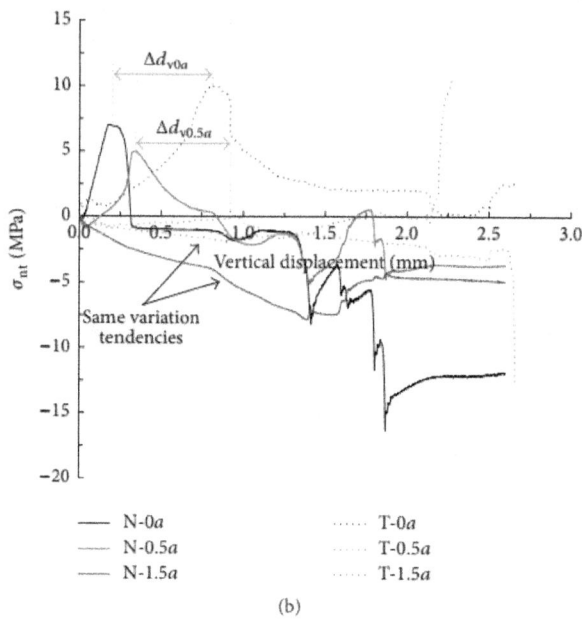

(b)

Figure 14: Normal stresses in the tangential direction of the opening profile: (a) the crown, (b) the floor.

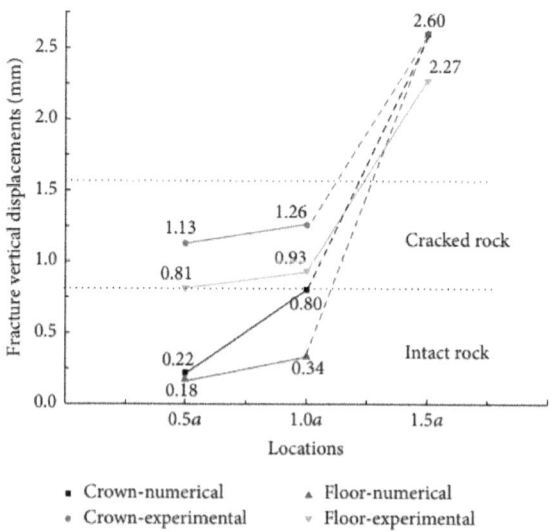

Figure 15: Comparison of numerical and experimental results.

Based on this comparison using the same geometrical dimensions and equivalent mechanical parameters, the numerical model of the perfectly intact specimen had a larger compression strength than those of naturally cracked specimen. However, under compression pressure, the cracks closed gradually and dissipated strain energy. They generated a much larger compression pressure at the initial fracture in the cracked surrounding rock. This is also why hysteretic fractures and weak rock bursts occurred often during the excavation of the Niba mountain tunnel.

CONCLUSION

In this study, uniaxial and biaxial compression tests were conducted on the naturally cracked rhyolite. The fracture modes and stress state were studied. The results of numerical simulation which was treated as intact rock were then compared with those of laboratory tests. According to the laboratory test and numerical simulation results, the following conclusions were drawn:(1)The initiation stress of rhyolite was 43% of the uniaxial compressive strength. The AE monitoring and ultimate failure of uniaxial compression test showed that there were numbers of original cracks in the natural rhyolite, which means that the rhyolite was naturally cracked and brittle.(2)Regarding the analysis of the normal stresses in the tangential direction of the opening profile distributed around the inverted U-shaped opening, the floor rock was tensile fractured prior to the crown, and the "V-shaped" compression-shear fractured region is formed at the sidewall near the arch springing point. As the initiation, propagation, and coalescence of cracks, strain energy accumulated during the loading was released.(3)Numerical model with intact rock had the same variations of surrounding rock stresses with laboratory test results. Compared to the perfectly intact rock numerical model, the crack initiation vertical displacement of laboratory test was larger, but the cracks propagate to deep surrounding rock faster. Meanwhile, the ultimate strength of the cracked biaxial specimen is lower, which means that the original cracks decreased the strength but postpone the initiation cracking of the surrounding rock. This behaviour of cracked surrounding rock had something to do with the strain energy releasing of original cracks. This process reveals the mechanism of hysteretic fractures of cracks surrounding the Niba mountain tunnel.

Moreover, this study also has some limitations. Although the application of the results presented in this study to practical projects is difficult, these results are significative to reveal the mechanism of hysteretic fractures in naturally cracked surrounding rock during the excavation of Niba mountain tunnel. It is also the basis for the systemic study of hysteretic fractures.

CONFLICT OF INTERESTS

The authors declare that there is no conflict of interests regarding the publication of this paper.

ACKNOWLEDGMENTS

This research was supported by the National Natural Science Foundation of China (no. 41272321) and the Doctorial Innovation Foundation of Southwest Jiaotong University. The authors wish to acknowledge Mr. Xiong Hun for his help during laboratory tests.

REFERENCES

1. L. N. Y. Wong and H. H. Einstein, "Crack coalescence in molded gypsum and carrara marble: part 1. macroscopic observations and interpretation," Rock Mechanics and Rock Engineering, vol. 42, no. 3, pp. 475–511, 2009. View at PublisherView at Google ScholarView at Scopus

2. R. H. C. Wong and K. T. Chau, "Crack coalescence in a rock-like material containing two cracks,"International Journal of Rock Mechanics and Mining Sciences, vol. 35, no. 2, pp. 147–164, 1998. View at PublisherView at Google ScholarView at Scopus

3. H. Lee and S. Jeon, "An experimental and numerical study of fracture coalescence in pre-cracked specimens under uniaxial compression," International Journal of Solids and Structures, vol. 48, no. 6, pp. 979–999, 2011. View at PublisherView at Google ScholarView at Scopus

4. L. Jia, M. Chen, W. Zhang et al., "Experimental study and numerical modeling of brittle fracture of carbonate rock under uniaxial compression," Mechanics Research Communications, vol. 50, pp. 58–62, 2013. View at PublisherView at Google ScholarView at Scopus

5. E. Eberhardt, D. Stead, B. Stimpson, and R. S. Read, "Identifying crack initiation and propagation thresholds in brittle rock," Canadian Geotechnical Journal, vol. 35, no. 2, pp. 222–233, 1998. View at PublisherView at Google ScholarView at Scopus

6. R. T. Ewy and N. G. W. Cook, "Deformation and fracture around cylindrical openings in rock—II. Initiation, growth and interaction of fractures," International Journal of Rock Mechanics and Mining Sciences & Geomechanics Abstracts, vol. 27, no. 5, pp. 409–427, 1990. View at Google Scholar

7. Fakhimi, F. Carvalho, T. Ishida, and J. F. Labuz, "Simulation of failure around a circular opening in rock," International Journal of Rock Mechanics and Mining Sciences, vol. 39, no. 4, pp. 507–515, 2002. View at PublisherView at Google ScholarView at Scopus

8. T.-T. Wang and T.-H. Huang, "Numerical simulation on anisotropic squeezing phenomenon of New Guanyin Tunnel," Journal of GeoEngineering, vol. 6, no. 3, pp. 125–133, 2011. View at PublisherView at Google ScholarView at Scopus

9. T.-T. Wang and T.-H. Huang, "Anisotropic deformation of a circular tunnel excavated in a rock mass containing sets of ubiquitous joints: theory analysis and numerical modeling," Rock Mechanics and Rock Engineering, vol. 47, no. 2, pp. 643–657, 2014. View at PublisherView at Google ScholarView at Scopus

10. X. P. Zhou, F. H. Wang, Q. H. Qian, and B. H. Zhang, "Zonal fracturing mechanism in deep crack-weakened rock masses," Theoretical and Applied Fracture Mechanics, vol. 50, no. 1, pp. 57–65, 2008. View at PublisherView at Google ScholarView at Scopus

11. X. P. Zhou, E. M. Xia, H. Q. Yang, and Q. H. Qian, "Different crack sizes analyzed for surrounding rock mass around underground caverns in Jinping I hydropower station," Theoretical and Applied Fracture Mechanics, vol. 57, no. 1, pp. 19–30, 2012. View at PublisherView at Google ScholarView at Scopus

12. M. Sagong, D. Park, J. Yoo, and J. S. Lee, "Experimental and numerical analyses of an opening in a jointed rock mass under biaxial compression," International Journal of Rock Mechanics and Mining Sciences, vol. 48, no. 7, pp. 1055–1067, 2011. View at PublisherView at Google ScholarView at Scopus

13. C. Zhang, Z. C. Wang, and Q. Wang, "Deformation and failure characteristics of the rock masses around deep underground caverns," Mathematical Problems in Engineering, vol. 2015, Article ID 230126, 13 pages, 2015. View at PublisherView at Google Scholar

14. M. Souley, F. Homand, and A. Thoraval, "The effect of joint constitutive laws on the modelling of an underground excavation and comparison with in situ measurements," International Journal of Rock Mechanics and Mining Sciences, vol. 34, no. 1, pp. 97–115, 1997. View at PublisherView at Google ScholarView at Scopus

15. Y. H. Hao and R. Azzam, "The plastic zones and displacements around underground openings in rock masses containing a fault," Tunnelling and Underground Space Technology, vol. 20, no. 1, pp. 49–61, 2005. View at

PublisherView at Google ScholarView at Scopus

16. Y. Jiang, B. Li, and Y. Yamashita, "Simulation of cracking near a large underground cavern in a discontinuous rock mass using the expanded distinct element method," International Journal of Rock Mechanics and Mining Sciences, vol. 46, no. 1, pp. 97–106, 2009. View at PublisherView at Google ScholarView at Scopus

17. W. C. Zhu, J. Liu, C. A. Tang, X. D. Zhao, and B. H. Brady, "Simulation of progressive fracturing processes around underground excavations under biaxial compression," Tunnelling and Underground Space Technology, vol. 20, no. 3, pp. 231–247, 2005. View at PublisherView at Google ScholarView at Scopus

18. P. Jia and C. A. Tang, "Numerical study on failure mechanism of tunnel in jointed rock mass,"Tunnelling and Underground Space Technology, vol. 23, no. 5, pp. 500–507, 2008. View at Publisher ·View at Google ScholarView at Scopus

19. T.-T. Wang and T.-H. Huang, "A constitutive model for the deformation of a rock mass containing sets of ubiquitous joints," International Journal of Rock Mechanics and Mining Sciences, vol. 46, no. 3, pp. 521–530, 2009. View at PublisherView at Google ScholarView at Scopus

20. R. G. Deng, X. M. Fu, J. Shao, and L. Deng, Study of the Rockmass Engineering Problems in Niba Mountain Deep and Long Tunnel, Southwest Jiaotong University Press, Chendu, China, 2010 (Chinese).

21. S. C. Li, X. D. Feng, and S. C. Li, "Numerical model for the zonal disintegration of the rock mass around deep underground workings," Theoretical and Applied Fracture Mechanics, vol. 66-67, pp. 65–73, 2013. View at PublisherView at Google ScholarView at Scopus

22. M. R. Ayatollahi, M. R. M. Aliha, and H. Saghafi, "An improved semi-circular bend specimen for investigating mixed mode brittle fracture," Engineering Fracture Mechanics, vol. 78, no. 1, pp. 110–123, 2011. View at PublisherView at Google ScholarView at Scopus

23. Z. B. Zhong, R. G. Deng, J. Li, and X. M. Fu, "Experimental study of triaxial mechanical properties of naturally fissured rhyolite," Chinese Journal of Rock Mechanics and Engineering, vol. 33, no. 6, pp. 1233–1240, 2014 (Chinese). View at Google ScholarView at Scopus

24. G. Barla and M. Barla, "Continuum and discontinuum modelling in tunnel engineering," Gradevinar, vol. 52, no. 10, pp. 567–576, 2000. View at Google Scholar

25. Y. Li, D. Zhang, Q. Fang, Q. Yu, and L. Xia, "A physical and numerical investigation of the failure mechanism of weak rocks surrounding

tunnels," Computers and Geotechnics, vol. 61, pp. 292–307, 2014. View at PublisherView at Google ScholarView at Scopus

26. R. G. Deng, Z. B. Zhong, X. M. Fu, W. M. Xiao, and J. Yin, "Experimental study on deformation and strength properties of microfissured rock," in Rock Characterisation, Modelling and Engineering Design Methods, X.-T. Feng, J. A. Hudson, and F. Tan, Eds., pp. 231–234, Taylor & Francis Group, London, UK, 2013. View at Google Scholar

27. Itasca, FLAC3D. Fast Lagrangian Analysis of Continua in 3-Dimensions. Version 4.0, Manual, Itasca, Minneapolis, Minn, USA, 2009.

Chapter 12

DEFORMATION AND FAILURE CHARACTERISTICS OF THE ROCK MASSES AROUND DEEP UNDERGROUND CAVERNS

Chong Zhang,[1] Zhechao Wang,[2] and Qi Wang[2]

[1]College of Electronic and Information Engineering, Henan University of Science and Technology, 263 Kaiyuandadao, Luolong, Luoyang, Henan 471023, China

[2]Geotechnical and Structural Engineering Research Center, Shandong University, 17923 Jingshi Road, Jinan, Shandong 250061, China

ABSTRACT

The deformation and failure characteristics of deep rock masses are the focus of this study on deep rock mass engineering. The study identifies the deformation and failure characteristics of a deep cavern under different ground stress conditions using model test and theoretical analysis methods. First, the similarity theory for model tests is introduced, and then the scale factors used in the present study are calculated according to the Froude criterion. Based on the study objectives, the details of the study methods (the similarity coefficient, the loading conditions, the test steps, etc.) are introduced. Finally, the failure characteristics of the deep cavern and the strain distribution characteristics surrounding the caverns under different ground stress conditions are identified using the model test. It was found that compared with shallow rock masses the rock masses of the deep cavern have a much greater tensile range, which reaches 1.5 times the diameter of the cavern under the conditions established in the present study. Under different ground stress conditions, there are differences in failure characteristics and the reasons of the differences were analyzed. The implication of the test results on the design of support system for deep caverns was presented.

INTRODUCTION

The construction of infrastructure such as hydropower stations and the deep exploitation of such energy resources as coal require us to understand the deformation and failure characteristics of deep underground caverns. The high ground stress states in deep rock masses have a great effect on the stability of the rocks surrounding caverns. After the excavation of a cavern is completed, readjustment of the high ground stress states in the rock masses will occur; stress redistribution often leads to stress concentration, resulting in the development of pressure on the rocks surrounding the lining or the support system of the deep underground engineering structure. When the pressure on the rocks surrounding a cavern exceeds the strength limit or the yield limit of the rock masses, plastic deformation or failure will occur in the rocks surrounding the cavern.

Currently, there are many theories regarding the relation between the depth of a cavern and the pressure of the rocks surrounding the cavern. In some equations, the pressure of the rocks surrounding a cavern is unrelated to the depth of the cavern; in other equations, however, the pressure of the rocks surrounding a cavern is related to the depth of the cavern. The current study shows that when the rocks surrounding a cavern are in a plastic deformation state, the embedded depth of the cavern is greater, and the pressure of the rocks surrounding the cavern is higher. The rocks surrounding a deep cavern are often in a high plastic state; thus, the pressure of the rocks surrounding the cavern increases with the increasing depth [1–3]. Rock masses with different properties will have different failure modes under high pressure from the surrounding rocks [4, 5].

Researchers from many countries have observed the failures of deep rock masses surrounding deep caverns [6], for example, the deep caverns in the metal mines in South Africa and Russia and the Jinchuan nickel mines in China. Based on the drilling data for the rocks surrounding a deep cavern and the comprehensive sectional view obtained from the drilling data, Shemyakin et al. developed the basic pattern of the failure phenomenon of deep surrounding rocks and noted that such a failure mode is quite different from the deformation and failure state of the original shallow surrounding rocks [7]. Guzev and Paroshin attempted to understand the nonmonotonic variation pattern of the stress related to the distance from the surface of the cavern using the non-Euclidean geometric model and discovered that such nonmonotonicity was related to the attenuation of the stress field along the cavern or the periodic variation of stress with time [8]. Metlov et al. revealed the physical basis of the alternating disintegration near the cavern from the perspective of equilibrium thermodynamics [9]. These authors described the instability evolution of the rocks from their elastic state when the cavern was excavated to their mature alternating disintegrated structure and simulated such a relation in a numerical experiment; however, Metlov et al. could not explain the mechanism and condition of the occurrence of such a relation. Reva analyzed the stability of zonally disintegrated surrounding rocks using the energy criterion [10]. Diederichs et al. analyzed the failure process of brittle rocks and concluded that tensile failure was the dominant failure mode during the damage and failure processes of brittle rocks and that the main cause for the final failure (splitting or the formation of a shear zone) was the development, accumulation, and interaction of tensile failure [11]. Zhou et al. analyzed the mechanism of the zonal disintegration of the rocks surrounding a deeply buried spherical cavern [12]. Li et al. conducted an analysis of the fracture form of the rocks surrounding a deeply buried tunnel during the excavation process [13]. Li et al. analyzed the mechanism of the occurrence of the zonal disintegration phenomenon of rock masses [14].

In terms of study methods, test studies on the failure of deep rock masses can be classified into two types: field test studies and model test studies. These two types of test studies have common characteristics: knowledge of patterns is derived from large amounts of test data, and the effects of some factors on the failure of deep rock masses are directly evaluated [4]. Field full-prototype-scale tests are advantageous in that they have visual results and phenomena and the test results can be used directly [6]. Physical simulation tests are an effective, economical, and important manner in which we study deep underground engineering structures. A physical model is a real physical entity. When the model satisfies the basic similarity principle, a physical model can preclude mathematical and mechanical difficulties and relatively truly, comprehensively, visually, and accurately reflect the overall mechanical characteristics, deformation trend, and stability characteristics of the rocks surrounding a deep underground engineering structure [4]. The results obtained from a physical model and the results obtained from numerical calculation or theoretical analysis can be mutually verified; a physical model can also be utilized to discover new mechanical phenomena and patterns to provide important bases for establishing new calculation theories and mathematical models.

In the present study, the deformation and failure characteristics of a deep cavern under different ground stress conditions are obtained using model test and theoretical analysis methods. The similarity theory for model tests is introduced first, and the scale factors used in the present study are calculated according to the Froude criterion. Then, based on the study objectives, the details of the study method (the similarity coefficient, the loading conditions, the test procedure, etc.) are introduced. Finally, the failure characteristics of the deep cavern as well as the strain distribution characteristics surrounding the cavern under different ground stress conditions are developed utilizing the model test. The research results can provide a basis for designing support systems for deep caverns.

SIMILARITY THEORY

Similarity theory is the theoretical basis for model tests. The results of a model test that satisfies similarity conditions can better reflect the actual conditions of the prototype. Whether the results of a model test can reflect the actual conditions of a prototype and provide a qualitative or quantitative design basis for the prototype depends on the degree of similarity between the model test and the physical process of the prototype [15].

Currently, there is no established mathematical equation for characterizing the entire failure process of the rocks surrounding a deep cavern. It is therefore necessary to first study the similarity criteria using the dimensional analysis method. According to dimensional analysis, for a rock mass medium, the possible physical phenomena (Q) of a model test are primarily related to the following physical quantities.

The following are the parameters of the rock mass medium: density (ρ), elastic modulus (E), compressive strength (RC), tensile strength (Rt), Poisson's ratio (v), cohesion (c), and internal friction angle (Φ). The geometric dimensional parameters of the cavern are as follows: span (D), height (H), and radius of the arch (R)

According to dimensional theory, among the physical quantities that describe physical phenomena, length, time, and mass can generally be selected as the basic physical quantities, and other physical quantities can be considered derived quantities (Table 1). For instance, length (l) is a physical quantity with a dimension of L, density (ρ) is a physical quantity with a dimension of ML^{-3}, and acceleration (a) is a physical quantity with a dimension of LT^{-2}

Table 1: Scale factors and dimensions of the main variables

Variables	Scale factors	Dimensions
Length	K_l	L
Density	K_ρ	ML^{-3}
Acceleration	K_a	LT^{-2}
Time	K_t	T
Stress	K_σ	$ML^{-1}T^{-2}$
Strain	K_ε	—
Poisson's ratio	K_μ	—
Friction angle	K_ϕ	—
Velocity	K_V	LT^{-1}
Force	K_F	MLT^{-2}
Specific weight	K_γ	$ML^{-2}T^{-2}$
Impulse	K_i	MLT^{-1}
Energy	K_E	ML^2T^{-2}

According to the basic scale law, the scale factor of each of the abovementioned physical quantities can be expressed using the scale factors of the quantities in its dimension instead of simply using the quantities in its dimension. Thus, the following relations can be derived

$$K_l = K_l, \tag{1}$$

$$K_\rho = K_m \cdot K_l^{-3}, \tag{2}$$

$$K_a = K_l \cdot K_t^{-2}, \tag{3}$$

where K_l, K_ρ, K_a, K_m, and K_t in the the above equations are all the scale factors of the quantities of the same type of model and the prototype. Based on (2) and (3), the following equation can be derived.

$$K_m = K_\rho \cdot K_l^3,$$

$$K_t = K_l^{1/2} \cdot K_a^{-1/2}. \tag{4}$$

The dimension of any variable can be expressed using the power product of M, L, and T. In addition, the scale factors of M, L, and T can be expressed using the scale factors $^{(K_l, K_\rho, \text{ and } K_a)}$ of the 3 selected independent variables (length, density, and acceleration). Therefore, the scale factor of any variable can also be expressed using the scale factors $^{(K_l, K_\rho, \text{ and } K_a)}$ of the 3 selected independent variables.

A geomechanical model test must comply with the related similarity relation criteria. The following criterion is the most important strength relation criterion that must be complied with

$$K_\sigma = K_l \cdot K_\rho \cdot K_g, \tag{5}$$

where $K\sigma = \sigma_m/\sigma_p$ represents the ratio of the compressive strength of the model to the compressive strength of the prototype, $K_l = l_m/l_p$ represents the ratio of the geometric dimension of the model to the geometric dimension of the prototype, $K\rho = \rho_m/\rho_p$ represents the ratio of the density of the material of the model to the density of the material of the prototype, and $K_g = g_m/g_p$ represents the ratio of the acceleration of the model body to the acceleration of the prototype body. Table 2 lists the scale factors of the variables.

Table 2: Scale factors of the variables

Variables	Scale factors	Froude scale factors
Length	K_l	K_l
Density	K_ρ	K_ρ
Acceleration	K_a	$K_a = 1$
Time	$K_t = K_l^{1/2} \cdot K_a^{-1/2}$	$K_t = K_l^{1/2}$
Stress	$K_\sigma = K_\rho \cdot K_a \cdot K_l$	$K_\sigma = K_\rho \cdot K_l$
Strain	K_ε	$K_\varepsilon = 1$
Poisson's ratio	K_μ	$K_\mu = 1$
Friction angle	K_ϕ	$K_\phi = 1$
Velocity	$K_V = K_a^{1/2} \cdot K_l^{1/2}$	$K_V = K_l^{1/2}$
Force	$K_F = K_\rho \cdot K_a \cdot K_l^3$	$K_F = K_\rho \cdot K_l^3$
Specific weight	$K_\gamma = K_\rho \cdot K_a$	$K_\gamma = K_\rho$
Impulse	$K_i = K_\rho \cdot K_a \cdot K_l^{7/2}$	$K_i = K_\rho \cdot K_l^{7/2}$
Energy	$K_w = K_\rho \cdot K_a \cdot K_l^4$	$K_w = K_\rho \cdot K_l^4$

It is extremely difficult to satisfy the numerous parameter similarity requirements in one test. Hence, it is necessary to perform proper simplification according to the test objective and the principles according to which the main issues should be stressed. According to the Froude scaling requirements, the material of a model should satisfy the following relations.

(i)   The material of a model should have the quantities of the dimension of the stress: $K_\sigma = K_\rho \cdot K_l$.

(ii)   Strain: . $K_\varepsilon = 1$.

(iii)   Poisson's ratio: $K_\mu = 1$.

(iv)   Friction angle: $K_\phi = 1$.

In addition, the actual engineering is extremely complex; thus, it is relatively difficult to completely, comprehensively, and truly simulate actual situations. The following simplification assumptions are made when designing the model test:

(1)   Only the effect of the ground stress field is considered; the effect of the dead weight of the surrounding rocks near the walls of the cavern is not considered.

(2)   Rock masses are homogeneous; the effect of the tectonic structures (e.g., bedding and joints) is not considered.

(3)   Considerations are not based strictly on the rheological similarity conditions of each material; however, errors caused by reading intervals are avoided.

RESEARCH METHOD

Similarity Coefficients of the Model Test

Geometric similarity conditions and stress similarity conditions are primarily considered based on the similarity issue of the model test of the present research project.

Geological Conditions

The prototype cavern has a buried depth (H) of 1,000 m and a rock mass density (ρ) of 2.4×10^3 kg/m^3 . The vertical initial ground stress load ($PV\ 0$) that is generated by the dead weight of the rock masses is 24.0 MPa. The side pressure coefficient (N) is 1/3. Therefore, the horizontal ground stress load ($Ph\ 0$) that is generated by the dead weight of the rock masses is 8.0 MPa. According to the national standard classification method for surrounding rocks, type II rock masses are selected. Type II rock masses have a uniaxial compressive strength of 30 MPa~60 MP$_a$. In the present study, type II rock masses with a uniaxial compressive strength of 40 MP$_a$ are selected. Caverns with straight walls and an arch vault, which are commonly used in underground engineering, have a span

(D) of 3,000 mm~5,000 mm. In the present study, a cavern with a span of 3,000 mm, a sidewall height of 1,500 mm, and a vault height of 1,500 mm were selected.

Dimension of the Model Cavern

The present study avoids the effect of the loading boundary of the model test device on the force on the inside of the model cavern and the effect of the boundary of the cavern on the force on the boundary of the model body. Theoretically, if the dimension of a model body is larger, the dimension of the cavern is smaller, and the interaction between the boundary of the model body and the boundary of the cavern is smaller. However, if the model body is too large, the test workload will be large, and the test period will be long. If the dimension of the cavern is too small, construction simulation will be difficult to conduct, and the requirements for testing and excavation techniques will be high. The test model used in the present study has a length of 1,000 mm, a width of 1,000 mm, and a thickness of 400 mm. Plane strain sections are perpendicular to the axial direction of the cavern. If the span of the model cavern is 200 mm, the length and width of the model body will be 5 times the span of the cavern; thus, the related theoretical requirements are met, and there will be no interaction between the boundary of the model body and the boundary of the cavern. Therefore, the geometric similarity coefficient is 1:15, and a span of 200 mm is selected as the span of the model cavern. Figure 1 shows the dimension of the model cavern.

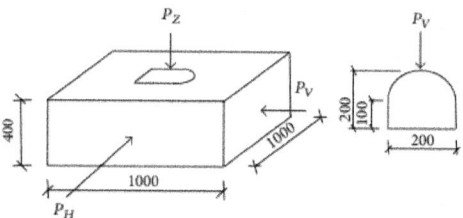

Figure 1: Dimensions of the section of the prototype cavern.

Selection of the Material of the Model

A stress similarity coefficient of 1:20 was selected. Low-grade cement mortar with a weight ratio of cement:sand:water of 1:14:1.4 was selected as the rock mass simulation material. The tamping method was used for molding the cement mortar. The material of the model has a compressive strength of 2.28 MPa. Table 3 lists the detailed mechanical parameters of the original rock, the required simulation material, and the selected simulation material.

Table 3: Mechanical parameters of the original rock masses and the material of the physical model

Types of surrounding rocks	Compressive strength (Rc) (MPa)	Tensile strength (Rt) (MPa)	Cohesion (C) (MPa)	Internal friction angle (φ) (°)	Deformation modulus (Em) (GPa)	Poisson's ratio (μ)	Density (ρ) (kg/m³)
Original rock II	40	2.7	2.0	50	20	0.25	2400
Required simulation material	2.0	0.14	0.1	50	1.0	0.25	1.8
Selected simulation material	2.28	0.3	0.8	54	0.63	0.25	1.8

Loading Conditions

The objective of this set of tests is to obtain the maximum test load and failure mode of the cavern with straight walls and an arch vault under different ground stress load conditions when there is plane strain and the side pressure coefficient is 1/3. The test cavern is a cavern with straight walls and an arch vault. The ground stress load for M1 is of the same order of magnitude as the uniaxial compressive strength of the material of the model; the ground stress load for M2 is 2.17 times the uniaxial compressive strength of the material of the model.

Test Steps

(1)   The ground stress load is evenly applied in 8 steps.

(2)   After the application of the ground stress load is completed, full-section excavation of the cavern begins; the excavation is completed in 4 steps. The failure of the cavern begins during the excavation process. To ensure the formation of the cavern, the rapid excavation method is used. The excavation of the next level begins immediately after each step of excavation is completed without carefully trimming the external shape of the cavern.

TEST RESULTS

M1 Test Results

Failure Conditions of the M1 Cavern

Figure 2 shows the macroscopic failure conditions of the cavern. The following are apparent in the figure:

- The failure of the cavern occurs at the sidewalls but not at the vault or the floor.

- All of the cracks in the sidewalls develop upwards from the junctions between the sidewalls and the floor in the forms of parabolas and clusters and basically disappear at locations that are level with the feet of the vault.

- There are 6 cracks in each of the left and right sidewalls; the degrees of damage to the left and right sidewalls are basically identical.

- With the exception of the cracked body between the first crack and the sidewall, which is relatively wide, all of the other cracked bodies between the cracks are relatively narrow. The width of the cracked body gradually increases as the distance from the sidewall increases.

- An obliquely downward crack from the foot of the vault in the left sidewall of the cavern intersects the 4 cracks that are adjacent to the wall of the cavern. An obliquely downward crack from the foot of the vault in the right sidewall of the cavern and an obliquely downward crack from the center of the right sidewall intersect the 5 cracks that are adjacent to the right sidewall.

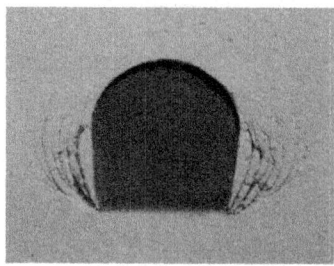

Figure 2: Image of the macroscopic failure of the M1 model.

Changes in the Dimension of the Section of the Cavern

The height of the cavern decreases from 200 mm to 194 mm (a 6 mm decrease), the span of the floor of the cavern decreases from 200 mm to 160 mm (a 40 mm decrease), the span between the centers of the two sidewalls decreases from 200 mm to 180 mm (a 20 mm decrease), and the span between the left foot of the vault and the right foot of the vault decreases from 200 mm to 185 mm (a 15 mm decrease) (Figure 3).

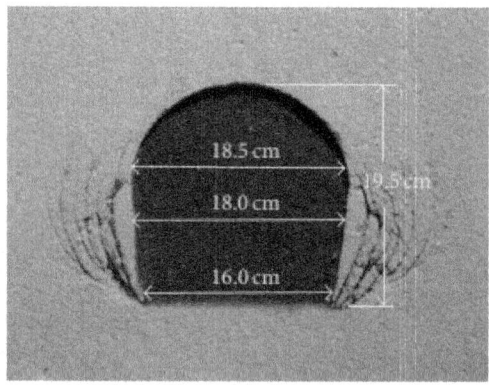

Figure 3: Change in the dimensions of the section of the M1 model cavern.

The following may also be observed from the above measurement data:

(i)   The entire cavern shrinks. Under the plane strain conditions along the axial direction of the cavern, the shrinkage of the dimension is relatively small in the direction in which the load is relatively large (the vault-floor direction); the shrinkage of the dimension is relatively large in the direction in which the load is relatively small (the left sidewall-right sidewall direction).

(ii)   No significant deformation of the vault and the floor of the cavern is observed. The left and right sidewalls are severely deformed. The shrinkage gradually increases from the junction between each sidewall and the respective foot of the vault to the junction between each sidewall and the floor. The lower portion of the section of the cavern exhibits an inverted trapezoid shape.

Crack Ranges

The widest section of the cracks in the left sidewall is level with the foot of the vault; the distribution width is 90 mm from the foot of the vault outwards; the cracks are 140 mm high; and the cracked bodies are 3 mm~20 mm wide.

The widest section of the cracks in the right sidewall is level with the foot of the vault; the distribution width is 90 mm from the foot of the vault outwards; the cracks have a height of 150 mm; and the cracked bodies have a width ranging from 3 mm to 20 mm (Figure 4).

Figure 4: Width and height of the cracks in the right sidewall of the M1 model.

The distribution width of the cracks is smaller than the distribution height of the cracks.

Strain Test Results

Figures 5 and 6 show the values of the radial and circumferential strains inside the medium above the vault, respectively, after the excavation is completed. In the present study, tensile strains are positive, and compressive strains are negative (the same is true for the M2 test).

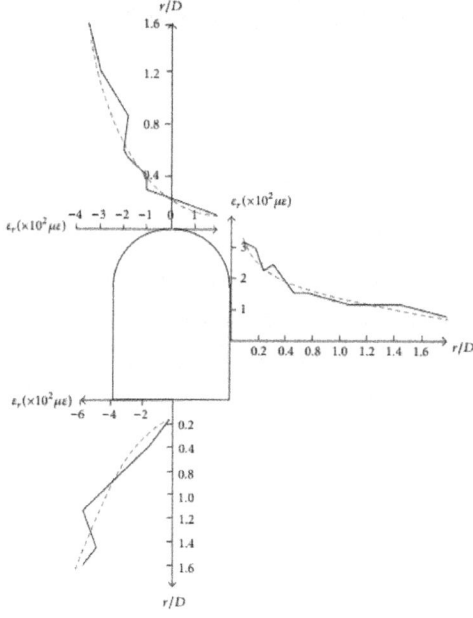

Figure 5: Radial strain curve of the M1 model cavern.

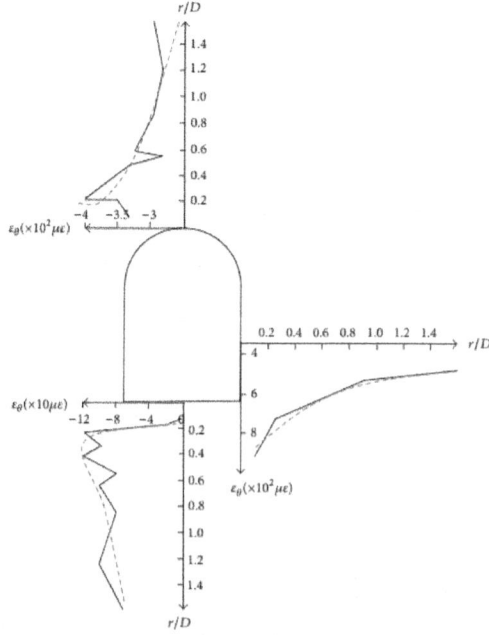

Figure 6: Circumferential strain curve of the M1 model cavern.

Radial Strains

(i) The radial support decreases with the radial strains on the vault, and the absolute values of the radial strains also decrease as the distance from the wall of the cavern increases. Compressive strains change to tensile strains at the locations near the wall of the cavern.

(ii) The radial strains on each of the two sidewalls are tensile strains and gradually increase as the distance from the sidewall decreases. After the excavation is completed, the value of the radial tensile strain on each of the two sidewalls is 3.6 times the value of the tensile strain at the measurement point that is farthest from the wall of the cavern.

(iii) The radial strains inside the model on the bottom of the floor are compressive strains and decrease as the distance from the floor decreases. The value of the radial compressive strain at the measurement point that is farthest from the floor is 12.2 times the value of the radial compressive strain at the measurement point closest to the floor.

Circumferential Strains

(i) The restricting effect of the surrounding rocks on the circumferential strains on the vault decreases as the distance from the walls of the cavern decreases, resulting in an increase in the values of the circumferential strains on the vault as the distance from the walls of the cavern decreases.

(ii) The value of the circumferential compressive strain on each of the sidewalls is 2.0 times the value of the compressive strain at the measurement point that is farthest from the sidewall, indicating that the degrees of the variations of the values of the radial tensile strains on each of the sidewalls are greater than the values of the circumferential compressive strains on each of the sidewalls as the distance from the sidewall decreases.

(iii) The circumferential strains inside the model on the floor are basically compressive strains and gradually increase as the distance from the floor decreases. For example, the value of the circumferential strain at the measurement point that is relatively close to the floor is 2.1 times the value of the circumferential strain at the measurement point that is farthest from the floor. The variation pattern of the circumferential strain begins changing at the location that is relatively close to the floor; the circumferential strain changes from a circumferential compressive strain to a circumferential tensile strain.

M2 Test Results

Failure Conditions of the M2 Cavern

Figure 7 shows the macroscopic failure of the cavern. The figure shows the following:

(i)   The failure of the cavern occurs at the sidewalls but not at the vault or the floor.

(ii)   All the cracks in the sidewalls develop upwards from the junctions between the sidewalls and the floor in the forms of parabolas and clusters and basically disappear at locations that are level with the foot of the vault.

(iii)   There are 6 cracks in each of the left and right sidewalls; the degrees of damage to the left and right sidewalls are basically identical.

(iv)   The cracked body between the second and the third cracks and the cracked body between the third and the fourth cracks in the left sidewall are relatively narrow; the cracked body between the second and the third cracks in the right sidewall is relatively narrow.

(v)   The 2 obliquely downward cracks from the foot of the vault in the left sidewall of the cavern intersect the cracks in the left sidewall. The 2 obliquely downward cracks from the foot of the vault in the right sidewall of the cavern and the 2 obliquely downward cracks from the center of the right sidewall intersect the cracks in the right sidewall. An obliquely downward crack from the top of the cracked body between the right sidewall and the first crack intersects the second cracked body.

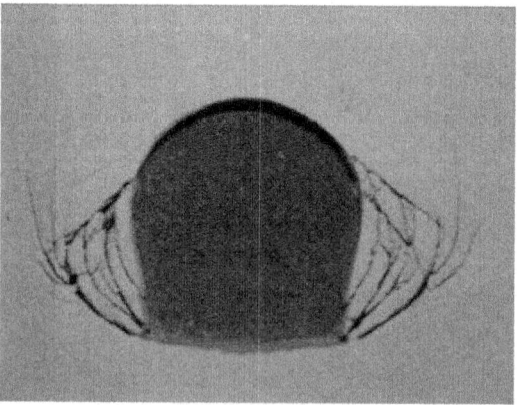

Figure 7: Image of the macroscopic failure of the M2 model.

Dimensions of the Section of the Cavern

The height of the cavern decreases from 200 mm to 195 mm (a 5 mm decrease), the span of the floor of the cavern decreases from 200 mm to 150 mm (a 50 mm decrease), the span between the centers of the two sidewalls decreases from 200 mm to 170 mm (a 30 mm decrease), and the span between the left and right feet of the vault decreases from 200 mm to 180 mm (a 20 mm decrease) (Figure 8).

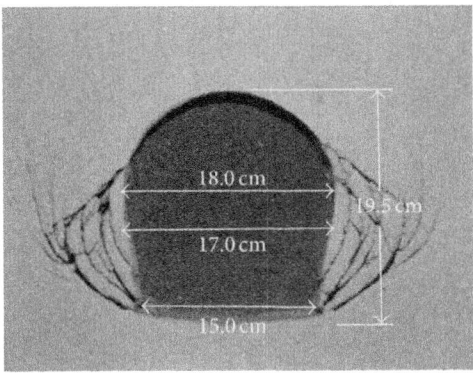

Figure 8: Change in the dimensions of the section of the M2 model cavern.

The above-mentioned measurement data also show the following:

(i)   The entire cavern shrinks. Under the plane strain conditions along the axial direction of the cavern, the shrinkage of the dimension is relatively small (5 mm) in the direction in which the load is relatively large (the vault-floor direction); the shrinkage of the dimension is relatively large (50 mm) in the direction in which the load is relatively small (the left sidewall-right sidewall direction).

(ii)   No significant deformation of the vault and the floor of the cavern was observed. The left and right sidewalls are severely deformed. The shrinkage gradually increases from the junction between each sidewall and the respective foot of the vault to the junction between each sidewall and the floor. The lower portion of the section of the cavern exhibits an inverted trapezoid shape.

Crack Ranges

(i)   The junctions between the cracks along the obliquely downward direction from each foot of the vault and the cracks along the obliquely upward direction from the bottom of the floor have an average width of 85 mm (Figure 9).

(ii)   The left crack that is on the same horizontal level as the left foot of the vault is 85 mm from the left foot of the vault horizontally; similarly, the right crack that is on the same horizontal level with the right foot of the vault is also 85 mm from the right foot of the vault horizontally.

(iii   )The widths of the cracked bodies range from 5 mm to 20 mm (Figure 10).

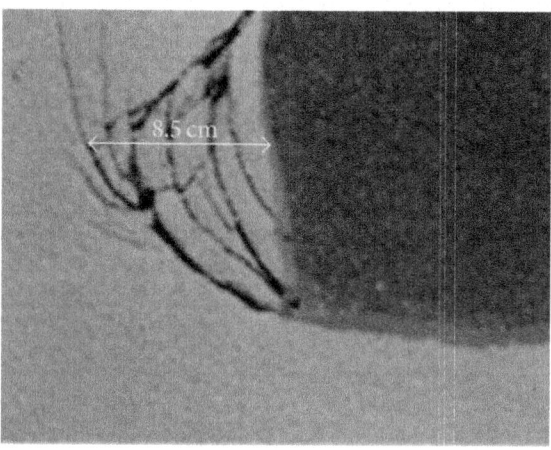

Figure 9: Width of the cracks in the left sidewall of the M2 model cavern.

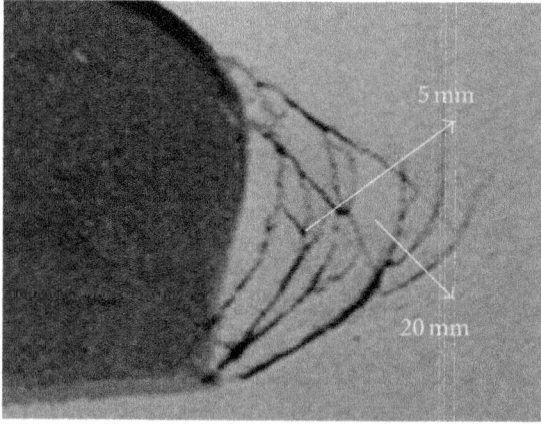

Figure 10: Range of the widths of the cracked bodies in the M2 model cavern.

Strain Test Results

Figures 11 and 12 show the radial and circumferential strain values inside the medium above the vault, respectively, after the excavation is completed.

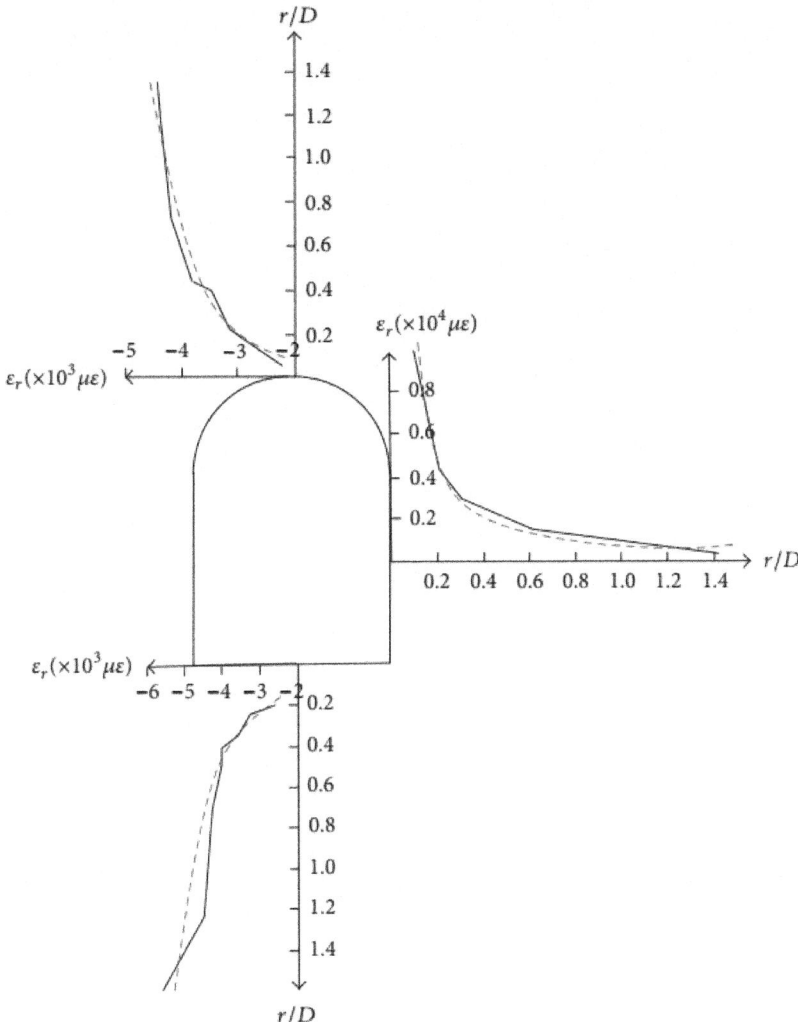

Figure 11: Radial strain curve of the M2 model cavern.

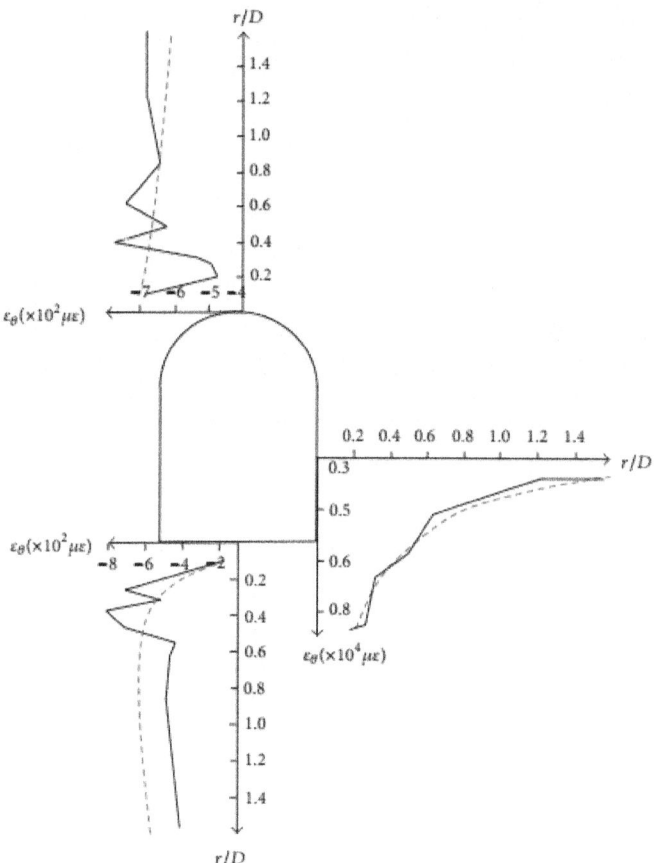

Figure 12: Circumferential strain curve of the M2 model cavern.

Radial Strains

(i)   The radial strains inside the model body on the top of the vault are all compressive strains; the values of the radial compressive strains increase as the distance from the vault increases; when compared to M1, the values of the radial compressive strains of M2 are much greater than the values of M1 at the corresponding locations.

(ii)   The radial strains on the sidewalls are all tensile strains; the values of the tensile strains decrease rapidly as the distance from the sidewall increases; when compared to M1, the values of the radial tensile strains of M2 are much greater than the values of M1 at the corresponding locations.

(iii)   The radial strains inside the model body on the bottom of the floor are all compressive strains; the values of the compressive strains increase as the distance from the floor increases; when compared to M1, the values of the radial strains of M2 are much greater than the values of M1 at the corresponding locations.

Circumferential Strains

(i)   The circumferential strains inside the model body on the top of the vault are also all compressive strains; the values of the circumferential compressive strains increase as the distance from the vault increases although the degrees of the increases in the values of the circumferential strains are significantly smaller than the values of the radial strains.

When compared to M1, the values of the circumferential strains of M2 are greater than the values of M1 at the corresponding locations although the degrees of the increases in the values of the circumferential strains are relatively small. The absolute values of the radial strains are greater than the values of the circumferential strains at the corresponding locations.

(ii)   The circumferential strains on the sidewalls are compressive strains; the values of the compressive strains decrease as the distance from the sidewall increases although the degrees of the decreases in the values of the circumferential strains are significantly smaller than the values of the radial strains; when compared to M1, the values of the circumferential strains of M2 are significantly greater than the values of M1 at the corresponding locations.

(iii)   The circumferential strains inside the model body on the bottom of the floor are compressive strains; however, there is no significant change in the values of the strains as the distance from the floor increases; when compared to M1, the values of the circumferential strains of M2 are significantly greater than the values of M1 at the corresponding locations. The values of the radial strains are greater than the values of the circumferential strains at the corresponding locations.

ANALYSIS OF THE TEST RESULTS

Deformation and Failure Characteristics of Deep Rock Masses

Based on the analysis of the test data, we know that the rock masses of the deep cavern have a tensile range of 1.5 times the diameter of the cavern and a failure range of 0.45 times the diameter of the cavern; however, the rock masses of a shallow cavern with the same shape as the deep cavern used in the present study have a tensile failure range of approximately 0.36 times the diameter of the cavern. When designing a support system for rock masses, the strengthening range of anchor rods and anchor cables should be greater than the tensile range of the cavern. Let us take a cavern with a span of 10 m as an example; the length of the support system for such a cavern should be greater than 15 m. Therefore, the strengthening range of the rock masses of a deep cavern should be far greater than the strengthening range of the rock masses of a shallow cavern.

Comparison of the Failure Modes

Overall, although there are differences in the ground stress load and the maximum test load between M1 and M2, the macroscopic failure phenomena of M1 and M2 are basically identical. By comparing the macroscopic failure phenomena of M1 and M2, the differences between M1 and M2 are exposed as follows:

(i)   The maximum test load of M1 is relatively small; the degree of damage to M1 is relatively small.

(ii)   The cracks along the obliquely downward directions of the left and right sidewalls of M2 are relatively well developed, which, combined with the cracks that have developed upwards from the bottoms of the sidewalls, form relatively significant wedge-shaped cracked bodies.

(iii) After the cracks originate from the bottoms of the left and right sidewalls of M1, the cracks continue developing upwards and move farther from the sidewalls as the distance from the floor increases; when the cracks have progressed to the locations that are on the same horizontal level with the centers of the sidewalls, the cracks are basically parallel to the sidewalls; in particular, the cracks in the right sidewall do not converge at the foot of the vault. After the cracks originate from the bottoms of the sidewalls of M2, the cracks also develop upwards as the distance from the floor increases; however, when the cracks encounter the cracks along the obliquely downward directions from the feet of the vault, the cracks that originated at the bottoms of the sidewalls stop developing; only the two outermost cracks do not intersect and continue to develop.

ANALYSIS OF THE FAILURE MECHANISM

Figure 13 shows the simplified diagram of the forces on a deeply buried circular cavern. Based on the equilibrium equation, the geometric equation, and the constitutive equation, the solutions of the radial stress, the circumferential stress, and the shear stress at an arbitrary point (such a point is at a distance r away from the center of the cavern; in addition, the angle between the line that connects such a point and the center of the cavern and the rightward horizontal axis is) are

$$\sigma_r = \frac{1}{2}(1+\lambda)P_0\left(1-\frac{R_0^2}{r^2}\right)$$

$$-\frac{1}{2}(1-\lambda)P_0\left(1-4\frac{R_0^2}{r^2}+3\frac{R_0^4}{r^4}\right)\cos 2\theta,$$

$$\sigma_\theta = \frac{1}{2}(1+\lambda)P_0\left(1+\frac{R_0^2}{r^2}\right)$$

$$+\frac{1}{2}(1-\lambda)P_0\left(1+3\frac{R_0^4}{r^4}\right)\cos 2\theta,$$

$$\tau_{r\theta} = \frac{1}{2}(1-\lambda)P_0\left(1+2\frac{R_0^2}{r^2}-3\frac{R_0^4}{r^4}\right)\sin 2\theta. \tag{6}$$

When $r=R_0$ that is, at the wall of the cavern,

$$\sigma_r = \tau_{r\theta} = 0,$$

$$\sigma_\theta = (1+\lambda)P_0 + 2(1-\lambda)P_0\cos 2\theta. \tag{7}$$

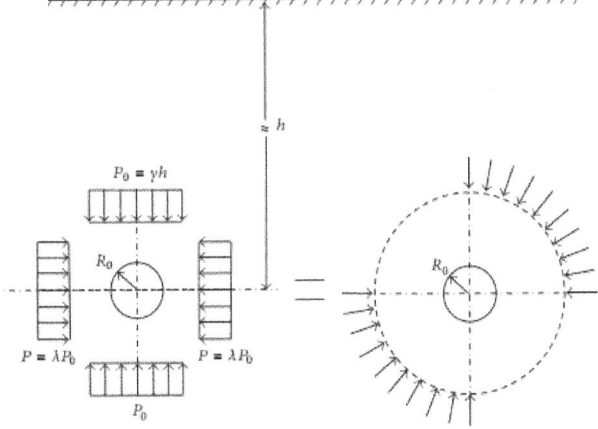

Figure 13: Simplified diagram of the forces on a deeply buried cavern.

Thus, in a circular cavern, there is only a circumferential stress along the periphery of the wall of the cavern; the radial stress and the shear stress are zero; the circumferential stress along the periphery of the wall of the cavern and the distribution of such circumferential stress primarily depend on the side pressure coefficient [λ]. Because the failure of the rocks surrounding a cavern begins from the wall of the cavern, the distribution of the circumferential stress along the periphery of the wall of the cavern determines the earliest failure point of the cavern; therefore, the circumferential stress along the wall of the M1 model is primarily analyzed in the present study.

The prototype of the present study is a semicircular cavern with straight walls and an arch vault; the prototype cavern has a buried depth of $H = 1000$ m, a rock mass density of $\rho = 2.4 \times 10^3$ kg/m^3, and a specific weight of $\gamma = 2.4 \times 10^4$ N/m^3. The key points of the semicircular cavern with straight walls and an arch vault are as follows: number 1 point (the top of the vault), number 2 point (the middle point between the top of the vault and foot of the vault), number 3 point (a foot of the vault), number 4 point (the center of a sidewall), and number 5 point (the bottom of a sidewall) (Figure 14). The calculation equation for the circumferential stresses on the wall of the cavern (σ_θ) is as follows

$$\sigma_\theta = \gamma (\alpha + \beta\lambda)(H' + kr_0),$$ (8)

Where α and β are calculation coefficients (see Table 4). Based on the aforementioned coefficients, the circumferential stresses at the key points on the walls of the cavern can be obtained (Table 5).

Table 4: Calculated coefficients for the circumferential stresses on the periphery of the semicircular cavern with straight walls and an arch vault

Point numbers	1	2	3	4	5
α	−0.9758	0.8131	2.1908	2.1704	3.4704
β	3.2138	1.2639	−0.9001	−0.7654	1.8451
k			1.1145		

Table 5: Values of the circumferential stresses at the key points of the semicircular cavern with straight walls and an arch vault

Circumferential stresses	1	2	3	4	5
σ_θ	−53.80	−49.93	−31.03	−33.78	−127.79

The values listed in the table indicate that after the excavation of the prototype cavern with straight walls and an arch vault is completed, the circumferential compressive stress on the bottom of each sidewall is greatest; the compressive shear failure begins at the bottom of each sidewall, and the cracks that originate from the bottom of each sidewall will gradually extend outwards, which can also be observed in the macroscopic failure of and the development of the cracks in M1, indicating that the failure mode of the cavern with straight walls and an arch vault under plane strain conditions that was obtained from the present study is correct and reasonable and in accordance with the actual stress state of a deep cavern.

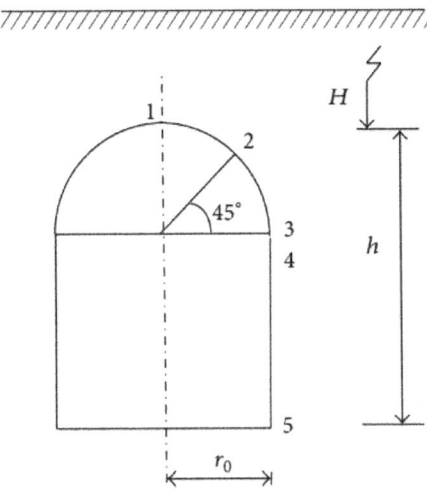

Figure 14: Key points of the semicircular cavern with straight walls and an arch vault.

CONCLUSIONS

The deformation and failure characteristics of deep rock masses are an important component of the research on deep rock mass engineering. In the present study, the deformation and failure characteristics of two deep caverns under different ground stress conditions are derived using the model test method. Using the model test, the failure characteristics of deep caverns and the strain distribution characteristics along the peripheries of the caverns under different ground stress conditions are derived. The test results show that the tensile range of the rock masses of a deep cavern is far greater than the tensile range of the rock masses of a shallow cavern. Under the conditions used in the present study, the rock masses have a tensile range of 1.5 times the diameter of the cavern and a failure range of approximately 0.45 times the diameter of the cavern. When designing a support system for rock masses, the strengthening range of anchor rods and anchor cables should be greater than the tensile range of the cavern.

The research results show that there are differences in the failure characteristics under different ground stress conditions: a cavern with a low cavern-excavation load has a slightly smaller failure range, and the main failure form of a cavern with a low cavern-excavation load is the outward development of fracture cracks from the feet of the walls, while the feet of the vault remain basically intact. Linear sliding failure occurs in both the feet of the vault and the feet of the walls of a cavern with a relatively large cavern-excavation load, and the linear sliding failure at the top and the bottom is approximately symmetrical.

The causes of the differences in the failure characteristics of a deep cavern under different ground stress conditions are analyzed; because of the difference between the pre- and poststress concentration states, the feet of the walls of a cavern with a relatively low cavern-excavation load enter a stress concentration state first; the subsequent load is applied to the cavern that is already in a stress concentration state, resulting in the occurrence of failure at the feet of the walls, which, in turn, results in the occurrence of stress transfer that reduces the damage to the other sections; thus, the failure of a cavern with a relatively small cavern-excavation load is relatively less severe.

CONFLICT OF INTERESTS

The authors declare no conflict of interests.

AUTHORS' CONTRIBUTION

Chong Zhang and Zhechao Wang conceived and designed the experiments; Qi Wang performed the experiments; Chong Zhang and Qi Wang analyzed the data; Zhechao Wang wrote the paper.

ACKNOWLEDGMENTS

The present study was supported by the general projects of the National Natural Science Foundation of China (Grant no. 51474095). The authors hereby wish to express their gratitude.

REFERENCES

1.   R. S. Read, "20 years of excavation response studies at AECL's underground research laboratory,"International Journal of Rock Mechanics and Mining Sciences, vol. 41, no. 8, pp. 1251–1275, 2004.

2.   Q. H. Qian, "The characteristic scientific phenomena of engineering response to deep Rock mass and the implication of deepness," Journal of East China Institute of Technology, vol. 27, no. 1, pp. 1–5, 2004 (Chinese).

3.   M. C. He, H. P. Xie, S. P. Peng, and Y. D. Jiang, "Study on rock mechanics in deep mining engineering,"Chinese Journal of Rock Mechanics and

Engineering, vol. 24, no. 16, pp. 2803–2813, 2005 (Chinese).

4. J. C. Gu, L. Y. Gu, A. M. Chen, J. Xu, and W. Chen, "Model test study on mechanism of layered fracture within surrounding rock of tunnels in deep stratum," Chinese Journal of Rock Mechanics and Engineering, vol. 27, no. 3, pp. 433–438, 2008 (Chinese).

5. X. P. Zhou and Q. H. Qian, "Zonal fracturing mechanism in deep tunnel," Chinese Journal of Rock Mechanics and Engineering, vol. 26, no. 5, pp. 877–885, 2007 (Chinese).

6. S. C. Li, H. P. Wang, Q. H. Qian, et al., "In-situ monitoring research on zonal disintegration of surrounding rock mass in deep mine roadways," Chinese Journal of Rock Mechanics and Engineering, vol. 27, no. 8, pp. 1545–1553, 2008 (Chinese).

7. E. I. Shemyakin, G. L. Fisenko, M. V. Kurlenya et al., "Zonal disintegration of rocks around underground workings, Part 1: data of in situ observations," Soviet Mining Science, vol. 22, no. 3, pp. 157–168, 1986.

8. M. A. Guzev and A. A. Paroshin, "Non-Euclidean model of the zonal disintegration of rocks around an underground working," Journal of Applied Mechanics and Technical Physics, vol. 42, no. 1, pp. 131–139, 2001. ·

9. L. S. Metlov, A. F. Morozov, and M. P. Zborshchik, "Physical foundations of mechanism of zonal rock failure in the vicinity of mine working," Journal of Mining Science, vol. 38, no. 2, pp. 150–155, 2002.

10. V. N. Reva, "Stability criteria of underground workings under zonal disintegration of rocks," Fiziko-Tekhnicheskie Problemy Razrabotki Poleznykh Iskopaemykh, no. 1, pp. 35–38, 2002.

11. M. S. Diederichs, P. K. Kaiser, and E. Eberhardt, "Damage initiation and propagation in hard rock during tunnelling and the influence of near-face stress rotation," International Journal of Rock Mechanics & Mining Sciences, vol. 41, no. 5, pp. 785–812, 2004.

12. X. P. Zhou, Q. H. Qian, B. H. Zhang, et al., "The mechanism of the zonal disintegration phenomenon around deep spherical tunnels," Engineering Mechanics, vol. 27, no. 1, pp. 69–75, 2010 (Chinese).

13. S. C. Li, Q. H. Qian, D. F. Zhang, et al., "Analysis of dynamic and fractured phenomena for excavation process of deep tunnel," Chinese Journal of Rock Mechanics and Engineering, vol. 28, no. 10, pp. 2104–2112, 2009 (Chinese).

14. Y. J. Li, Y. S. Pan, and Z. H. Li, "Analysis of mechanism of zonal disintegration of rocks," Chinese Journal of Geotechnical Engineering, vol. 28, no. 9, pp. 1124–1128, 2006 (Chinese).

15. M. C. Liao, Z. K. Guo, and F. Liu, "Determination of the specimen size for model experiment on zonal fracturing of deep rock mass," Journal of Disaster Prevention and Mitigation Engineering, vol. 26, no. 1, pp. 58–62, 2006 (Chinese).

Chapter 13

APPLICATION OF BASE FORCE ELEMENT METHOD ON COMPLEMENTARY ENERGY PRINCIPLE TO ROCK MECHANICS PROBLEMS

Yijiang Peng, Qing Guo, Zhaofeng, Zhang, and Yanyan Shan

The Key Laboratory of Urban Security and Disaster Engineering, Ministry of Education, Beijing University of Technology, Beijing 100124, China

ABSTRACT

The four-mid-node plane model of base force element method (BFEM) on complementary energy principle is used to analyze the rock mechanics problems. The method to simulate the crack propagation using the BFEM is proposed. And the calculation method of safety factor for rock mass stability was presented for the BFEM on complementary energy principle. The numerical researches show that the results of the BFEM are consistent with the results of conventional quadrilateral isoparametric element and quadrilateral reduced integration element, and the nonlinear BFEM has some advantages in dealing crack propagation and calculating safety factor of stability.

INTRODUCTION

The finite element method (FEM) has been playing a very important role in solving various problems in engineering and science. However, the conventional finite element method (FEM) based on the displacement model has some shortcomings, such as large deformation, treatment of incompressible materials, bending of thin plates, and moving boundary problems. In the past decades, numerous efforts techniques have been proposed for developing finite element models which are robust and insensitive to mesh distortion, such as the hybrid stress method [1–4], the equilibrium models [5, 6], the mixed approach [7], the integrated force method [8–11], the incompatible displacement modes [12, 13], the assumed strain method [14–17], the enhanced strain modes [18, 19], the selectively reduced integration scheme [20], the quasiconforming element method [21], the generalized conforming method [22], the Alpha finite element method [23], the new spline finite element method [24, 25], the unsymmetric method [26–29], the new natural coordinate methods [30–33], the smoothed finite element method [34], and the base force element method [35–43].

In recent years, some scholars are studying other types of numerical analysis methods, such as boundary element method [44, 45] and meshless method [46, 47]. And some scholars still adhere to explore the finite element method based on complementary energy principle [48–51]. However, these methods have not been widely applied in engineering.

In this paper, the base force element method (BFEM) on complementary energy principle is used to analyze the engineering problems of rock mechanics. The "base forces" was introduced by Gao [52], who used the concept to replace various stress tensors for the description of the stress state at a point. These base forces can be directly obtained from the strain energy. For large deformation problems, when the base forces were adopted, the derivation of basic formulae was simplified by Gao [53] and Gao et al. [54–56]. Based on the concept of the base forces, precise expressions for stiffness and compliance matrices for the FEM were obtained by Gao [52]. The applications of the stiffness matrix to the plane problems of elasticity using the plane quadrilateral element and the polygonal element were researched by Peng et al. [37]. Using the concept of base forces as state variables, a three-dimensional formulation of base force element method (BFEM) on complementary energy principle was proposed by Peng and Liu [35] for geometrically nonlinear problems. And the new finite element method based on the concept of base forces was called as the Base Force Element Method (BFEM) by Peng and Liu [35]. A three-dimensional model of base force element method (BFEM) on complementary energy principle was proposed by Liu and Peng [36] for elasticity problems. A 4-mid-node plane element model of the BFEM on complementary energy principle was proposed by Peng et al. [38] for geometrically nonlinear problem, which is derived by assuming that the stress is uniformly distributed on each edges of a plane element. In the paper [39], an arbitrary convex polygonal element model of the BFEM on complementary energy principle was proposed for geometrically nonlinear problem. In the paper [43], a 4-mid-node plane model of BFEM on complementary energy principle was researched, and its computational performance was studied. The convex polygonal element model of BFEM on complementary energy principle was given by Peng et al. [40] for arbitrary mesh problems. In the paper [41], the concave polygonal element model of BFEM on complementary energy principle was proposed for the concave polygonal mesh problems. In the paper [42], the BFEM on potential energy principle was used to analyze recycled aggregate concrete (RAC) on mesolevel, in which the model of BFEM with triangular element was derived, and the simulation results of the BFEM agree with the test results of recycled aggregate concrete. In recently, the BFEM on damage mechanics has been used to analyze the compressive strength, the size effects of compressive strength, and fracture process of concrete at mesolevel, and the analysis method is the new way for investigating fracture mechanism and numerical simulation of mechanical properties for concrete.

The purpose of this paper is to survey the base forces element method on complementary energy principle for large-scale computing problems in rock engineering problems.

MODEL OF THE BFEM

Compliance Matrix

Compliance Matrix. Consider a 4-mid-node plane element as shown in Figure 1; the compliance matrix of a base force element can be obtained as [43]

$$C_{IJ} = \frac{1+\nu}{EA}\left(r_{IJ}U - \frac{\nu}{1+\nu}r_I \otimes r_J\right), \quad (I,J=1,2,3,4) \tag{1}$$

in which E is Young's modulus,] is Poisson's ratio, A is the area of an element, U is the unit tensor, and rIJ is the dot product of radius vectors r_I and r_J at points I and J.

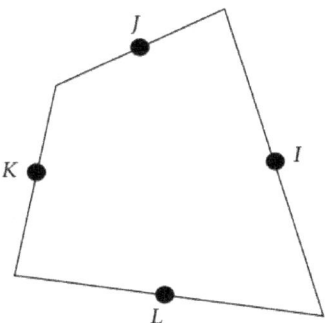

Figure 1: Four-mid-node plane element.

For a plane rectangular coordinate system, the radius vectors rI and rJ of points I and J can be written as

$$\mathbf{r}_I = x_I\mathbf{e}_x + y_I\mathbf{e}_y, \qquad \mathbf{r}_J = x_J\mathbf{e}_x + y_J\mathbf{e}_y \tag{2}$$

in which $\mathbf{e}_x$, $\mathbf{e}_y$ are the unit vectors. Further, the compliance matrix of an element can be reduced as follows:

$$\begin{aligned}
C_{IJ} = \frac{1+\nu}{EA}\Bigg[& \left(\frac{1}{1+\nu}x_Ix_J + y_Iy_J\right)\mathbf{e}_x \otimes \mathbf{e}_x \\
& - \frac{\nu}{1+\nu}x_Iy_J\mathbf{e}_x \otimes \mathbf{e}_y - \frac{\nu}{1+\nu}y_Ix_J\mathbf{e}_y \otimes \mathbf{e}_x \\
& + \left(x_Ix_J + \frac{1}{1+\nu}y_Iy_J\right)\mathbf{e}_y \otimes \mathbf{e}_y\Bigg].
\end{aligned} \tag{3}$$

For a plane strain problem, it is necessary to replace E by $E/(1-\nu^2)$ and ν by $\nu/(1-\nu)$ in (1) and (3).

The characteristics of the BFEM on complementary energy principle are that the model does not introduce an interpolating function and is not necessary to introduce the Gauss integral for calculating the compliance coefficient at a point.

GOVERNING EQUATIONS

The total complementary energy of the elastic system which has n elements can be written as

$$\Pi_C = \sum_n \left(W_C^e - \bar{\mathbf{u}}_I \cdot \mathbf{T}^I\right), \tag{4}$$

where W_C^e is the complementary energy of an element and $\mathbf{T}^I$ and $\bar{\mathbf{u}}_I$ are the resultant force vectors and the given displacement acting on the center node I of the edge I, respectively. The equilibrium conditions can be released by the Lagrange multiplier method, and a new complementary energy function for an element can be introduced as follows:

$$\Pi_C^{e,*}(\mathbf{T},\lambda,\lambda_\theta) = \Pi_C^e(\mathbf{T}) + \lambda\left(\sum_{I=1}^{4}\mathbf{T}^I\right) + \lambda_\theta\left(\mathbf{T}^I \times \mathbf{r}_I\right), \tag{5}$$

where $\lambda = \lambda_x\mathbf{e}_x + \lambda_y\mathbf{e}_y$ and λ_θ are the Lagrange multipliers. For the elastic system, the modified total complementary energy function of the elastic system which contains n elements should meet the following equation by means of the modified complementary energy principle:

$$\delta\Pi_C^* = \sum_n \left[\delta\Pi_C^{e,*}(\mathbf{T},\lambda,\lambda_\theta)\right] = 0. \tag{6}$$

Further, (6) can be expressed as

$$\frac{\partial \Pi_C^* \left(\mathbf{T}, \lambda, \lambda_\theta \right)}{\partial \mathbf{T}} = 0, \qquad \frac{\partial \Pi_C^* \left(\mathbf{T}, \lambda, \lambda_\theta \right)}{\partial \lambda} = 0,$$

$$\frac{\partial \Pi_C^* \left(\mathbf{T}, \lambda, \lambda_\theta \right)}{\partial \lambda_\theta} = 0.$$

$$(7)$$

The first of (7) is the compatibility equations and displacement boundary conditions for the elastic system.

According to this equation, the displacement boundary conditions in this paper can be implemented in the BFEM. The second of (7) is the force equilibrium equation of each element. The third of (7) is the moment equilibrium equation of each element. These are the governing equations of the BFEM. From the equations, we can obtain the resultant forces acting on the center points of the edges of all elements.

Stress Tensor of an Element.

Consider the 4-mid-node plane element as shown in Figure 1; the real stress σ of the element can be replaced by the average stress $\bar{\sigma}$ if the element is small enough. According to the definitions of Cauchy stress tensors, the stress expressions of an element can be obtained as

$$\sigma = \frac{1}{A} \mathbf{T}^I \otimes \mathbf{r}_I,$$

$$(8)$$

where $\otimes$ is the dyadic symbol, $\mathbf{T}^I$ and $\mathbf{r}I$ are the resultant force vectors acting on the center node I of the edge I and the radius vector of the node I, respectively, and the summation rule is implied. For a plane rectangular coordinate system, the force vectors $\mathbf{T}^I$ of the node I can be written as

$$\mathbf{T}^I = T_x^I \mathbf{e}_x + T_y^I \mathbf{e}_y,$$

$$(9)$$

where $T^I{}_x$ and $T^I{}_y$ are the components of the force vector $\mathbf{T}I$ along coordinates x and y, respectively. Further, the stress tensors of an element can be reduced as follows:

$$\sigma = \frac{1}{A} \sum_{I=1}^{4} \left[T_x^I x_I \mathbf{e}_x \otimes \mathbf{e}_x + T_x^I y_I \mathbf{e}_x \otimes \mathbf{e}_y \right.$$

$$\left. + T_y^I x_I \mathbf{e}_y \otimes \mathbf{e}_x + T_y^I y_I \mathbf{e}_y \otimes \mathbf{e}_y \right].$$

$$(10)$$

Displacement Vector of Nodes. According to the governing equation of an element of the BFEM, the explicit expression of displacement can be obtained as

$$\boldsymbol{\delta}_I = \mathbf{C}_{IJ} \cdot \mathbf{T}^J + \boldsymbol{\lambda} + \lambda_\theta \boldsymbol{\varepsilon} \cdot \mathbf{r}_I,$$

(11)

in which ε is the alternating tensor, λ and λ_θ are the Lagrange multipliers [43], and ε and λ can be expressed as

$$\boldsymbol{\varepsilon} = \mathbf{e}_x \otimes \mathbf{e}_y - \mathbf{e}_y \otimes \mathbf{e}_x,$$

$$\boldsymbol{\lambda} = \lambda_x \mathbf{e}_x + \lambda_y \mathbf{e}_y.$$

(12)

Further, the displacement vectors of an element can be reduced as follows [43]:

$$\boldsymbol{\delta}_I = \left(C_{IxJx} T_x^J + C_{IxJy} T_y^J + \lambda_x + \lambda_\theta y_I \right) \mathbf{e}_x$$

$$+ \left(C_{IyJx} T_x^J + C_{IyJy} T_y^J + \lambda_y - \lambda_\theta x_I \right) \mathbf{e}_y.$$

(13)

SIMULATION OF GRAVITY AND MATERIAL

In engineering problems, regardless of the dam, rock slope, or other structure of rock mass, the gravity of structure should be considered in the numerical calculation. We did not take into account the gravity problem when we previously prepared the program of BFEM on complementary energy principle. In order to consider the gravity of structure, three problems must be solved, including the calculation problem of equivalent node loads in the BFEM on complementary energy principle, the problem of exerting gravity in the software of the BFEM and the calculation problem of stress tensor in an element when the gravity is added.

Equivalent Node Loads of Gravity

For the BFEM on complementary energy principle, the equivalent node loads of gravity in an element can be calculated according to the principle of virtual work, as shown in Figure 2, and the expression can be given as

$$\{Q\}^e = -\frac{\rho g A t}{4} \begin{bmatrix} 0 & 1 & 0 & 1 & 0 & 1 & 0 & 1 \end{bmatrix}^T$$

(14)

$$Q^I = -\frac{\rho g A t}{4} \mathbf{e}_y, \quad (I = 1, 2, 3, 4),$$

(15)

where t is the thickness of an element, ρ is the density of material, and g is acceleration of gravity

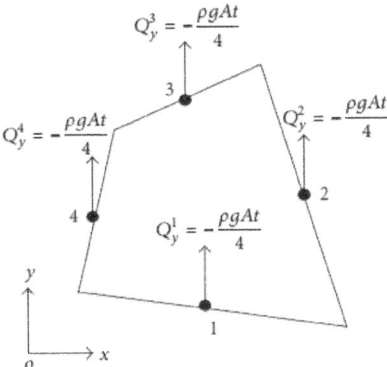

Figure 2: Equivalent node loads of gravity in an element.

Stress Calculation of an Element

When the gravity of an element is not taken into account, as shown in Figure 3, we calculate the force acting on the edges of an element first. Then, the stress tensor of the element can be calculated by (8) or (10).

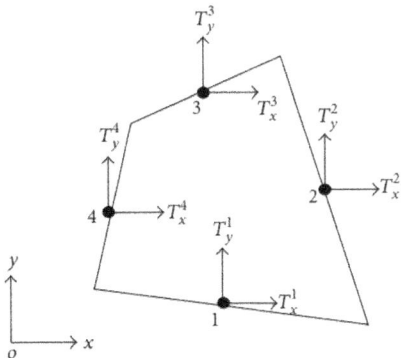

Figure 3: An element without gravity.

Further, the stress tensors of an element can also be reduced as follows:

$$\bar{\sigma} = \frac{1}{A}\left(\mathbf{T}^1 \otimes \mathbf{r}_1 + \mathbf{T}^2 \otimes \mathbf{r}_2 + \mathbf{T}^3 \otimes \mathbf{r}_3 + \mathbf{T}^4 \otimes \mathbf{r}_4\right),$$

(16)

Where

$\mathbf{T}^I = T_x^I \mathbf{e}_x + T_y^I \mathbf{e}_y$, $\mathbf{r}^I = x_I \mathbf{e}_x + y_I \mathbf{e}_y$, $(I = 1, 2, 3, 4)$, x_I and y_I are the coordinates of node I, respectively.

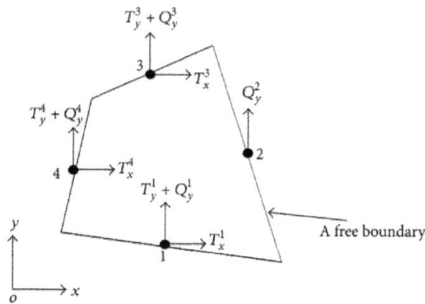

Figure 4: Considering gravity and a free boundary.

When the gravity of an element is taken into account and there is a free boundary, as shown in Figure 4, the stress tensor of the element can be calculated as

$$\bar{\sigma} = \frac{1}{A} \left[(\mathbf{T}^1 + \mathbf{Q}^1) \otimes \mathbf{r}_1 + \mathbf{Q}^2 \otimes \mathbf{r}_2 \right. $$
$$\left. + (\mathbf{T}^3 + \mathbf{Q}^3) \otimes \mathbf{r}_3 + (\mathbf{T}^4 + \mathbf{Q}^4) \otimes \mathbf{r}_4 \right]. \tag{17}$$

When the gravity of an element is taken into account and there is a force boundary condition, as shown in Figure 5, the stress tensor of the element can be calculated as

$$\bar{\sigma} = \frac{1}{A} \left[(\mathbf{T}^1 + \mathbf{Q}^1) \otimes \mathbf{r}_1 + (\mathbf{F} + \mathbf{Q}^2) \otimes \mathbf{r}_2 \right. $$
$$\left. + (\mathbf{T}^3 + \mathbf{Q}^3) \otimes \mathbf{r}_3 + (\mathbf{T}^4 + \mathbf{Q}^4) \otimes \mathbf{r}_4 \right] \tag{18}$$

In which $\mathbf{F} = F_x \mathbf{e}_x + F_y \mathbf{e}_y$.

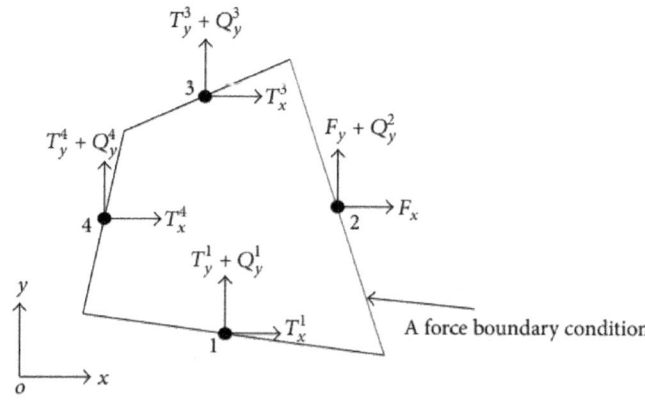

Figure 5: Considering gravity and a force boundary condition.

When the gravity of an element is not taken into account and there is a force boundary condition, as shown in Figure 6, the stress tensor of the element can be calculated as

$$\bar{\sigma} = \frac{1}{A}\left(\mathbf{T}^1 \otimes \mathbf{r}_1 + \mathbf{F} \otimes \mathbf{r}_2 + \mathbf{T}^3 \otimes \mathbf{r}_3 + \mathbf{T}^4 \otimes \mathbf{r}_4\right).$$

$$(19)$$

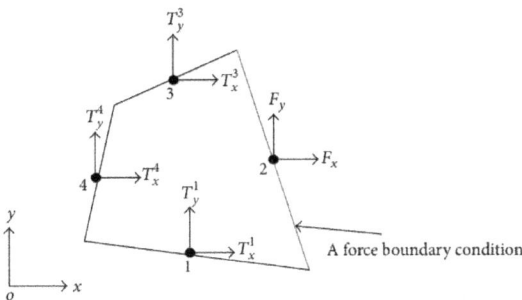

Figure 6: Considering a force boundary condition and no gravity.

Simulation of Different Materials

We adopt two one dimensional array variables to reflect the change of elastic modulus and Poisson's ratio of different materials, respectively.

The Nonlinear Model of BFEM for Crack Propagation Problems

The conventional displacement model of FEM requires the mesh reconstruction for the crack propagation problems. Therefore, the conventional FEM has deficiencies for the crack propagation problems. Because the contact forces between each element are used, the BFEM on complementary energy principle has advantage in the simulation of crack propagation problems. In the BFEM on complementary energy principle, we only need to deal with the constraint conditions and compatibility equations of displacement

Failure Criteria of the Contact Interface between Two Elements

According to the control equations of the BFEM on complementary energy principle, the forces acting on the edge on an element can easily be calculated. According to the failure criteria expressed by the forces acting on the edge of an element, we can judge whether the interface of elements is cracking. If there is cracking of element interface, the nonlinear processing must be carried out.

Condition of Elastic State at Contact Interface. When $T_n < 0$ and $|T_s| < -f \cdot T_n + c \cdot l$ or $0 < T_n < t_0$ and $|T_s| < c \cdot l$, the contact interface is not cracked and in the elastic state. Here, T_n and T_s are the normal interface force and tangential interface force at contact interface between two elements,

respectively. c and f are the cohesion and the internal friction coefficient of the contact interface, respectively. l is the length of an element. t_0 is the tensile strength of an element, and it is positive in tension.

Condition of Positive Slip at Contact Interface.

When $T_n < 0$ and $T_s \geq -f \cdot T_n + c \cdot l$, the contact interface of the two elements began to slip. We need to get rid of the tangential displacement constraint at the contact interface between the two elements that it has been cracked and put the frictional force as load that is used to an initial force condition. The new load acting on the contact interface between the two elements in the second loop of calculation can be written as follows:

$$T_{s0} = f \cdot T_s. \tag{20}$$

When $0 < T_n < t_0$ and $T_s \geq c \cdot l$, the contact interface of the two elements begins to slip. We need to get rid of the tangential displacement constraint and the normal displacement constraint at the contact interface of the two elements. And the contact interface will crack after slipping

Condition of Negative Slip at Contact Interface

When $T_n < 0$ and $T_s \leq f \cdot T_n - c \cdot l$, the contact interface of the two elements began to slip. We need to get rid of the tangential displacement constraint at the contact interface between the two elements that it has been cracked, and put the frictional force as load that is an initial force condition. The new load acting on the contact interface between the two elements in the second loop of calculation is

$$T_{s0} = -f \cdot T_s. \tag{21}$$

When $0 < T_n < t_0$ and $T_s \leq -c \cdot l$, the contact interface of the two elements begins to slip. We need to get rid of the tangential displacement constraint and the normal displacement constraint at the contact interfaces of elements. And the contact interface will crack after slipping.

Condition of Pull Cracking at Contact Interface. When $T_n \geq t_0$, the contact interface of the two elements begins to crack. We need to get rid of the tangential displacement constraint and the normal displacement constraint at the contact interfaces of elements.

After the above checks, the computer program of BFEM uses the new loads and constraint conditions to solve the governing equations of BFEM on complementary energy principle and obtains the new forces acting on the contact interfaces of elements. The program repeats the above checks until there are no new cracking at the contact interfaces or the solutions of nonlinear equations cannot be convergence since the interface cracks are too long.

Flow Chart of the Nonlinear BEFM for Crack Propagation Problems.

The flow chart of the nonlinear BEFM of crack propagation problems can be shown in Figure 7.

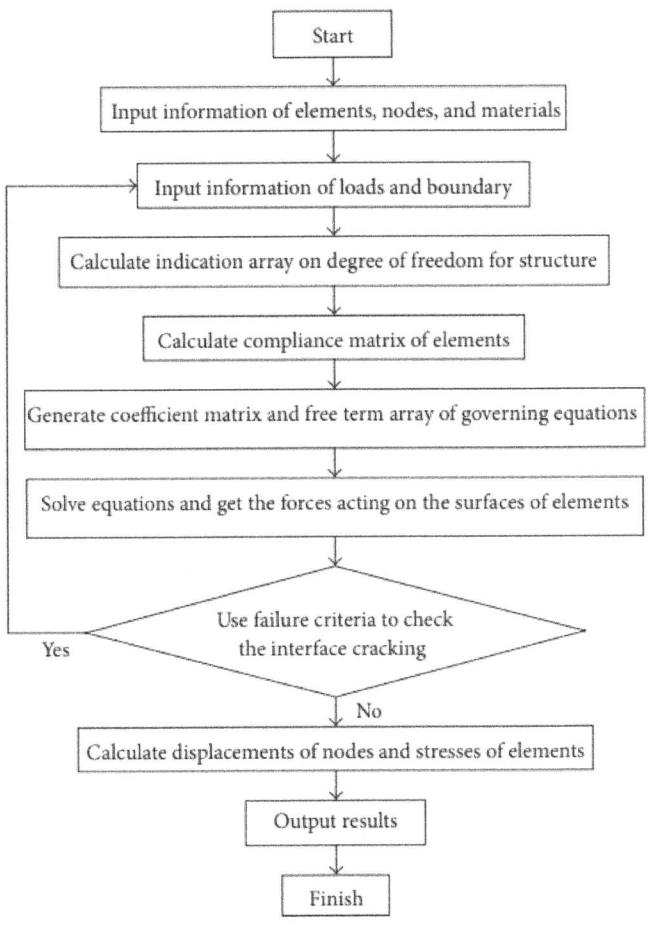

Figure 7: Flow chart of the nonlinear BEFM for crack propagation problems.

CALCULATION METHOD ON SAFETY FACTOR OF STABILITY IN THE BFEM

There are many methods to calculate the safety factor in engineering. The traditional finite element method usually used the rock joint elements to calculate factor of safety along the joint path. When the base force element method

is used to analyze the stability of the rock mass in order to get the safety factor along the joint path, it is very easy. First, we calculate the surface forces of all elements according to the different load combinations. Then, we accumulate the sliding resistances and the sliding forces along the sliding path, respectively. Further, the safety factor of stability along the sliding path can obtained by the following equation:

$$K = \frac{\sum_{i=1}^{n}(-T_{ni}f_i + c_i l_i)}{\sum_{i=1}^{n} T_{si}},$$

(22)

where T_{ni} and T_{si} are the normal interface force and tangential interface force at contact interface between two elements along the sliding path, respectively. c_i and f_i are the cohesion and the internal friction coefficient of the contact interface between two elements along the sliding path, respectively. l_i is the length of the interface in the element along the sliding path

NUMERICAL EXAMPLES

Example 1: A Rock Pillar under the Action of Gravity. Consider a thick pillar of rock under the action of gravity shown in Figure 8. And its width is 5 m, its height is 10 m, modulus of elasticity $E = 1 \times 10^8$ Pa, Poisson ratio $v = 0.3$, density of rock $\rho = 2.45$ t/m³ , and acceleration of gravity $g = 9.8$ m/s² . The calculation is considered into the plane stress problem.

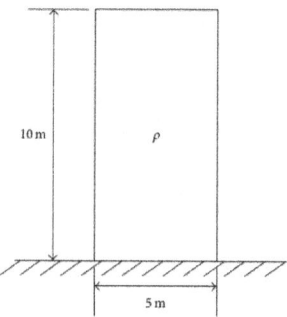

Figure 8: A rock pillar under the action of gravity.

The calculation is done using three different element meshes with the center nodes of edges of elements as shown in Figure 9.

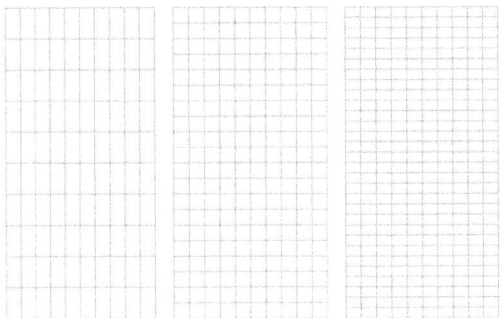

Figure 9: Three kinds of meshes for a rock pillar.

The values of stress components and displacement components of the rock pillar are listed in Tables 1–4, respectively. Comparisons of the results from the conventional quadrilateral isoparametric element (Q4 model) and quadrilateral reduced integration element (Q4R model) are also given in Tables 1–4, respectively. The numerical results of the present model are consistent with those of the Q4 model and Q4R model and have shown good computational stability.

Table 1: Displacement u_y at the top of the pillar

Meshes	BFEM model $(\times 10^{-2}\,\text{m})$	Q4 model $(\times 10^{-2}\,\text{m})$	Q4R model $(\times 10^{-2}\,\text{m})$
10×10	−1.1939	−1.1917	−1.2048
10×20	−1.1940	−1.1931	−1.1964
10×30	−1.1940	−1.1934	−1.1940

Table 2: Displacement u_y at $h = 5$ of the pillar

Meshes	BFEM $(\times 10^{-3}\,\text{m})$	Q4 model $(\times 10^{-3}\,\text{m})$	Q4R model $(\times 10^{-3}\,\text{m})$
10×10	−8.7592	−8.6975	−8.8045
10×20	−8.7630	−8.7118	−8.6726
10×30	−8.7637	−8.7149	−8.7363

Table 3: Stress σ_x at center of the elements at top of the pillar

Meshes	BFEM (kPa)	Q4 model (kPa)	Q4R model (kPa)
10 × 10	−12.016	−12.244	−12.062
10 × 20	−6.005	−6.0773	−6.013
10 × 30	−4.003	−4.0382	−4.002

Table 4: Stress σ_y at center of the elements at bottom of the pillar

Meshes	BFEM (kPa)	Q4 model (kPa)	Q4R model (kPa)
10 × 10	−244.194	−242.945	−244.349
10 × 20	−262.976	−260.116	−263.387
10 × 30	−272.102	−268.002	−272.552

Example 2: Analysis for a Rock Pillar with Four Materials under Pure Shear Effect

Consider a rock pillar with four materials under pure shear effect shown in Figure 10. And its modulus of elasticity $E_1 = 10^9$, $E_2 = 10^8$, $E_3 = 10^7$, and $E_4 = 10^6$, respectively. And Poisson ratio $v = 0.3$, the shear stress on surface of the structure $\tau = 1$. The calculation is considered into the plane stress problem and the dimensionless values.

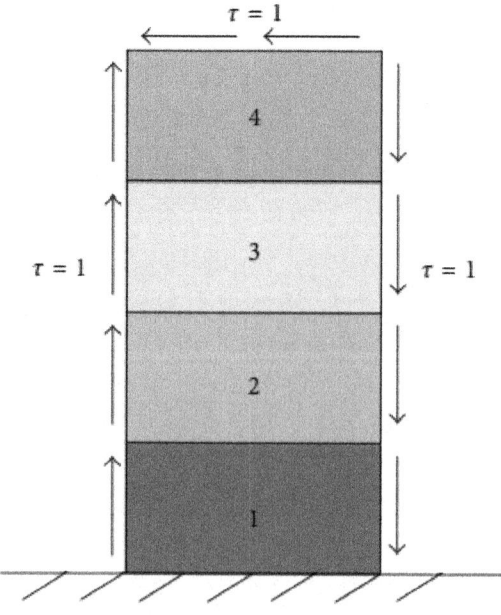

Figure 10: A rock pillar with four kinds of materials.

The calculation is done using the mesh with the center nodes of edges of elements as shown in Figure 11.

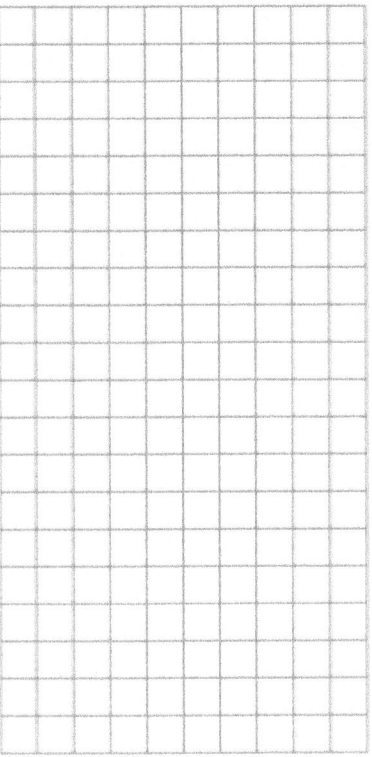

Figure 11: Mesh with 200 elements, 430 nodes.

The values of stress components of the rock pillar are listed in Table 5, respectively. Comparisons of the results from the theoretical analysis are also given in Table 5, respectively.

Example 3: A Rock Pillar under Water Pressure and Gravity

Consider a rock pillar under water pressure and gravity shown in Figure 12. And its modulus of elasticity $E = 108$ Pa, Poisson' ratio $v = 0.3$, density of rock $\rho_1 = 2.4$ t/m^3 , density of water $\rho_2 = 1.0$ t/m^3 , and acceleration of gravity $g = 9.8$ m/s^2 . The calculation is considered into the plane strain problem.

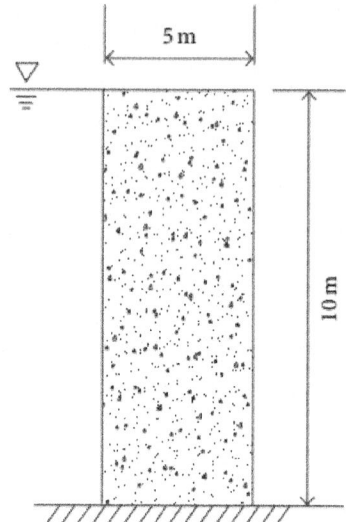

Figure 12: A rock pillar subjected to water pressure and gravity.

The calculation is done using four different element meshes with the center nodes of edges of elements as shown in Figure 13.

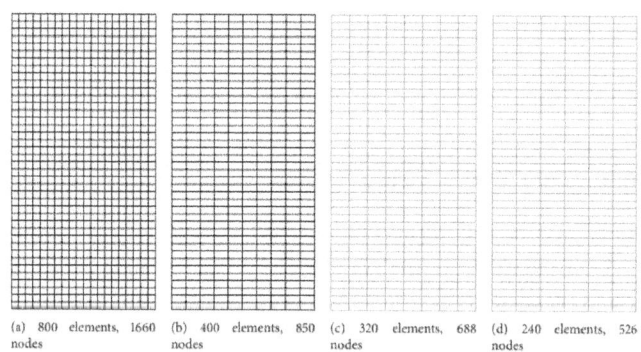

(a) 800 elements, 1660 nodes   (b) 400 elements, 850 nodes   (c) 320 elements, 688 nodes   (d) 240 elements, 526 nodes

Figure 13: Four kinds of meshes for a rock pillar.

The values of stress components and displacement components of the rock pillar are listed in Tables 6–12, respectively. Comparisons of the results from the conventional quadrilateral is oparametric element (Q4 model) and quadrilateral reduced integration element (Q4R model) are also given in Tables 6–12, respectively.

Table 5: Stress solution of the pillar with four kinds of materials under uniform shearing forces

	σ_x	τ_{xy}	σ_y
BFEM	0.000	1.000	0.000
Exact	0.000	1.000	0.000

The numerical results of the present model are consistent with those of Q4R model, and the 4-mid-node element of BFEM has given good performance compared with Q4 model for the large aspect ratio of elements. Due to the different method calculating the equivalent node loads of water pressure between the base force elements and the element of traditional FEM, there is a slight error about the calculation results of the horizontal stresses σ_x of the element in lower right corner of rock pillar between the BFEM and the Q4R model, as shown in Table 10.

Table 6: Displacement u_x at $x = 2.5$ m of the pillar top

Meshes	BFEM ($\times 10^{-2}$ m)	Q4 model ($\times 10^{-2}$ m)	Q4R model ($\times 10^{-2}$ m)
20 × 40	4.0219	4.0056	4.0190
10 × 40	4.0415	4.0094	4.0398
8 × 40	4.0563	4.0135	4.0532
6 × 40	4.0891	4.0239	4.0863

Table 7: Displacement u_x at $h = 5$ m on the right of the pillar

Meshes	BFEM ($\times 10^{-2}$ m)	Q4 model ($\times 10^{-2}$ m)	Q4R model ($\times 10^{-2}$ m)
20 × 40	2.0837	2.0758	2.0830
10 × 40	2.0816	2.0667	2.0815
8 × 40	2.0820	2.0626	2.0823
6 × 40	2.0849	2.0566	2.0862

Table 8: Displacement u_y at $x = 2.5$ m of the pillar top

Meshes	BFEM ($\times 10^{-3}$ m)	Q4 model ($\times 10^{-3}$ m)	Q4R model ($\times 10^{-3}$ m)
20 × 40	−9.8891	−9.8869	−9.8895
10 × 40	−9.8888	−9.8856	−9.8898
8 × 40	−9.8889	−9.8850	−9.8943
6 × 40	−9.8889	−9.8840	−9.8926

Table 9: Displacement u_y at $h = 5\,\text{m}$ on the right of the pillar

Meshes	BFEM model ($\times 10^{-2}$ m)	Q4 model ($\times 10^{-2}$ m)	Q4R model ($\times 10^{-2}$ m)
20 × 40	−1.5918	−1.5878	−1.5913
10 × 40	−1.5488	−1.5412	−1.5483
8 × 40	−1.5290	−1.51894	−1.5284
6 × 40	−1.4983	−1.4832	−1.4975

Table 10: Stress σ_x at center of the element at lower right of pillar

Meshes	BFEM (kPa)	Q4 model (kPa)	Q4R model (kPa)
20 × 40	−128.963	−179.706	−132.974
10 × 40	−142.729	−191.962	−150.839
8 × 40	−141.996	−193.919	−152.815
6 × 40	−134.954	−194.249	−150.941

Table 11: Stress τ_{xy} at center of the element at lower right of pillar

Meshes	BFEM (kPa)	Q4 model (kPa)	Q4R model (kPa)
20 × 40	152.442	172.480	153.366
10 × 40	153.439	167.708	154.637
8 × 40	151.262	163.966	152.612
6 × 40	146.664	157.624	148.288

Table 12: Stress σ_y at center of the element at lower right of pillar

Meshes	BFEM (kPa)	Q4 model (kPa)	Q4R model (kPa)
20 × 40	−802.328	−796.173	−802.741
10 × 40	−707.932	−700.016	−708.582
8 × 40	−675.867	−666.975	−676.616
6 × 40	−634.179	−623.753	−635.039

Example 4: Stress Analysis of Concrete Gravity Dam

Consider a concrete gravity dam shown in Figure 14. And height of the dam is 65 m, bottom width is 49 m, the water level is 60 m, the elastic modulus of concrete $E_1 = 15$ GPa, Poisson ratio of rock $v_1 = 0.2$, the elastic modulus of rock $E_2 = 30$ GPa, Poisson ratio of rock $V_2 = 0.3$, density of concrete is 2.45 t/m^3, density of water is 1 t/m^3, and acceleration of gravity $g = 9.8$ m/s^2. The calculation is considered into the plane strain problem and considered the effect of rock foundation of the dam.

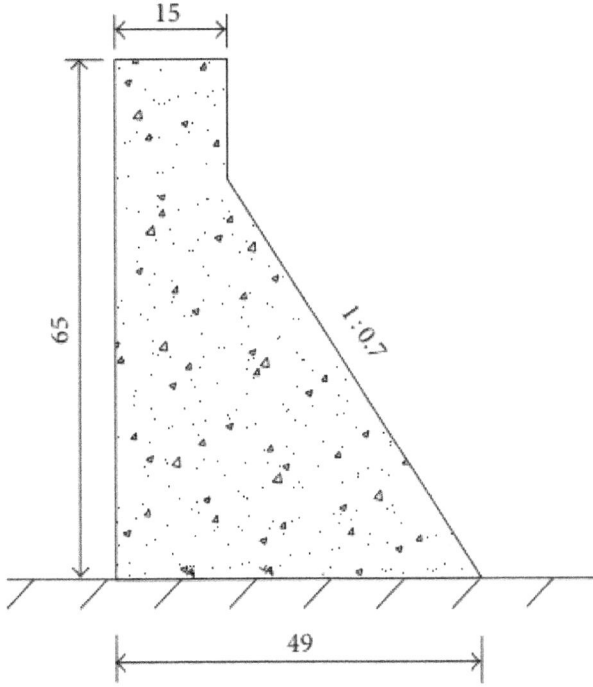

Figure 14: A concrete gravity dam.

The calculation is done using the mesh with the center nodes of edges of elements as shown in Figure 15. In this calculation, we do not consider the initial geostress field. The boundaries of the foundation are used the fixed constraint. The origin of coordinates is located at the bottom of the dam, and is 25 meters away from the dam heel.

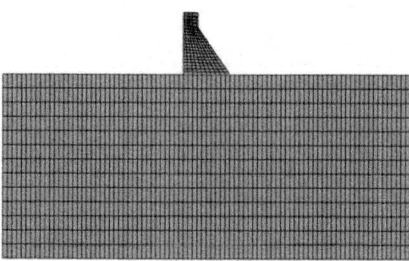

Figure 15: Meshes of the dam and its foundation (1344 elements, 2809 nodes).

The values of stress components and displacement components of the dam are plotted in Figures 16–25, respectively. Comparisons of the results from the conventional quadrilateral iso parametric element (Q4 model) and quadrilateral reduced integration element (Q4R model) are also given in Figures 16–25, respectively. The numerical results of the present model are consistent with those of the Q4 model and Q4R model and have shown good computational stability

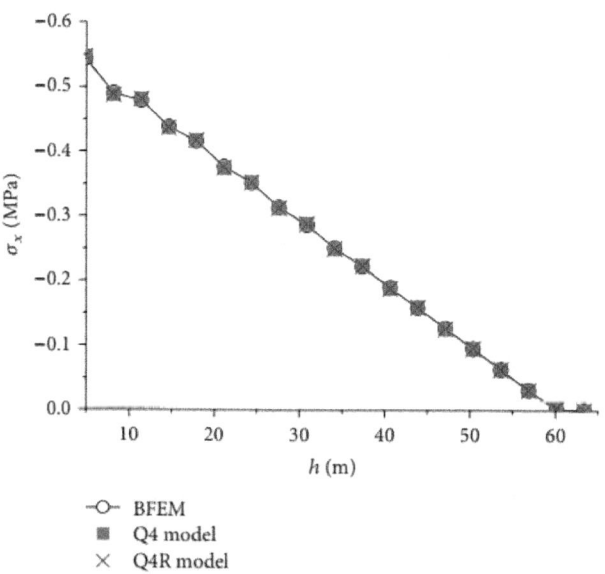

Figure 16: The $^{h\text{-}\sigma_x}$ curves at the upstream face of the dam.

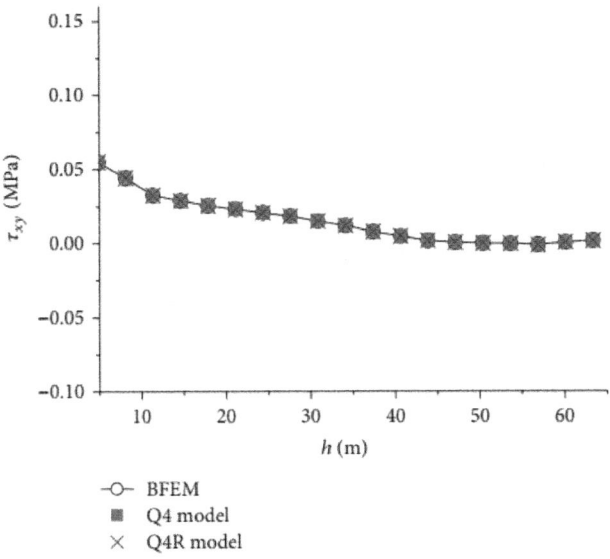

Figure 17: The $^{h-\tau_{xy}}$ curves at the upstream face of the dam

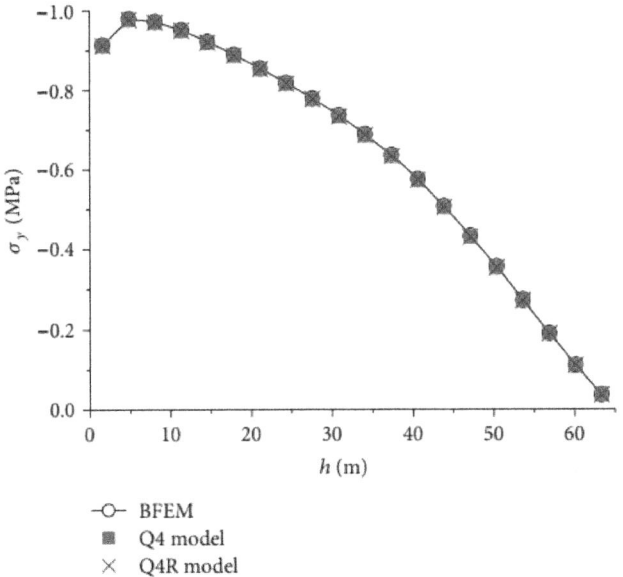

Figure 18: The $^{h-\sigma_y}$ curves at the upstream face of the dam.

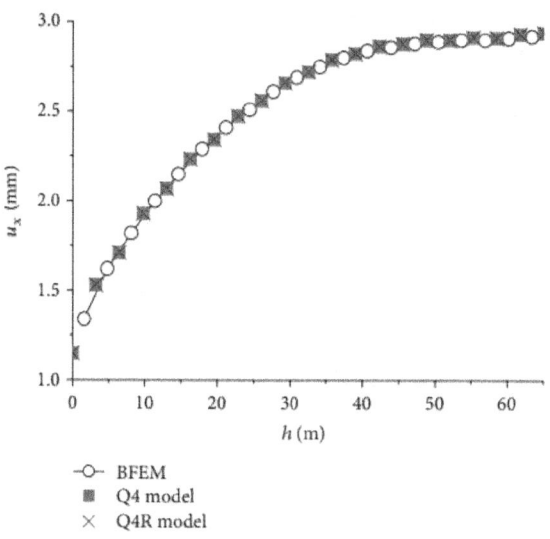

Figure 19: The $h\text{-}u_x$ curves at the upstream face of the dam.

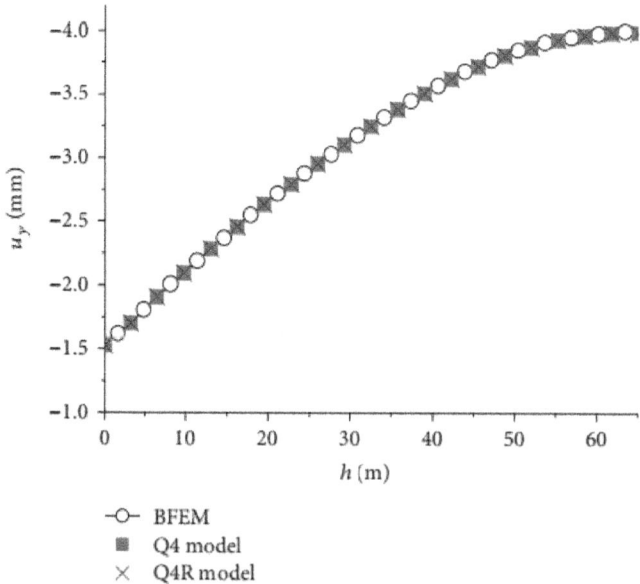

Figure 20: The $h\text{-}u_y$ curves at the upstream face of the dam.

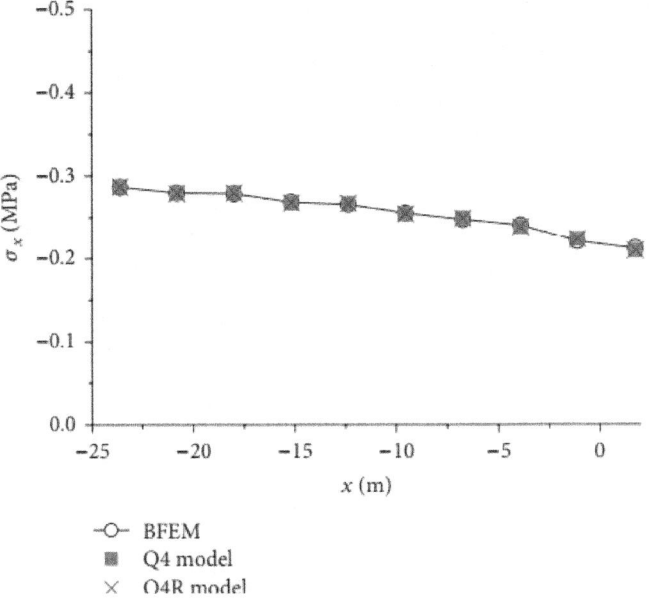

Figure 21: The $h_{-\sigma x}$ curves at the half height of the dam

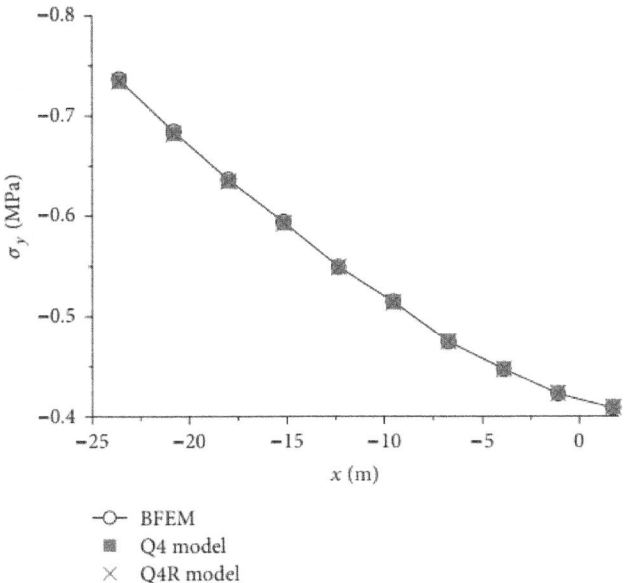

Figure 22: The h-σy curves at the half height of the dam.

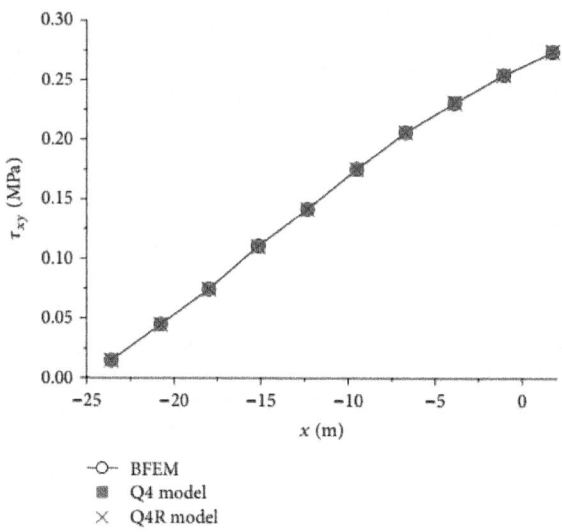

Figure 23: The h-τ_{xy} curves at the half height of the dam.

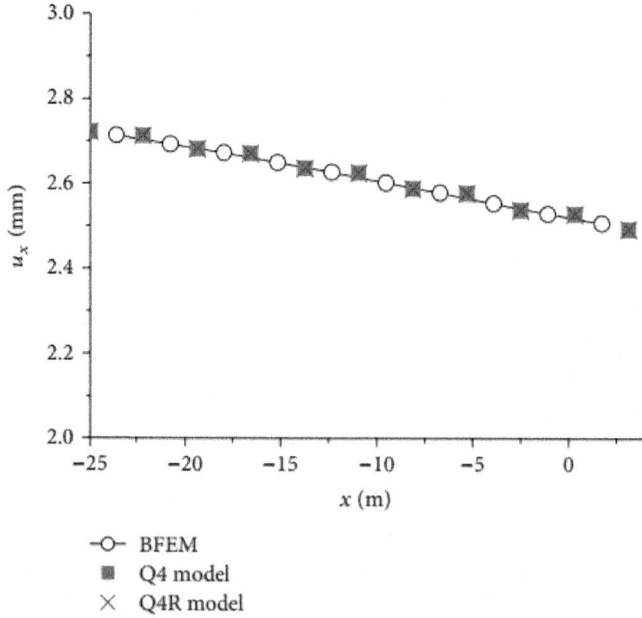

Figure 24: The h-$_{ux}$ curves at the half height of the dam.

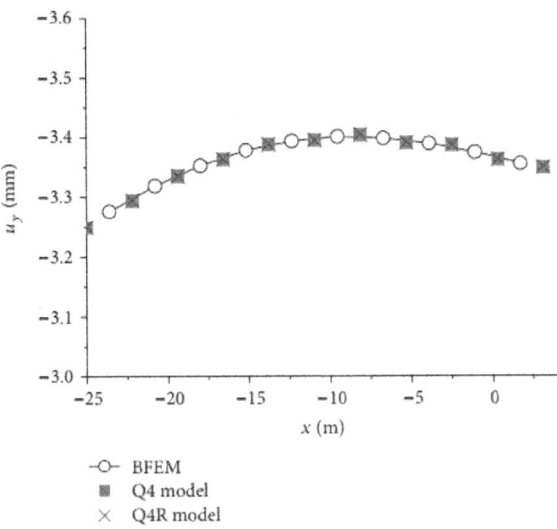

Figure 25: The h-u_y curves at the half height of the dam

Example 5: Simulation and Analysis on the Horizontal Crack Propagation of Rock Block

Consider a rock block subjected by the horizontal thrust and vertical pressure shown in Figure 26. For the convenience of study, we do not consider the weight. And we use the dimensionless numerical analysis

Assuming elastic modulus $E=1$, Poisson ratio $V = 0.3$, tensile strength of a large number, and the uniform load $p_1 = 1$ and $p_2 = 1$. The calculation is considered into the plane stress problem and the dimensionless values.

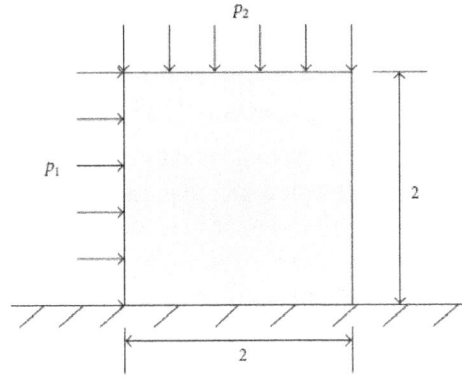

Figure 26: A rock block subjected by the horizontal thrust and vertical pressure.

The calculation is done using the mesh with the center nodes of edges of elements as shown in Figure 27.

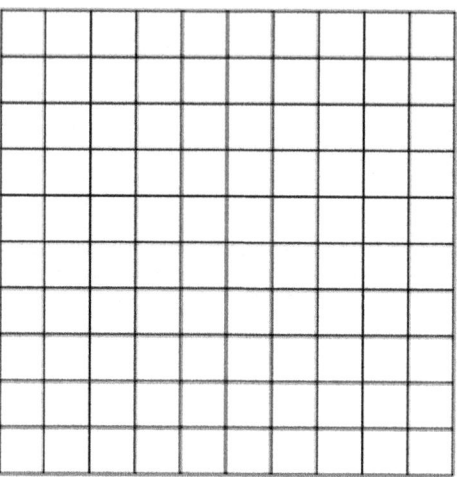

Figure 27: Meshes of the rock block.

In order to check whether the interface between the rock block and the ground will crack, we assume the friction coefficient of interface $f = 0.5$ and change the value of interface cohesion c. The results of calculation using the computer program of the nonlinear BFEM shown in Section 4 and the failure criteria in Section 4.1 are as follows. (1) When $c = 10$, there is no cracks. (2) When $c = 2.8$, there is one element interface crack.

(1)   When $c = 10$, there is no cracks.

(2)   When $c = 2.8$, there is one element interface crack

(3)   When $c=2$, there are three elements' interfaces cracks.

(4)   When $c = 1.5$, there are five elements' interfaces cracks.

(5)   When $c = 1.1$, there are seven elements' interfaces cracks.

(6)   When $c = 1.0$, the cracks are too long, and too little structural constraints have been insufficient to solve the equations. When the interface cohesion c is 1.5, the case of crack propagation of rock block is shown in Figure 28, and the safety factor of stability is $K=2$ which is consistent with the result of the theoretical analysis using the rigid limit equilibrium method.

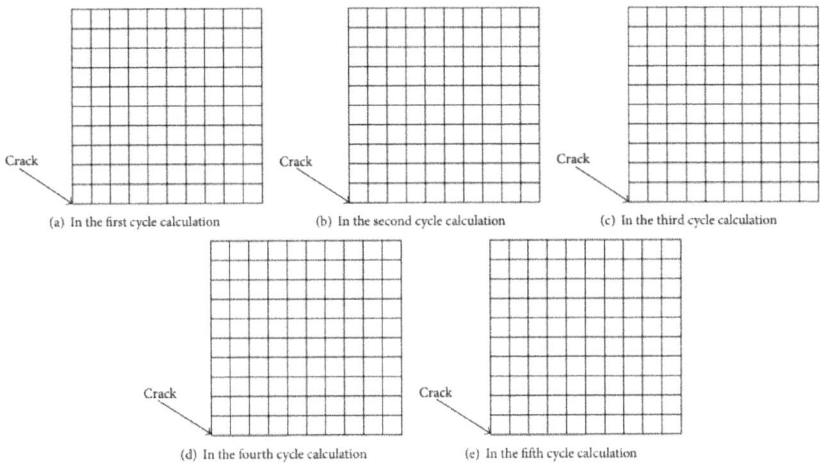

(a) In the first cycle calculation   (b) In the second cycle calculation   (c) In the third cycle calculation

(d) In the fourth cycle calculation   (e) In the fifth cycle calculation

Figure 28: Crack propagation path of the rock block.

Example 6: Analysis on the Crack Propagation of Concrete Gravity Dam. Consider a concrete gravity dam shown in Figure 14. And height of the dam is 65 m, bottom width is 49 m, the water level is 60 m, The elastic modulus of concrete E = 15 GPa, Poisson ratio of rock v = 0.2, density of concrete is 2.45 t/m³ , density of water is 1 t/m³ , and acceleration of gravity g = 9.8 m/s² . The calculation is considered into the plane strain problem and is not considered the effect of rock foundation of the dam. The calculation is done using the mesh with the center nodes of edges of elements as shown in Figure 29.

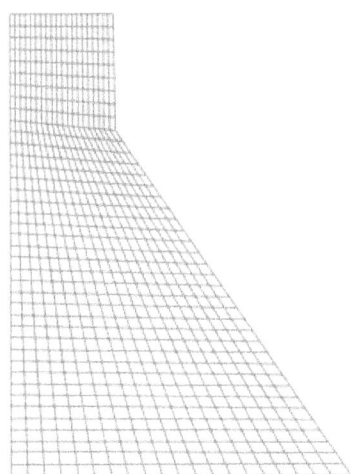

Figure 29: Mesh of gravity dam (800 elements, 1660 nodes).

No Initial Crack in Dam Foundation

We assume that the foundation surface of dam is weak structural interface and analyze the crack propagation and the safety factor by adjusting the value of the friction coefficient and the cohesion. The results of calculation using the computer program of the nonlinear BFEM shown in Section 4 and the failure criteria in Section 4.1 are as follows.

(1)   When $c = 10^6$ Pa and $f = 0.95$, there is no cracks.

(2)   When $c = 0.5 \times 10^6$ Pa and $f = 0.5$, there are two elements' interface cracks.

(3)   When $c = 0.1 \times 10^6$ Pa and $f = 0.4$, there are eight elements' interfaces cracks. The case of crack propagation of dam is shown in Figure 30. When $c = 10^6$ Pa and $f = 0.5$, the safety factor of stability $K = 4.0$ which is consistent with the result of the theoretical analysis. When $c = 0.5 \times 10^6$ Pa and $f = 0.5$,

the safety factor of stability $K = 2.55$ which is consistent with the results of the theoretical analysis using the rigid limit equilibrium method.

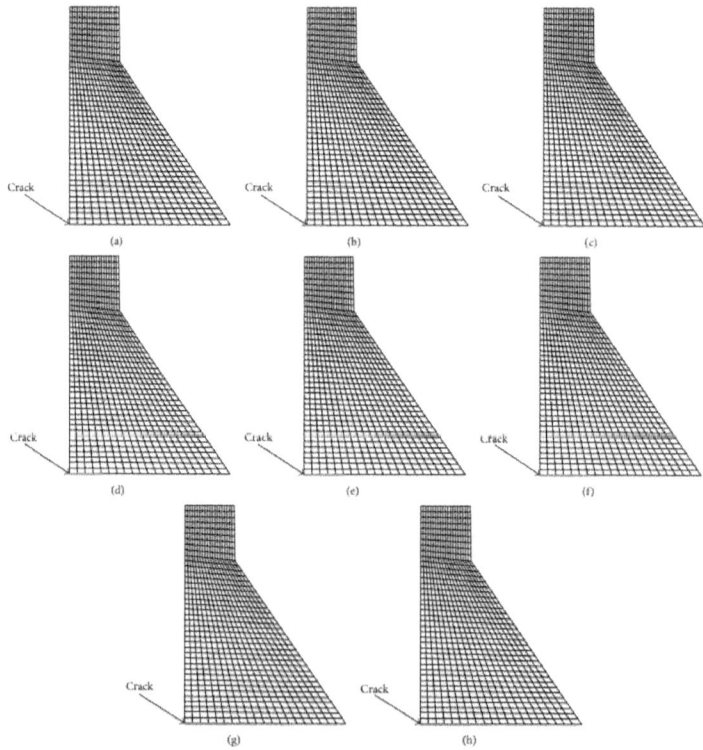

Figure 30: Crack propagation path of the gravity dam.

Existing an Initial Crack in Dam Foundation

There is an initial crack in the dam heel as shown in Figure 31. We assume that the dam foundation surface is weak structural interface and analyze the crack propagation and the safety factor by adjusting the value of the friction coefficient and the cohesion. The results of calculation using the computer linear BFEM shown in Section 4 and the failure criteria in Section 4.1 are as follows.

(1)   When $c = 10^6$ Pa and $f = 0.95$, there is no cracks.

(2)   When $c = 0.5 \times 10^6$ Pa and $f = 0.5$, there is one element interface cracks.

(3)   When $c = 0.1 \times 10^6$ Pa and $f = 0.5$, there are three element' interface cracks.

(4)   When $c = 0.1 \times 10^6$ Pa and $f = 0.4$, there are seven elements' interfaces cracks.

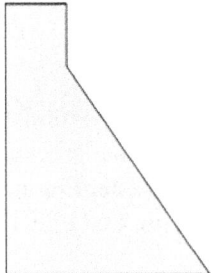

Figure 31: A gravity dam with a crack at the dam heel.

CONCLUSIONS

In this paper, the base force element method (BFEM) on complementary energy principle is used to analyze the rock mechanics problems. The methods to simulate the gravity of an element, the crack propagation and the safety factor of stability are proposed for the BFEM on complementary energy principle. The following conclusions can be drawn.(1)The calculation results of the BFEM on complementary energy principle show that the numerical results of the present method coincide with the theoretical solution, the results of conventional quadrilateral isoparametric element (Q4 model) and quadrilateral reduced integration element (Q4R model). The correctness of the present method and its computer program is verified.(2)The research results show that the BFEM on complementary energy principle has a good computational precision and stability is not sensitive to the effects on the aspect ratio of element and can be used for large-scale scientific and engineering computing.(3)The results of the BFEM for crack propagation problems show that the nonlinear BFEM can solve the cracking problem and simulate the crack propagation of the interface in rock mechanics engineering.(4)The BFEM on complementary energy principle was applied to analyze the stability of rock mass and dam, and the results of safety factor are consistent with the results of the theoretical solutions using the rigid limit equilibrium method. The research results show the present method can be easily used to calculate the safety factor in rock engineering. (5)This paper researched only the cracking problems with horizontal crack in rock mass and calculated only the safety factor of a single slip channel in rock mass.(6)The cracking problems of inclined cracks and the safety factor of multiple sliding channels in rock mass are studying, and the further research results will be published in the future.

CONFLICT OF INTERESTS

The authors declare that there is no conflict of interests.

ACKNOWLEDGMENTS

This work is supported by the National Natural Science Foundation of China, nos. 10972015 and 11172015 and the preexploration project of the Key Laboratory of Urban Security and Disaster Engineering, Ministry of Education, Beijing University of Technology, no. USDE201404.

REFERENCES

1.   T. H. H. Pian, "Derivation of element stiffness matrices by assumed stress distributions," AIAA Journal, vol. 2, no. 7, pp. 1333–1336, 1964.

2.   T. H. Pian and D. P. Chen, "Alternative ways for formulation of hybrid stress elements," International Journal for Numerical Methods in Engineering, vol. 18, no. 11, pp. 1679–1684, 1982.

3.   T. H. H. Pian and K. Sumihara, "Rational approach for assumed stress finite elements," International Journal for Numerical Methods in Engineering, vol. 20, no. 9, pp. 1685–1695, 1984.·

4.   C. Zhang, D. Wang, J. Zhang, W. Feng, and Q. Huang, "On the equivalence of various hybrid finite elements and a new orthogonalization method for explicit element stiffness formulation," Finite Elements in Analysis and Design, vol. 43, no. 4, pp. 321–332, 2007.

5.   B. Fraeijs de Veubeke, "Displacement and equilibrium models in the finite element method," in Stress Analysis, O. C. Zienkiewicz and G. S. Holister, Eds., pp. 145–197, John Wiley & Sons, New York, NY, USA, 1965.

6. B. F. de Veubeke, "A new variational principle for finite elastic displacements," International Journal of Engineering Science, vol. 10, no. 9, pp. 745–763, 1972.

7. R. L. Taylor and O. C. Zienkiewicz, "Complementary energy with penalty functions in finite element analysis," in Energy Methods in Finite Element Analysis, R. Glowinski, Ed., pp. 153–174, John Wiley & Sons, New York, NY, USA, 1979.

8. S. N. Patniak, "An integrated force method for discrete analysis," International Journal for Numerical Methods in Engineering, vol. 6, no. 2, pp. 237–251, 1973.

9. S. N. Patnaik, "The integrated force method versus the standard force method," Computers and Structures, vol. 22, no. 2, pp. 151–163, 1986.

10. S. N. Patnaik, "The variational energy formulation for the integrated force method," AIAA Journal, vol. 24, no. 1, pp. 129–137, 1986.

11. S. N. Patnaik, L. Berke, and R. H. Gallagher, "Integrated force method versus displacement method for finite element analysis," Computers and Structures, vol. 38, no. 4, pp. 377–407, 1991.·

12. E. L. Wilson, R. L. Tayler, W. P. Doherty, and J. Ghaboussi, "Incompatible displacement models," inNumerical and Computational Methods in Structural Mechanics, S. J. Fenves, N. Perrone, A. R. Robinson, and W. C. Schnobrich, Eds., pp. 43–57, Academic Press, New York, NY, USA, 1973.

13. R. L. Taylor, P. J. Beresford, and E. L. Wilson, "A non-conforming element for stress analysis,"International Journal for Numerical Methods in Engineering, vol. 10, no. 6, pp. 1211–1219, 1976.

14. J. C. Simo and T. J. R. Hughes, "On the variational foundations of assumed strain methods," Journal of Applied Mechanics, vol. 53, no. 1, pp. 51–54, 1986.

15. J. C. Simo and M. S. Rifai, "Class of mixed assumed strain methods and the method of incompatible modes," International Journal for Numerical Methods in Engineering, vol. 29, no. 8, pp. 1595–1638, 1990. ·

16. R. H. Macneal, "Derivation of element stiffness matrices by assumed strain distributions," Nuclear Engineering and Design, vol. 70, no. 1, pp. 3–12, 1982.

17. T. Belytschko and L. P. Bindeman, "Assumed strain stabilization of the 4-node quadrilateral with 1-point quadrature for nonlinear problems," Computer Methods in Applied Mechanics and Engineering, vol. 88, no. 3, pp. 311–340, 1991.

18. R. Piltner and R. L. Taylor, "A quadrilateral mixed finite element with two enhanced strain modes,"International Journal for Numerical Methods in Engineering, vol. 38, no. 11, pp. 1783–1808, 1995.

19. R. Piltner and R. L. Taylor, "A systematic constructions of B-bar functions for linear and nonlinear mixed-enhanced finite elements for plane elasticity problems," International Journal for Numerical Methods in Engineering, vol. 44, no. 5, pp. 615–639, 1997.

20. T. J. R. Hughes, "Generalization of selective integration procedures to anisotropic and nonlinear media,"International Journal for Numerical Methods in Engineering, vol. 15, no. 9, pp. 1413–1418, 1980.

21. T. Limin, C. Wanji, and L. Yingxi, "Formulation of quasi-conforming element and Hu-Washizu principle," Computers and Structures, vol. 19, no. 1-2, pp. 247–250, 1984.

22. L. Yu-qiu and H. Min-feng, "A generalized conforming isoparametric element," Applied Mathematics and Mechanics, vol. 9, no. 10, pp. 929–936, 1988.

23. G. R. Liu, T. Nguyen-Thoi, and K. Y. Lam, "A novel alpha finite element method (αFEM) for exact solution to mechanics problems using triangular and tetrahedral elements," Computer Methods in Applied Mechanics and Engineering, vol. 197, no. 45–48, pp. 3883–3897, 2008.

24. J. Chen, C.-J. Li, and W.-J. Chen, "A 17-node quadrilateral spline finite element using the triangular area coordinates," Applied Mathematics and Mechanics (English Edition), vol. 31, no. 1, pp. 125–134, 2010. ·

25. J. Chen, C.-J. Li, and W.-J. Chen, "A family of spline finite elements," Computers and Structures, vol. 88, no. 11-12, pp. 718–727, 2010.

26. S. Rajendran and K. M. Liew, "A novel unsymmetric 8-node plane element immune to mesh distortion under a quadratic displacement field," International Journal for Numerical Methods in Engineering, vol. 58, no. 11, pp. 1713–1748, 2003.

27. S. Rajendran, "A technique to develop mesh-distortion immune finite elements," Computer Methods in Applied Mechanics and Engineering, vol. 199, no. 17–20, pp. 1044–1063, 2010.

28. E. T. Ooi, S. Rajendran, and J. H. Yeo, "A 20-node hexahedron element with enhanced distortion tolerance," International Journal for Numerical Methods in Engineering, vol. 60, no. 15, pp. 2501–2530, 2004.

29. E. T. Ooi, S. Rajendran, and J. H. Yeo, "Remedies to rotational frame dependence and interpolation failure of US-QUAD8

element," Communications in Numerical Methods in Engineering, vol. 24, no. 11, pp. 1203–1217, 2008.

30. Y. Long, L. Juxuan, Z. Long, and C. Song, "Area co-ordinates used in quadrilateral elements,"Communications in Numerical Methods in Engineering, vol. 15, no. 8, pp. 533–543, 1999.

31. Y. Q. Long, S. Cen, and Z. F. Long, Advanced Finite Element Method in Structural Engineering, Springer/Tsinghua University Press, Berlin, Germany, 2009.

32. Z. F. Long, J. X. Li, S. Cen, and Y. Q. Long, "Some basic formulae for area coordinates used in quadrilateral elements," Communications in Numerical Methods in Engineering, vol. 15, no. 12, pp. 841–852, 1999.

33. Z.-F. Long, S. Cen, L. Wang, X.-R. Fu, and Y.-Q. Long, "The third form of the quadrilateral area coordinate method (QACM-III): theory, application, and scheme of composite coordinate interpolation," Finite Elements in Analysis and Design, vol. 46, no. 10, pp. 805–818, 2010.

34. G. R. Liu, K. Y. Dai, and T. T. Nguyen, "A smoothed finite element method for mechanics problems,"Computational Mechanics, vol. 39, no. 6, pp. 859–877, 2007. ·

35. Y. Peng and Y. Liu, "Base force element method of complementary energy principle for large rotation problems," Acta Mechanica Sinica, vol. 25, no. 4, pp. 507–515, 2009.

36. Y. Liu and Y. Peng, "Base force element method (BFEM) on complementary energy principle for linear elasticity problem," Science China: Physics, Mechanics and Astronomy, vol. 54, no. 11, pp. 2025–2032, 2011.

37. Y. Peng, Z. Dong, B. Peng, and Y. Liu, "Base force element method (BFEM) on potential energy principle for elasticity problems," International Journal of Mechanics and Materials in Design, vol. 7, no. 3, pp. 245–251, 2011.

38. Y. Peng, Z. Dong, B. Peng, and N. Zong, "The application of 2D base force element method (BFEM) to geometrically non-linear analysis," International Journal of Non-Linear Mechanics, vol. 47, no. 3, pp. 153–161, 2012.

39. Y.-J. Peng, J.-W. Pu, B. Peng, and L.-J. Zhang, "Two-dimensional model of base force element method (BFEM) on complementary energy principle for geometrically nonlinear problems," Finite Elements in Analysis and Design, vol. 75, pp. 78–84, 2013.

40. Y. J. Peng, N. N. Zong, L. J. Zhang, and J. W. Pu, "Application of 2D base

force element method with complementary energy principle for arbitrary meshes," Engineering Computations, vol. 31, no. 4, pp. 1–15, 2014.

41. Y. Peng, L. Zhang, J. Pu, and Q. Guo, "A two-dimensional base force element method using concave polygonal mesh," Engineering Analysis with Boundary Elements, vol. 42, pp. 45–50, 2014.

42. Y. Peng, Y. Liu, J. Pu, and L. Zhang, "Application of base force element method to mesomechanics analysis for recycled aggregate concrete," Mathematical Problems in Engineering, vol. 2013, Article ID 292801, 8 pages, 2013.

43. Y. Liu, Y. Peng, L. Zhang, and Q. Guo, "A 4-mid-node plane model of base force element method on complementary energy principle," Mathematical Problems in Engineering, vol. 2013, Article ID 706759, 8 pages, 2013.

44. C. Y. Dong and G. L. Zhang, "Boundary element analysis of three dimensional nanoscale inhomogeneities," International Journal of Solids and Structures, vol. 50, no. 1, pp. 201–208, 2013.

45. C. Y. Dong and E. Pan, "Boundary element analysis of nanoinhomogeneities of arbitrary shapes with surface and interface effects," Engineering Analysis with Boundary Elements, vol. 35, no. 8, pp. 996–1002, 2011.

46. S. S. Chen, Q. H. Li, Y. H. Liu, and Z. Q. Xue, "A meshless local natural neighbour interpolation method for analysis of two-dimensional piezoelectric structures," Engineering Analysis with Boundary Elements, vol. 37, no. 2, pp. 273–279, 2013.

47. S. Chen, Y. Liu, J. Li, and Z. Cen, "Performance of the MLPG method for static shakedown analysis for bounded kinematic hardening structures," European Journal of Mechanics, A/Solids, vol. 30, no. 2, pp. 183–194, 2011.

48. S. Cen, X.-R. Fu, and M.-J. Zhou, "8- and 12-node plane hybrid stress-function elements immune to severely distorted mesh containing elements with concave shapes," Computer Methods in Applied Mechanics and Engineering, vol. 200, no. 29–32, pp. 2321–2336, 2011.

49. S. Cen, G.-H. Zhou, and X.-R. Fu, "A shape-free 8-node plane element unsymmetric analytical trial function method," International Journal for Numerical Methods in Engineering, vol. 91, no. 2, pp. 158–185, 2012.

50. H. A. F. A. Santos, "Complementary-energy methods for geometrically non-linear structural models: an overview and recent developments in the analysis of frames," Archives of Computational Methods in Engineering, vol. 18, no. 4, pp. 405–440, 2011.

51. H. A. F. A. Santos and C. I. Almeida Paulo, "On a pure complementary

energy principle and a force-based finite element formulation for non-linear elastic cables," International Journal of Non-Linear Mechanics, vol. 46, no. 2, pp. 395–406, 2011.

52. Y. C. Gao, "A new description of the stress state at a point with applications," Archive of Applied Mechanics, vol. 73, no. 3-4, pp. 171–183, 2003.

53. Y. C. Gao, "Asymptotic analysis of the nonlinear Boussinesq problem for a kind of incompressible rubber material (compression case)," Journal of Elasticity, vol. 64, no. 2-3, pp. 111–130, 2001.

54. Y. C. Gao and T. J. Gao, "Large deformation contact of a rubber notch with a rigid wedge," International Journal of Solids and Structures, vol. 37, no. 32, pp. 4319–4334, 2000.

55. Y. C. Gao and S. H. Chen, "Analysis of a rubber cone tensioned by a concentrated force," Mechanics Research Communications, vol. 28, no. 1, pp. 49–54, 2001. ·

56. Y.-C. Gao, M. Jin, and G.-S. Dui, "Stresses, singularities, and a complementary energy principle for large strain elasticity," Applied Mechanics Reviews, vol. 61, no. 3, Article ID 030801, 16 pages, 2008.

Chapter 14

GEOCHEMISTRY OF HYDROTHERMAL ALTERATION IN VOLCANIC ROCKS

Silvina Marfil and Pedro Maiza

Universidad Nacional del Sur – INGEOSUR- CIC de la Provincia de Buenos Aires – CONICET Argentina

INTRODUCTION

Hydrothermal alteration is a chemical replacement of the original minerals in a rock by new minerals where a hydrothermal fluid delivers the chemical reactants and removes the aqueous reaction products. An understanding of hydrothermal alteration is of value because it provides insights into the chemical attributes and origins of ore fluids and the physical conditions of ore formation (Reed M. 1997).

Within a mineral deposit, the solution channelways are usually obvious because precipitated minerals and altered wallrocks remain as evidence. The direction in which the solutions flowed, especially in flat-lying deposits, is usually less obvious but in many cases can be enferred from mineral zoning or similar evidence (Skinner B. 1997). The mobility of major, minor and rare-earth elements (REE) during alteration processes in different environments has been documented by numerous authors and has been used to discriminate the origin of kaolin deposits. (Sturchio et al., 1986; De Groot & Baker, 1992; Gouveia et al., 1993; van der Weijden & van der Weijden, 1995; Condie et al., 1995; Dill et al.,1997, 2000; Galán et al., 1998, 2007; Pandarinath et al., 2008, among others). Terakado & Fujitani (1998) studied the REE and other trace elements in silicastones, alunites and related rocks in order to examine the behaviour of trace elements in the acidic hydrothermal alteration of silicic volcanic rocks. They found that most of the elements such as Na, Fe, Ba and LREE were leached from the silicastones, while HREE, Th, Hf and Zr were retained in the rocks, even under strongly acidic hydrothermal processes.

Alunite samples have LREEenriched and HREE-depleted features. According to Dill et al. (1997), the ratios $TiO_2 + Fe_2O_3$ vs. $Cr + Nb$, Zr vs. TiO_2 and $Ba + Sr$ vs. $Ce + Y + La$ in kaolinites allow discrimination between hypogene and supergene kaolinization processes. The APS-bearing argillaceous zones that formed during supergene processes are significantly enriched in REE relative to hypogene equivalents (Dill, 2000).

Pandarinath et al. (2008) studied the effects of hydrothermal alteration on major, rare-earth, and other trace-element concentrations in rhyolitic rocks of the Los Azufres geothermal field, Mexico. They concluded that the hydrothermal alteration resulted in a decrease in MnO, P_2O_5, Ta, Rb/Zr and Rb/Nb, and an increase in Zr, Nb and Nb/Y. The greater variances of Y, Ce, Pr, Nd, Sm, Lu and Pb in altered rocks are probably due to hydrothermal processes, whereas smaller variances of CaO, Sr, Rb/Sr and Rb/Ba in altered rocks suggest that these processes led to more uniform chemical rock compositions. The concentrations of REE were not significantly different between fresh and altered rhyolitic rocks, which implies that either these elements remained immobile or were reincorporated into secondary minerals during the hydrothermal alteration of the rhyolitic rocks. Papoulis & Tsolis-Katagas (2008) studied kaolin deposits in the western and southern parts of Limnos Island, northeast Aegean Sea, Greece, and they found two types of hydrothermal alteration zones: smectite-illite-halloysite and kaolinite-dickite-rich zones.

Mineral assemblages reveal that temperatures ranged from <100°C (smectite-rich and halloysite rich zones) to ~270°C (kaolinite-dickite-rich zones). Limited supergene alteration was observed in the less hydrothermally altered rocks of the illite-rich zones as suggested by the presence of jarosite and pyrite. The development of the various assemblages depends not only on the temperature and composition of the hydrothermal fluids but also on the distance of the rock from the fault or the channel of the ascending hydrothermal fluids. Papoulis et al. (2004) used the K and Na content in a kaolin deposit from Greece to measure the degree of alteration of primary rocks. The positive correlation between Al_2O_3 and LOI and the large negative loadings between SiO_2-Al_2O_3 and SiO_2-LOI indicate that LOI and Al_2O_3 contents increase in the more altered samples. Ece et al. (2008) studied acid-sulphate hydrothermal alteration in andesitic tuffs in the Biga Peninsula (Turkey), and concluded that changes in the chemical composition of geothermal waters through time and the chemistry of the intermediate products in clay deposits also control the formation of alunite. The main geothermal activity and acidsulphate alterations for the mineralizations of alunite and halloysite occurred after the onset of NAF-related faulting in the Biga Peninsula. The P_2O_5 enrichment in alunite nodules suggests a deep magmatic source for the geothermal waters that passed through the shallow-level granodiorite intrusions. The minerals that constitute various alteration assemblages depend on: temperature, pressure, primary rock composition, primary fluid composition and the ratio of fluid to rock in the reaction that produced the alteration. (Reed M. 1997). The aim of the study described in this chapter was to evaluate the relation between the chemical composition of major, minor and trace elements, the stable isotopes in kaolin (O and D) and the mineralogical alteration zonation to confirm the hydrothermal genesis of the deposits, working with volcanic rocks from Patagonia Argentina. Knowing the genesis of such deposits is crucial in order to establish exploration criteria and evaluate reserves. These Patagonian deposits are derived from Mesozoic rhyolitic and andesitic rocks. There have been several studies aimed at discovering the origin of the primary deposits (Domínguez & Murray, 1995, 1997; Cravero et al., 1991, 2001; Domínguez et al., 2008, Marfil et al., 2005). Most of the studies deal with the mineralogy, chemical composition, the deposit structure and oxygen-deuterium isotopic data. The deposits examined in this study are located in different areas of Patagonia in the Provinces of Río Negro and Chubut and they are studied by Cravero et al. (2010), Maiza et al. (2009) and Marfil et al. (2005, 2010).

Geologically, the area is characterized by the presence of a set of volcanic rocks and tuffs with minor clastic sediments that overlie a basement of Mesozoic age, essentially constituted by granites. At the base, the volcanic complex is composed of andesites, which is known asVera formation. This complex is followed by a succession of sandstone tuffs, rich in fossil plants (Dicrodium Flora formation) and finally by a suite of ignimbritic tuffs and flows known as the Sierra Colorada formation, deposited at the top of the series. The age of this volcano-sedimetary complex ranges from Triassic to Middle Jurassic. Kaolinite deposits are enclosed in rhyolitic tuffs of the Sierra Colorada formation. An important silicification, developed at the top of the formation, seems to have protected the altered areas from erosive processes giving rise to smooth elevations in the landscape.

The occurrence of hydrothermal events in the region is also proven by the presence of fluorite and Pb-Cu-Zn veins (Labudia & Hayase, 1975). Fluorite veins are hosted by rhyolites presenting an alteration mineralogy with sericite, carbonates, silica and kaolinite (Labudia & Hayase, 1975). According to Manera (1972) and Hayase & Manera (1973), homogenization temperatures of fluid inclusions in fluorite are between 150 and 240 °C (without correction for pressure). The presence of kaolinite and the formation temperatures of these veins suggest a possible relationship between fluorite-base metal veins and the studied kaolinite deposits of the area. In this case, kaolinite occurrences could be used as a prospecting tool for other types of mineral deposits in the region.

GEOLOGICAL SETTING

The deposits examined in this study are hydrothermal origin, located in the provinces of Río Negro (Blanquita, Equivocada and Loma Blanca mines) and Chubut (Estrella Gaucha mine) (Patagonia, Argentina) (Fig. 1).

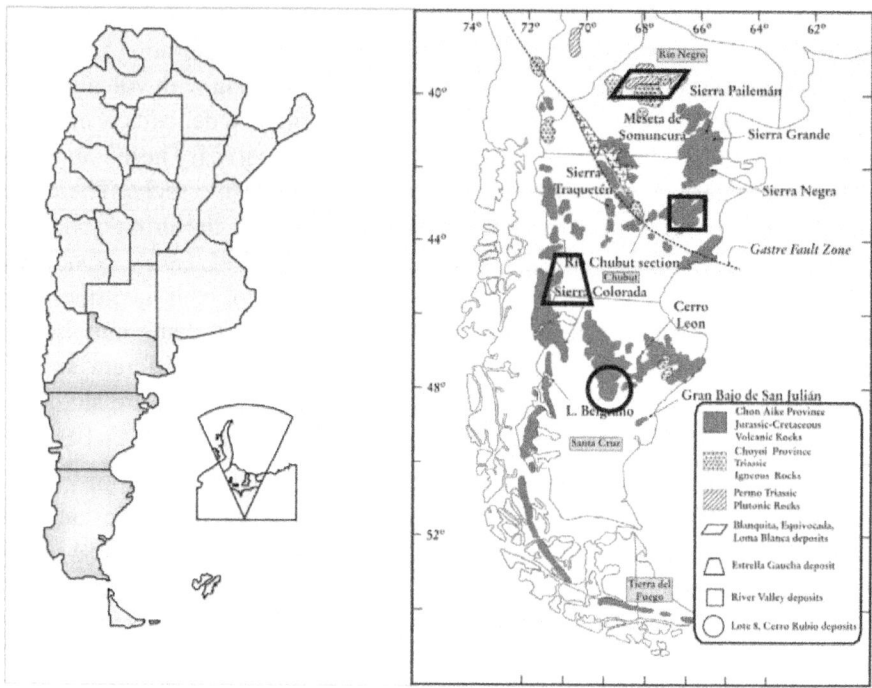

Figure 1: Localization of the kaolin deposits in a sketch map of Patagonia (Cravero et al. 2010, modified).

In the provinces of Chubut and Santa Cruz there are many kaolin deposits, some of them very extensive, locates in the Chubut river valley and lote 8, 18, 19 and Cerro Rubio respectively. These deposits are sedimentary and residual origin, formed from the alteration of volcanic materials, especially Jurassic rhyolitic tuffs. In the province of Chubut there is another group of kaolin deposit related with hydrothermal alteration. Estrella Gaucha mine is one of them. It is situated in western Chubut Province, 70 km from Alto Río Senguer and 20 km from Aldea Apeleg. It is located at SW from Cerro Bayo at 1620 meters above sea level. Ploszkiewicz & Ramos (1977), found andesites and dacites of Ñirehuao Formation and dacitic ignimbrites of Gato Formation. These volcanic rocks are Cretaceous edge. The outcropping rocks in the area studied are conglomeratic sandstones with moderate alteration and stratified. In their composition are common rhyolites, andesites and siliceous sandstones.

These rocks are the base of the deposit. Iron oxides are abundant and they are chloritized. The lithic components are selective altered related with the textures and mineralogical composition. It is possible to observe particles replaced pseudomorphologicaly by chlorite, kaolinite, alunite and quartz. Some grains conserve the original textures while in others the reemplacement is massive forming monomineral particles. The mineralization is developed from the alteration of rhyolitic tuffs of the Lower Cretaceous Payaniyeu Formation in contact with tuff sandstone of Apeleg Formation. The relation between both Formations is not clear because the alteration processes developed in the transition zone (Ploszkiewicz & Ramos 1977). The mineralogy consists mainly of dickite with rare alunite and variable amounts of quartz distributed in four alteration zones, from a silicified upper zone, grading downwards to an alunite zone, then a kaolinized zone and, finally, a sericite-chlorite zone (Hayase et al. 1971, Maiza 1972, 1981, Maiza & Hayase 1975 and Maiza et al. 2009). (Fig 2a) The Chubut River Valley deposits and those of the Santa Cruz province are developed on Jurassic volcanic rocks which are predominantly rhyolitic and form one of the world's most voluminous silicic provinces (Pankurst & Rapela, 1995). It is called Chon-Aike Province (Pankurst et al., 1998), which ranges in age from Early Jurassic to Early Cretaceous. This province comprises several formations that, depending on the geographical locality, receive different names (Marifiland Chon-Aike, among others). In Chubut, the deposits were formed on the Marifil Formation (Malvicini & Llambias, 1974) and are found along the Chubut River in an area of ~60 km². In Santa Cruz, the primary deposits came from the alteration of rhyolitic ash-fall tuffs and ignimbrites of the Chon-Aike Formation which formerly belonged to the Bahía Laura Group) (Domínguez & Murray, 1997; Cravero et al., 2001). In both the Marifil and Chon-Aike rocks, the kaolinized areas have a wide horizontal extension, limited thickness (8-12 m at most) and a downward decreasing degree of alteration. The mineralogical composition is kaolinite with minor halloysite. illite and relict quartz and feldspars (Dominguez & Murray, 1997; Cravero et al., 2001; Dominguez et al., 2008). Table 1 synthesizes the geology and mineralogy of the deposits (Cravero et al. 2010). Blanquita and Equivocada mines are situated 30 km SE of Los Menucos in the province of Río Negro (Fig. 1). The mineralized zone is distributed along a belt 5 to 8 km wide and 20 km long with an approximate area of 110 km². Kaolinite deposits are enclosed in rhyolitic tuffs of the Sierra Colorada Formation, Triassic to Middle Jurassic in age. Blanquita mine is characterized by the presence of kaolinite and alunite with scarce dickite and pyrophyllite, whereas in the Equivocada mine, kaolinite is associated with dickite and traces of alunite (without pyrophyllite) (Marfil et al., 2005). The rhyolitic tuffs, which host most of the kaolinized bodies, were deposited on top of the porphyritic rhyolites. The texture of the tuff is variable, from fine-grained up to agglomerate levels, and shows a microcrystalline matrix with abundant quartz, sanidine, plagioclase, hornblende and biotite. Aleration is moderate with sericitizacion and kaolinization of feldspars and chloritization of amphiboles and biotites (Marfil et al. 2005) (Fig. 2b Blanquita Mine and Fig. 2c Equivocada Mine).

Figure 2: a. Estrella Gaucha mine (kaolinized zone). b. Blanquita mine. c. Equivocada mine. d. Loma Blanca mine (Advanced argillic alteration zone).

Loma Blanca is situated 70 km NW of Los Menucos (Río Negro Province, Argentina) (Fig. 1). The parent rocks are andesites and their tuffs from the Vera Formation (Lower Triassic) (Los Menucos Group), which lie unconformably on the La Esperanza granite and Colo Niyeu metamorphites. 150 m to the northeast of the deposit there is an outcrop of granite intruded by aplitic dykes. Its geological importance arises from the mineralogy, structure and setting of the strongly mineralized area and its relationship with the different lithologic units (Fig 2d). Hayase & Maiza (1972a) studied the mineralogy of the Loma Blanca deposit using X-ray diffraction (XRD), thermogravimetric (TG) and differential thermal analysis (DTA), and microscopy. They concluded that the deposit was formed by the activity of an acid hydrothermal solution or superheated solfatara.

Table 1: Host-rock composition, alteration mineralogy, age and origin of the Patagonia kaolin deposits (Cravero et al. 2010, modified)

Deposit	Host rocks	Mineral association	Age	Origin
Blanquita	Rhyolitic tuffs	Kaolinite*** Alunite ** Dickite* Pyrophyllite *	Triassic to Middle Jurassic	Hypogene
Equivocada	Rhyolitic tuffs	Kaolinite*** Alunite* Dickite** Natroalunite**	Triassic to Middle Jurassic	Hypogene
Loma Blanca	Andesites and Andesitic tuffs	Kaolinite*** Alunite** Dickite** Pyrophyllite** Diaspore* Illite** Montmorillonite* Chlorite*	Lower Triassic	Hypogene
Estrella Gaucha	Rhyolitic tuffs	Dickite*** Alunite*	Lower Cretaceous	Hypogene
Chubut River Valley	Rhyolitic tuffs	Kaolinite*** Quartz-Fd** Illite-halloysite*	Lower Jurassic- Lower Cretaceous	Supergene
Santa Cruz	Rhyolitic tuffs	Kaolinite *** Quartz-Fd** Illite-halloysite*	Lower Jurassic- Lower Cretaceous	Supergene

*** Abundant
** Scarce
* Very rare

They proposed a concentric zonation model. From the parent rock outward, different alteration patterns were recognized: a zone with sericite, chlorite and montmorillonite; a zone with kaolinite and dickite; a zone with dickite, pyrophyllite and alunite-natroalunite; and a zone with quartz, disseminated sulphides and diaspore (Fig. 3). (Marfil et al. 2010). The structure is not clearly exposed due to the scarce outcrops and smooth geomorphical topography covered by recent sediments. Geological evidence from the area indicates the development of tectonic fracturing and differential block movements affecting the heterogeneous volcanosedimentary pile that channelled the hydrothermal solutions responsible for the mineralization. The mineral assemblage, with high-temperature minerals (diaspore, pyrophyllite and dickite), the abovementioned sulphides, the areal structure and favourable lithology related to mineralization processes identify this zone as suitable for metallic mineral prospecting. The floor of the Loma Blanca mine consists of andesite that shows progressive alteration and development of a propylitic zone. Tuffaceous levels are transitionally interlayered; they are affected by tectonics and have developed breccias zones that, owing to their greater permeability, provide pathways for mineralizing solutions. Towards the top of the mineralized zone, the texture is obliterated due to the intense silicification that has erased the lithological characteristics of the affected rocks (Maiza et al., 2009). The oldest formation in the area is Colo Niyeu, which is a low-grade metasedimentary basement. According to Labudia & Bjerg (1994), it could only be assigned a pre-Permian age. This unit was intruded by different plutonites of the so-called Complejo Plutónico La Esperanza. Close to the deposit, outcrops attributed to the Donosa granite, due to their mineralogical and textural characteristics, were recognized; the age of the intrusion, according to Pankurst et al. (1992), will be Late Permian. Over these formations lie the lithological components of the Los Menucos Group, of Triassic age; they are intruded by andesitic dykes of the Taquetren Formation. The field relations of these units have been obscured by tectonism, covered by recent sediments and masked by the mineralizing process (Marfil et al. 2010).

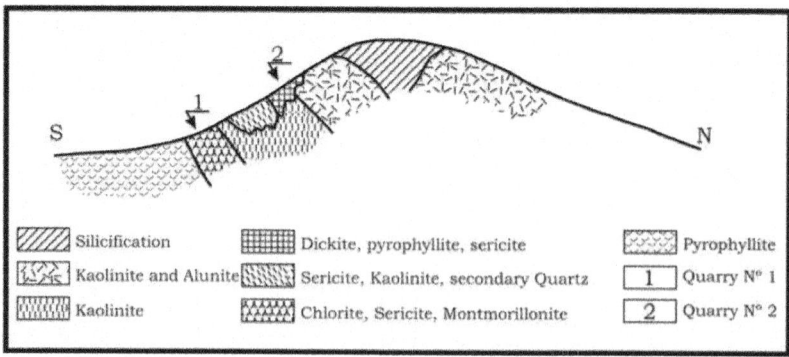

Figure 3: Scheme of the Loma Blanca mine (Marfil et al. 2010).

METHODS AND INSTRUMENTATION

The petrographic studies were carried out with an Olympus B2-UMA trinocular petrographic microscope with a built-in Sony 151A video camera, a high-resolution monitor and Image-Pro Plus image processing software were used. The mineralogical composition of bulk samples was determined by XRD, using a Rigaku D-Max III, with Cu-Kα radiation and a graphite monochromator operated at 35 kV and 15 mA. The XRD patterns were recorded from 2° to 60° 2θ. Chemical analyses of bulk samples for major, minor and trace elements were carried out by inductively-coupled plasma mass-spectrometry (ICP). Kaolin samples were selected for H and O isotope, SEM, IR and DTA-TG analyses. Isotope results are presented as % deviation with respect to SMOW. The reproducibility of results is better than ±0.5% for O and ±10% for H.

MINERALOGICAL COMPOSITION

Estrella Gaucha Mine

The mineralogy of the kaolinized zone in this deposit is very simple. The mineral more abundant is dickite, with minor amount of alunite and quartz and scarce pyrophillite and diaspore. The presence of dickite was mentioned in other kaolin deposits: Tres Picos Mines (Province of Neuquen) (Hayase & Maiza 1970, Losada et al. 1975), Loma Blanca Mine (Province of Río Negro) (Hayase & Maiza 1972a) and Adelita Mine (Province of Río Negro) (Hayase & Maiza 1972b) buy in any case is the more abundant.

In Estrella Gaucha mine dickite is associated with fine size quartz. To the top of the mineralization the more abundant mineral is alunite, with relictic inclusions of dickte, in a 20-30 meters thickness zone. The deposit culminates with a silicified zone with quartz with saccharoidal texture. Massive dickite, constitutes tabular crystals of 200 microns (Fig. 4a). It has parallel extinction (to 4°) and very low birefringence (0.006 to 0.008).

Blanquita Mine

The kaolinization processes have completely obliterated the original textures of the rock, leaving only quartz (Fig. 4b). The mineralogy includes dickite, kaolinite, pyrophyllite, variable amounts of quartz and scarce alunite (Marfil et al. 2005).

Equivocada Mine

The alteration processes is similar to that mentioned in Blanquita mine. It has almost erased the original texture and mineralogy of the tuffs. Only quartz and some biotite remnants are still visible (Fig. 4c). The lithic particles have been pseudomorphically replaced by kaolinite and dickite. The alteration mineralogy consist of a core of alunite grading outwards to an association of kaolinite-illite, illite-zeolite and finally fresh rock (Hayase & Maiza 1970, Marfil et al. 2005). The whole mineralization was later discordantly overprinted by natroalunite (Maiza & Mas 1981).

Loma Blanca Mine

In the propylitic alteration zone the original texture is preserved. Plagioclase (andesine) crystals are altered, mainly to illite and calcite. Chlorite pseudomorphically replaces the mafic minerals, and goethite is formed from the remaining Fe (Fig. 4d). Saccharoidal quartz is found as patches in the groundmass. In sericitic alteration zone the groundmass is completely argillized with secondary quartz. Iron oxides stain the whole rock.

Sericite is the main alteration product. Quartz forms saccharoidal textures where three or more anhedral crystals are grouped. No original minerals are preserved. In the intermediate argillic alteration the andesitic tuff is completely kaolinized. There is a stockwork of pure dickite veins, where the area between the veins is composed of a quartzdickite association with variable amounts of illite, and scarce diaspore (Fig. 4e). There are veins of dickite associated with bohemite crystals. Diaspore was identified in silicified zone and is closely related to dickite.

Advanced argillic alteration zone coincides with the greatest mineralization. The texture resembles that of tuff with lithic clasts replaced by natroalunite. There is no relict quartz. The groundmass is an association of quartz-kaolin and metallic minerals. Hydroxides of Al can be recognized in certain areas. Natroalunite is abundant in this area and is associated with dickite and pyrophyllite (Fig. 4f).

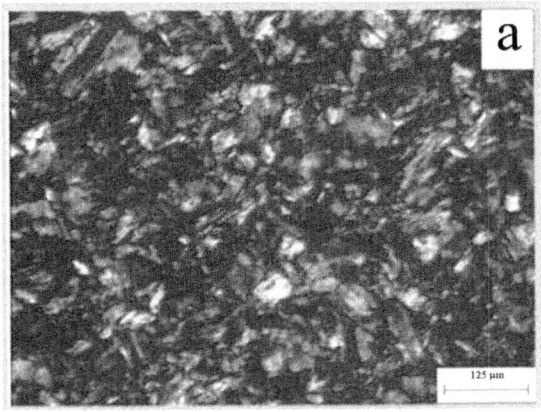

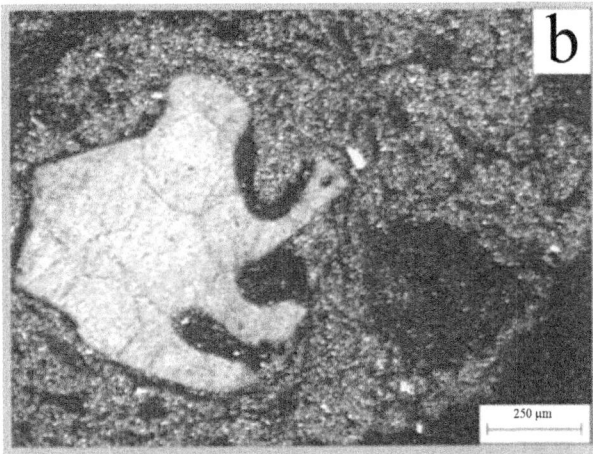

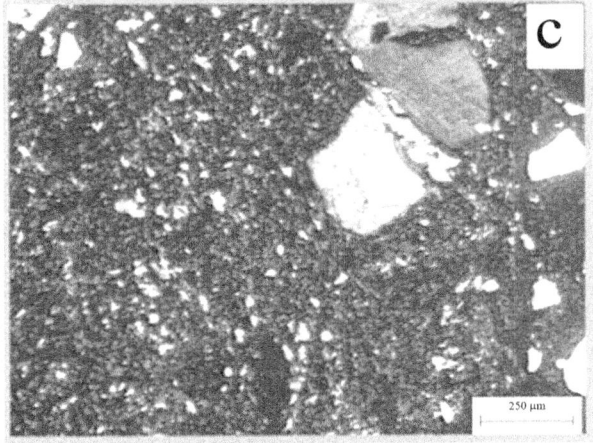

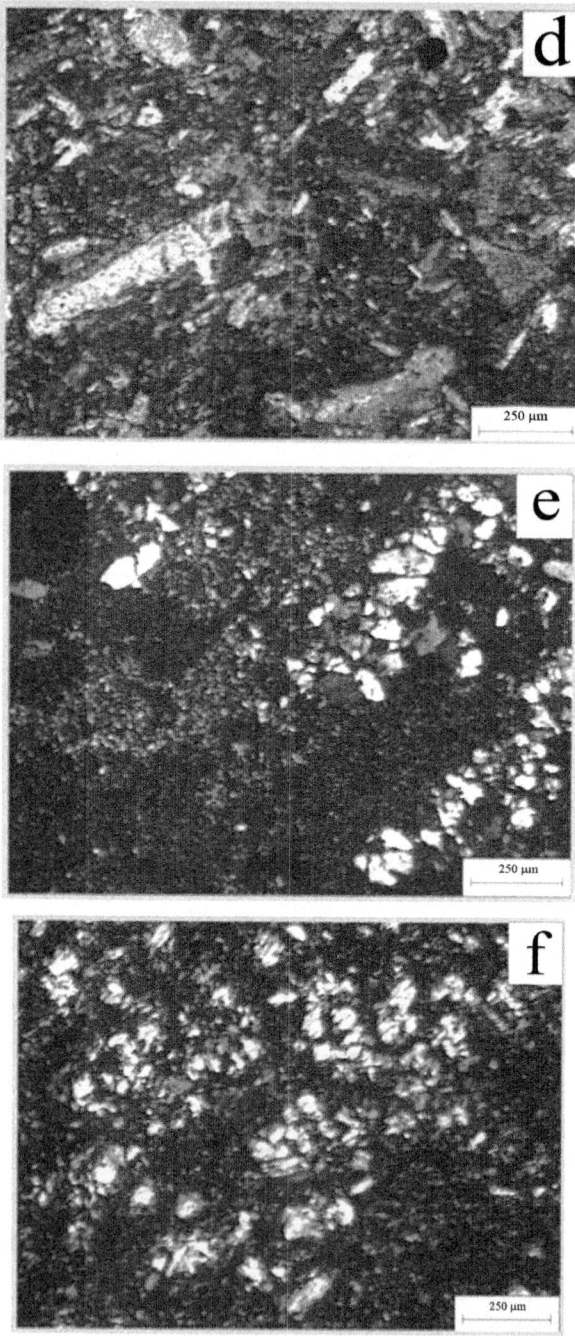

Figure 4: a. Tabular crystals of massive dickite from Estrella Gaucha mine. b. Blanquita mine.: The kaolinization processes have completely obliterated the original

textures of the rock, leaving only quartz. c. Equivocada mine: The original texture and mineralogy of the tuffs have almost erased. d. Propylitic alteration zone in Loma Blanca mine: The original texture is preserved. e. Intermediate argillic alteration: Andesitic tuff completely kaolinized. f. Advanced argillic alteration zone coincides with the greatest mineralization.

XRD

When kaolinite and dickite are associated, it is very difficult to differentiate them by XRD. The two triplets located between 34° and 40°2, were studied in detail. Dickite appears in very small proportions compared with kaolinite in the samples Loma Blanca, Blanquita and Equivocada mines while in Estrella Gaucha mine dickite is the predominant mineral. In the last one, dickite with variable amounts of quartz (between 0 to 25 %) was determined. In some samples scarce alunite was identified.

In "Loma Blanca" in the advanced argillic alteration zone, besides alunite-natroalunite, diaspore at the top and pyrophyllite in the highest area of the profile were identified. These minerals are found in isolated masses within the area of greatest hydrothermal activity as relics of a first stage of higher pressure and temperature. In the propylitic alteration zone, chlorite and scarce illite were identified. Feldspar from the andesitic rock is still present. The predominant alteration mineral is illite together with abundant quartz. In the intermediate argillic alteration zone, kaolinite and quartz reflections were identified. Natroalunite composition was determined by XRD using basal reflections (003, 006 and 009). Maiza & Mas (1980), synthesized the minerals of the series from sericite alteration in a H_2SO_4 solution with varying amounts of KCl and NaCl. They determined the c_o values in each of the products obtained and checked them by chemical analysis. With these values they fitted a straight line by the least-squares method. The calculated value for natroalunite from Loma Blanca is 16.83 Å, which corresponds to 87% Na. The formula calculated on this basis is $(Na0.13K_{0.87})Al_3(SO_4)_2(OH)_6$. (Marfil et al. 2010).

CHEMICAL COMPOSITION

The kaolinte content in Blanquita, calculated from de mineralogical composition and major element analysis ranges between 27 and 73 % and between 40 and 97% in Equivocada. Samples with a greater silica content correspond to the silicified zone of the deposits. The alunite content, estimated from S concentration, varied from 0.3 to 2.1 % in both deposits, the larger contents being related to the more intense kaolinization. On the other hand, an increase in the alumina content can be related to an increase in the degree of alteration of the fresh rock, reaching values close to theoretical ones for pure kaolinite. In these samples, the alkali content is also small (Table 2). The concentration of MnO and MgO in the samples analysed are < 0.01 wt % (Marfil et al. 2005). The trace element contents of kaolinite helps to understand the origin. Kaolinites formed from hydrothermal alteration of acid-type igneous rock show enrichments in S, Ba and Sr, whereas Cr, Nb, Ti and the lanthanides tend to concentrate in kaolinites formed from meteoric processes (Dill et al. 1997). The S contents was determined to ranges from 0.03 to 0.36 wt %, this may reflect the presence of small amounts of alunite in the samples analysed. The Zr and Ti contents show a positive linear correlation with greater values of both elements found in samples from Equivocada mine. The (Ti + Fe) contens are less than 1 wt% and the (Cr + Nb) contents range from 0 and 174 ppm with no significant difference between samples from the two deposits. The (Sr + Ba) concentrations vary between 1000 and 10000 ppm, with the larger values probably related to the presence of trace amount of barite. The Ce + Y + La concentraton is also variable (from 3 to 323 ppm). (Fig 5a, b, c and d) (Table 3). (Marfil et al. 2005). The chondrite-normalized REE diagrams are shown in figure 6a and 6b corresponding to Blanquita and Equivocada respectively.

In Loma Blanca mine Fe_2O_3, CaO, Na_2O and K_2O contents decrease from propylitic to advanced argillic alteration zones, whereas Al_2O_3 and LOI increase in the kaolinite natroalunite zone (Table 2). The percentage of alunite, calculated from the S content, ranges between 2% and 25%. The sulphur content in some samples is attributed to the sulphides present (covellite, sphalerite and pyrite). Sulphides were identified with a petrographic microscope on polished sections in almost all the samples, so the total percentage of natroalunite is considered to be slightly smaller than the calculated value. Relatively large Ba, Sr, V and Zr contents were observed, mainly in intermediate argillic and advanced argillic zones. Co, Ni, Cu, Zn and Rb are more common in the propylitic zone. Be, Ge, In, Sn, Mo, Nb and Ag are insignificant in all the samples (Table 3). Ba + Sr vs. Ce + Y + La, Fe_2O_3 + TiO_2 vs. C r + Nb and Zr vs. TiO_2 plots are shown in Fig. 5.

Table 2: Chemical analysis of major elements on whole rock samples (average weight %)

Deposit/samples		SiO$_2$	Al$_2$O$_3$	Fe$_2$O$_3$	CaO	Na$_2$O	K$_2$O	TiO$_2$	P$_2$O$_5$	S	LOI
Blanquita / 18	Average	66.08	22.74	0.28	0.73	0.15	0.06	0.16	0.20	0.16	9.39
	SD	10.31	7.21	0.23	1.78	0.09	0.03	0.07	0.14	0.09	2.97
	Minimum	46.28	10.84	0.11	0.02	0.00	0.01	0.25	0.05	0.03	14.60
	Maximum	84.78	37.21	0.80	6.28	0.23	0.12	0.05	0.64	0.36	4.22
Equivocada/13	Average	60.58	28.19	0.11	0.06	0.02	0.05	0.37	0.25	0.14	10.65
	SD	11.18	8.84	0.07	0.04	0.02	0.04	0.21	0.15	0.09	2.83
	Minimum	45.36	16.13	0.02	0.03	0.00	0.18	0.01	0.01	0.09	14.15
	Maximum	76.25	40.17	0.10	0.18	0.02	0.00	0.58	0.52	0.36	5.98
Loma Blanca/20	Average	58.49	22.20	2.54	0.50	1.22	0.69	0.80	0.28	1.13	11.28
	SD	8.67	7.57	3.02	0.75	1.88	0.76	0.20	0.15	1.49	4.73
	Minimum	34.02	16.25	0.04	0.05	0.09	0.00	0.10	0.17	0.04	5.16
	Maximum	69.18	37.82	9.45	2.86	5.42	3.01	1.12	0.80	3.68	23.03
Estrella Gaucha/9	Average	50.82	42.67	0.05	0.05	0.07	0.02	0.47	0.25	0.12	13.30
	SD	10.23	12.53	0.07	0.01	0.02	0.02	0.34	0.17	0.11	2.66
	Minimum	44.50	20.94	0.01	0.01	0.02	0	0.13	0.07	0.03	8.49
	Maximum	69.06	38.34	0.24	0.06	0.09	0.04	1.08	0.48	0.21	14.92

Table 3: Chemical analysis of trace elements (average ppm)

Deposit/samples		Ba	Sr	Y	Sc	Zr	V	Cr	Ga	Ge	As	Nb	Mo	Ag	La	Ce
Blanquita/18	Average	919	1660	4.8	8.9	74.4	107.4	80.3	18	4.7	390.9	5.7	5.8	7.1	35.7	62.4
	SD	2049	906	2.5	3.0	24.6	66.6	43.7	9	2.5	216.4	2.9	4.9	6.5	25.5	57.4
	Minimum	100	335	2	5	33	41	0	10	2	140	0	0	1	9	13
	Maximum	9041	4457	11	17	110	274	170	34	11	707	10	17	21	93	225
Equivocada/13	Average	610.4	1540	3.5	6	133.2	83.3	32.3	60.0	2.6	30.8	7.3	2.2	2.1	38.7	67.4
	SD	491.6	1056	1.9	2.2	73.6	59.2	28.9	29.7	1.6	36.4	5.0	2.6	2.0	16.3	32.3
	Minimum	56	32	0	3	11	20	0	0	1	0	0	0	0	1	2
	Maximum	1840	3160	6	11	235	212	90	77	7	107	15	6	6.9	60	118
Loma Blanca/20	Average	1122	1867	5.8	7.4	158.4	180.1	26.5	21.2	0.5	10.8	5.1	2.2	2.1	33.3	65.6
	SD	1282	1166	11.2	3.4	46.0	74.55	30.0	5.12	0.6	8.97	1.6	2.6	2.0	18.4	37.6
	Minimum	292	92	0	5	58	109	0	12	0	0	0	0	0	19	203
	Maximum	5990	4765	47	17	272	391	120	32	2	45	7	7	6.9	103	30.0
Estrella Gaucha/9	Average	282	1585	2.4	5.7	80.9	285.1	71.1	53.0	4.5	0	3.3	2.0	0	29.2	47.7
	SD	169	1200	2.0	3.1	50.3	128.1	76.7	25.9	1.7	0	2.4	2.4	0	16.5	33.9
	Minimum	72	447	0	2	27	102	0	17	2	0	0	0	0	5	10
	Maximum	563	3845	4	11	166	454	210	90	7	0	6	6	0	58	108

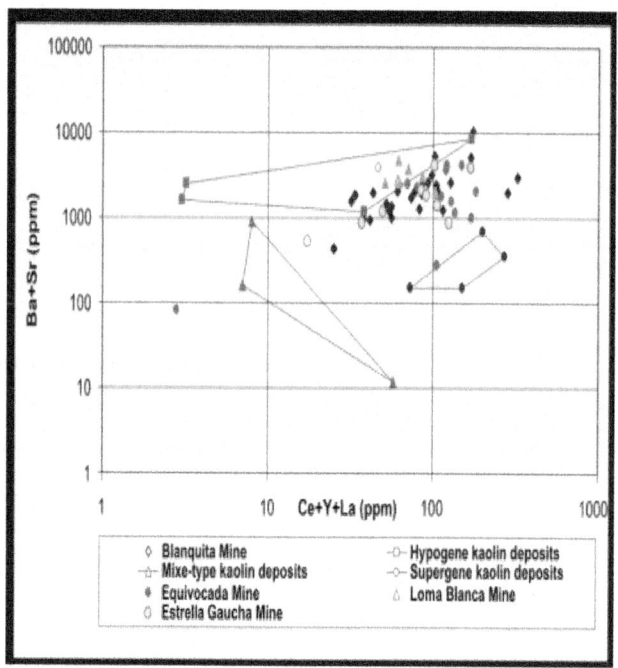

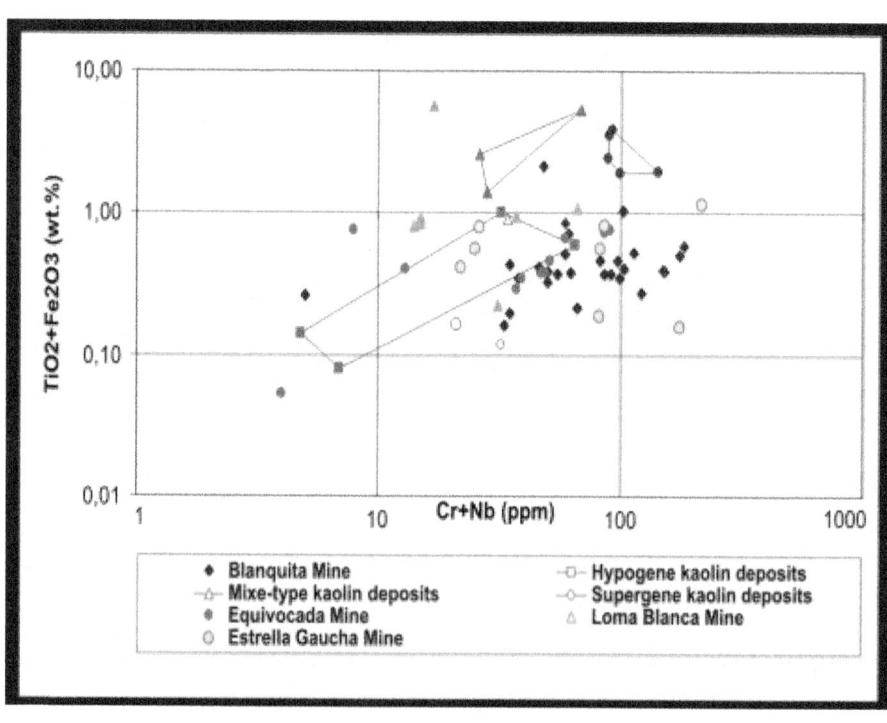

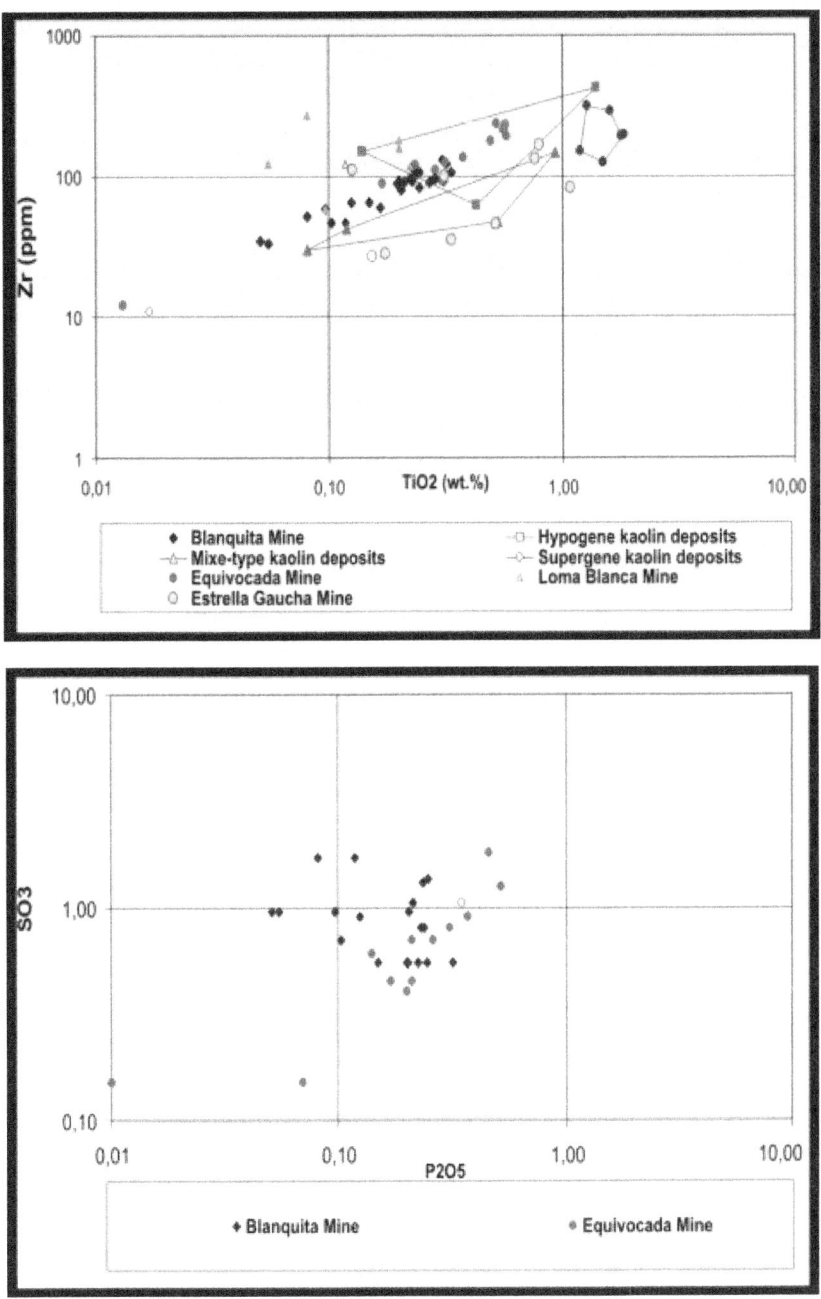

Figure 5: Data plot from Blanquita, Equivocada, Loma Blanca and Estrella Gaucha mines. a. (Ba + Sr) vs. (Ce + Y + La). b. (TiO_2 + Fe_2O_3) vs. (Cr. + Nb) c. Zr vs. TiO_2. d. SO_3 vs. P_2O_5 from Blanquita and Equivocada mines.

The chondrite-normalized REE diagram according to Boynton, 1984 (in Rollinson, 1992), for the samples of the propylitic and sericitic alteration zones are shown in Fig. 6c and 6d for the samples of intermediate and advanced argillic alteration zones. In the latter, especially in pure kaolin samples, there is marked LREE impoverishment with respect to HREE, especially when compared with the less altered samples. There is no evidence of a positive Ce anomaly, which is typical of the deposits of residual and/or meteoric origin (Cravero et al., 2001). LREE are more abundant than HREE in the intermediate argillic alteration and advanced argillic alteration zones (Marfil et al. 2010).

The results of the chemical analysis of major elements in Estrella Gaucha mine, show that SiO_2 ranges from 44.50 to 69.06 % and Al_2O_3 38.34 to 21.70 %. The SO_3 content (between 0.08 and 0.93 %) was adjudicated to alunite. The alumina amount allowed determining kaolin content between 75 and 100 %. The MnO is less than 0.01 % (Table 2) (Maiza et al. 2009). The results of trace element contents are shown in Table 3 and plotted in Fig. 5. The content of TiO_2 +Fe_2O_3 are lower than 1%wt, Cr+Nb range between 1 and 214 ppm, Sr+Ba between 1.000 and 10.000 while Ce+Y+La is low (15 – 160 ppm). These results are closed with that determined in Blanquita, Equivocada and Loma Blanca mines. The content of As, Rb, Ag, In, Sn, Cs and Bi are insignificant.

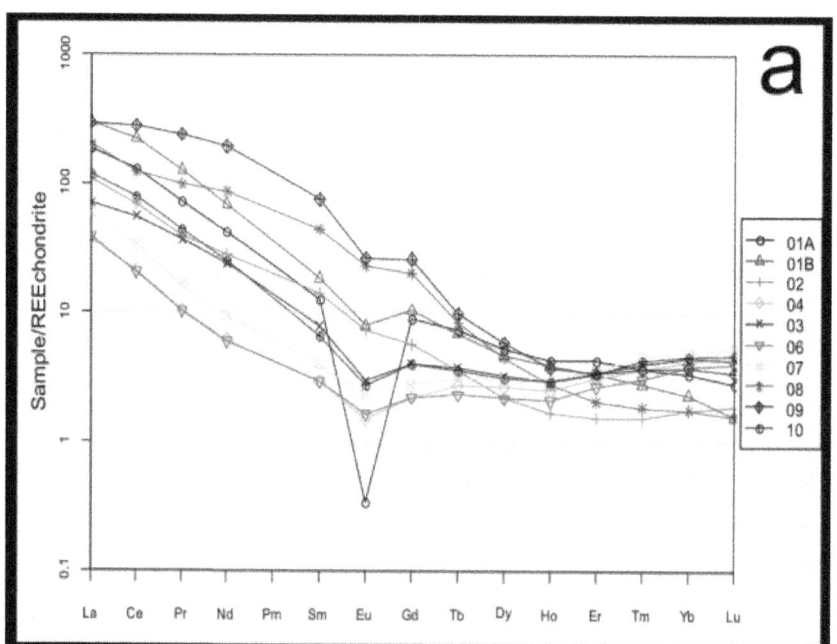

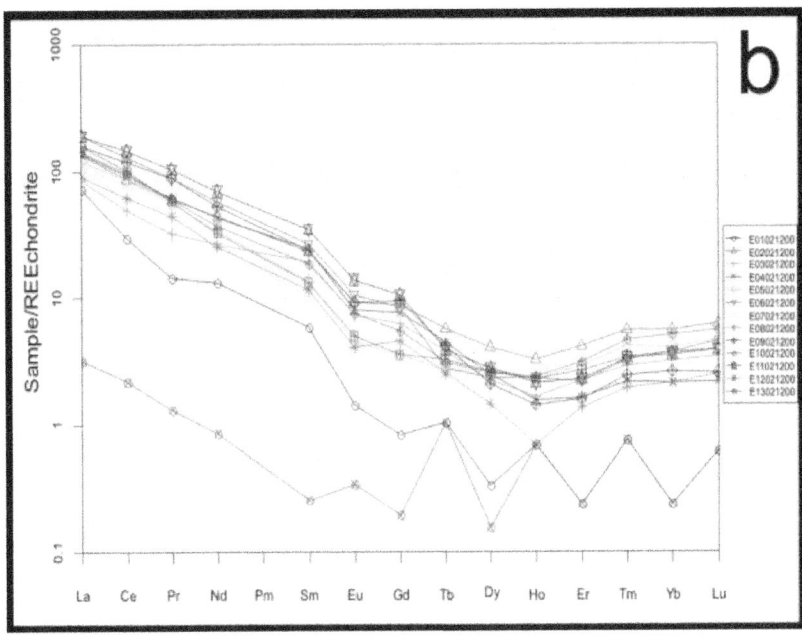

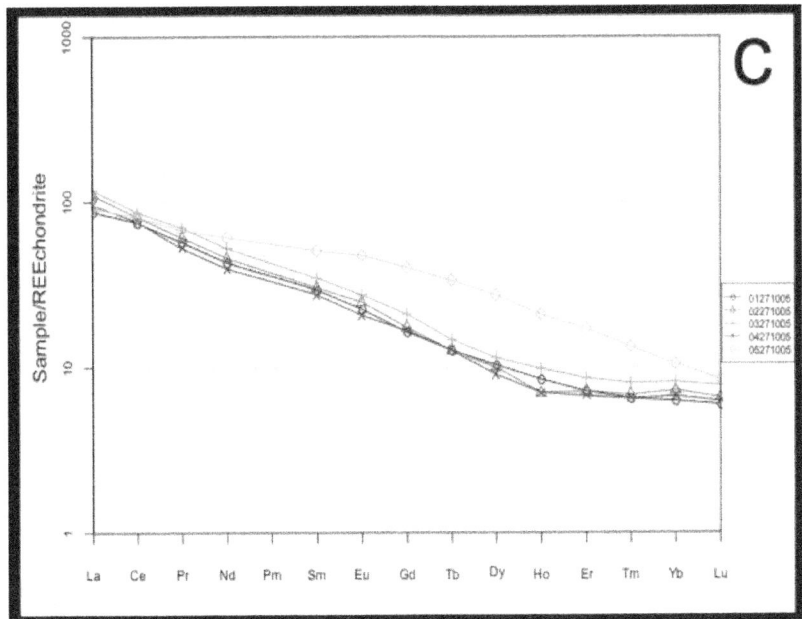

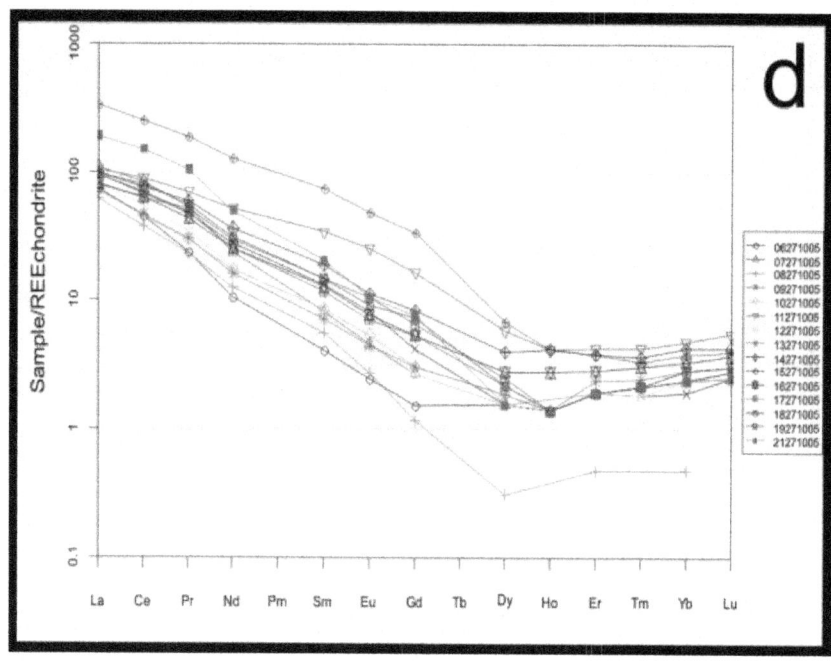

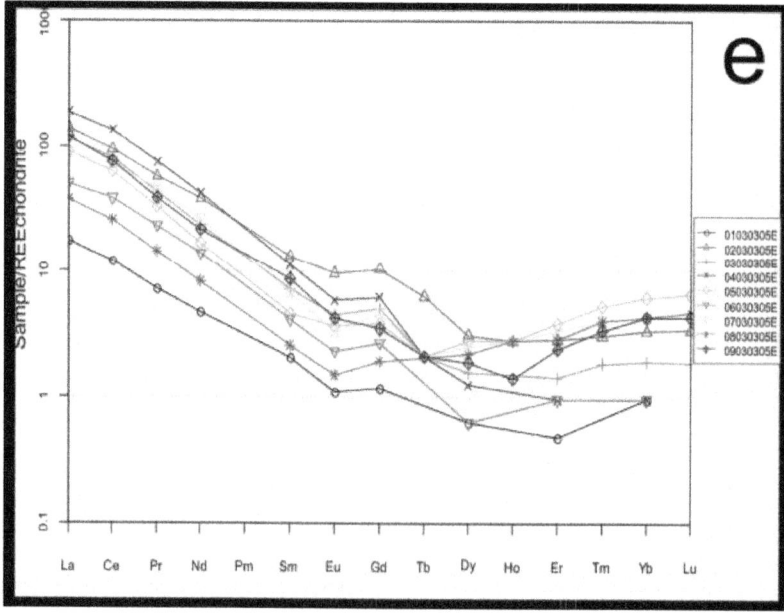

Figure 6: Chondrite-normalized REE. a. Blanquita mine. b. Equivocada mine. c. Loma Blanca mine: Propylitic and sericitic alteration zones. (d) Loma Blanca mine: Intermediate argillic alteration and advanced argillic alteration zones. e. Estrella Gaucha mine.

In Fig. 6e the chondrite-normalized REE are shown for the samples from Estrella Gaucha. It is possible to observe a LREE impoverishment with respect to HREE, similar to that observed in Blanquita, Equivocada and Loma Blanca mines. Negative Eu anomaly was identified and Ce anomaly was no observed.

O AND H ISOTOPES

The O and H stable isotope composition of kaolinites samples from the four mines studied are shown in Fig. 7. According to Savin & Lee (1988), the isotopic composition of kaolinite may reflect the geological conditions during its formation, provide the mineral did not suffer isotopic changes after its deposition. Thus, the O isotope composition in kaolinites of sedimentary origin usually varies from +19 to 23 ‰ and kaolinites from residual deposits has a $\delta^{18}O$ between +15 and +19‰ (Murray & Janssen, 1984). These values are compatible with a meteoric origin at temperatures between 20 and 25 °C. It is important to note than the $\delta^{18}O$ values of the analyzed samples from the four mines, are much lower than those assumed for kaolinite deposit formed under superficial conditions, thus suggesting a different origin for kaolinization fluids (Marfil et al. 2005).

Table 4: $\delta^{18}O$ and dD ‰ average of kaolin samples from Blanquita, Equivocada, Loma Blanca and Estrella Gaucha mines

Deposit/samples		$\delta^{18}O_{SMOW}$ ‰	δD_{SMOW} ‰
Blanquita / 6	Average	7.3	-102.8
	SD	2.0	14.5
	Minimum	4.8	-116.6
	Maximum	9.6	-88
Equivocada/6	Average	8.1	-95.9
	SD	2.2	5.0
	Minimum	5.1	-105.1
	Maximum	10.3	-90.0
Loma Blanca/4	Average	11.6	-84
	SD	1.3	0.8
	Minimum	10.4	-85
	Maximum	13.2	-83
Estrella Gaucha/3	Average	15.9	-84.3
	SD	0.3	2.1
	Minimum	15.6	-86
	Maximum	16.1	-82

$\delta^{18}O$ and δD values of kaolinites from Blanquita and Equivocada ranges from +4.8 to +10.3 ‰ and from -88 to -116‰ respectively (Table 4). While the O isotope composition is similar in both deposits, the δD values of Blanquite mine are slightly more negative. (Martil et al 2005). Both $\delta^{18}O$ and δD values are clearly different from those reported by Cravero et al (1991) in kaolinites from deposits of residual origin in the Santa Cruz and Chubut provinces ($\delta^{18}O$ from +16.5 to +18.8‰ and δD from -57.5 to -86.5‰). In a later paper, Cravero et al. (2001) presented additional data from Cerro Rubio and La Esperanza deposit in the province of Santa Cruz with kaolinites having $\delta^{18}O$ and δD values of +24‰ and -98‰ respectively.

In Loma Blanca mine $\delta^{18}O$ values in kaolinites range from 10.8 ‰ to 13.2 ‰ and δD from - 83‰ to -85‰ (Table 4). Although these values are within the range of variation for hydrothermal kaolins (Murray & Janssen, 1984), they are larger than those determined for Blanquita and Equivocada mines.

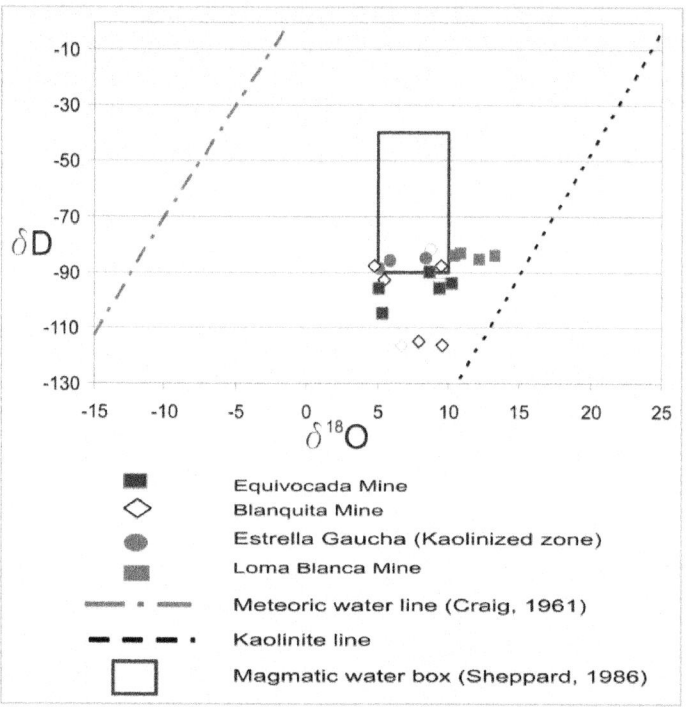

Figure 7: δ 18O and δD values in Loma Blanca, Estrella Gaucha, Blanquita and Equivocada kaolins.

$\delta^{18}O$ and δD values of kaolinites from Estrella Gaucha mine range from +5.1 to 8.8 ‰ and from -82 to -89‰ respectively (Table 4). They are similar to that obtained in Blanquita and Equivocada mines. The absence of primary minerals with fluid inclusions, not allowed determining the formation temperature of the deposit. However the presence of diaspore and pyrophyllite in the alteration minerals allows narrow the conditions between 250 and 350 °C, because they are the stability limits of those minerals.

DISCUSSION

Dill et al.(1997) used the relationships between SO_3 vs. P_2O_3, Zr vs. TiO_2, Ti + Fe vs. Cr + Nb, and Ba + Sr vs. Ce + Y + La to discriminate between kaolin deposits of different origins in Peru. They found that S, Ba and Sr are enriched during hydrothermal alteration, whereas Cr, Nb, Ti and lanthanide elements are concentrated mainly during weathering.

The trace element content and P vs. S, Zr vs. Ti, (Cr + Nb) vs. (Ti + Fe) and (Ce + Y + La) vs. (Ba + Sr) ratios of kaolin samples from the four mines studied do not differ significantly. This suggests that all the deposits might have formed by the same genetic process. However, although the Ti + Fe and Sr + Ba contents indicate a hypogenic origin (Dill at al. 1997), the presence of large Cr + Nb and Ce + Y + La concentrations in some samples could indicate the presence of kaolinite of supergenic origin as well. Titanium may be released from a primary mineral in the parent rock (e.g. biotite) during either hypogenic or supergenic kaolinization. However, as supergenic alteration seems to be more efficient, the Ti content in kaolinite has been used to discriminate between the two processes (Dill et al 1997). Because it behaves a geochemically immobile element in superficial conditions, Zr is also a good indicator of the degree of meteorization of the parent rock. Consequently, kaolin samples with high Ti and Zr contents point to a superficial environment of formation (Marfil et al. 2005).

Cravero et al. (2010) compared kaolin deposit from hypogene and supergene origin from Patagonia Argentina. In the hypogene deposits, S, Sr, Pb, V, P_2O_5, and LREE are more abundant, whereas Fe_2O_3, Y, Rb, U, Zr and HREE predominate in the supergene (weathered) deposits. It is important to consider that the samples are altered and the degree of alteration is not the same in all of them. The main alteration products in all the samples are clay minerals of the kaolin group. The approach adopted is to consider that the greater the LOI, the greater the resulting alteration degree. The concentrations of S, Pb and Sr, present only in the hypogene deposits, vary with the degree of alteration, indicating that these elements are mainly concentrated in the clay minerals. Zr does not show a clear behaviour; in the hypogene deposits it appears to be mobile during alteration, whereas under weathering conditions no clear relationship can be established. The V and U contents increase in both types of deposits with the degree of alteration; hence they can be considered to be immobile in both environments. P_2O_5 and S are only present in hypogene deposits and their contents increase with the alteration degree. Y shows no relation with the degree of alteration.

The greater content of HREE in the supergene deposits is more easily explained by considering that these elements are fractionated during alteration in the hydrothermal deposits and remain unchanged during weathering, thereby giving greater values in the supergene deposits, where no fractionation has occurred. Regarding the mobility of elements during hydrothermal alteration, other authors have found different behaviours. In rhyolitic rocks from Yellowstone drill cores, Ti, Al, Fe, Sc, Co, Y, Zr, REE, Hf, Ta and Th remained relatively immobile (Sturchio et al., 1986).

During the hydrothermal alteration of rhyolitic rocks of the Los Azufres geothermal field, Mexico (Pandarinath et al., 2008), REE concentrations were not significantly different between fresh and altered rocks. The immobile role of REE during hydrothermal processes in rhyolitic rocks has been reported by De Groot & Baker (1992). Nevertheless, in the advanced argillic alteration zone of Rangan area (Central Iran), composed mainly of alunitejarosite and pyrophylite, LREE are relatively immobile in the rocks of this zone and depleted in LREE and HREE (Parsapoor et al., 2009). These authors considered that this behaviour may be due to the entry of these elements into the lattice of alunite-jarosite. LREE can in fact substitute for K in the large-radius cations (A) of the alunite-jarosite formula. Alunite is a common component of the hypogene deposits studied here.

The samples from Patagonia studied by Cravero et al. 2010 have been taken as representative of the whole altered area, so the element contents do not come from a specific part of the weathering profile; therefore it can be considered that the conservative behaviour of the REE elements arises from the fact that they have fractionation could have occurred within the profile. Another characteristic of the REE that supports a supergene origin is the presence of a Ce anomaly, produced when Ce^{3+} is oxidized to Ce^{4+}. When these data are plotted in the graphical style of Dill et al. (1997, 2000), some of the relationships are not as clear as in their work, except for SO_3 vs. P_2O_3. Their supergene deposits are characterized by much greater contents of Ce + Y + La than those formed under hypogene conditions, whereas in Patagonia, these values are dispersed. In Peru, supergene kaolins are characterized by large values of both TiO_2 and Zr, whereas in Patagonia only Zr shows the same behaviour. Cr + Nb also show the opposite trend, while in Peru the greatest contents are related to a supergene origin; in Patagonia they are associated with hydrothermal deposits. In both areas, the same behaviour is found for S, P_2O_5, Ba and Sr, with the greater values being found in the deposits formed in the hypogene environment. The greater amounts of S and V in the hypogene deposits are explained by the presence of minerals such as alunite (sulphate) and mottramite (vanadate).

More information about the formation conditions of the studied deposits can be obtained from their mineralogy. Kaolin deposits of hydrothermal origin are constituted by kaolinite with dickite, alunite and pyrophyllite (Murray, 1984). However, in kaolin deposits of sedimentary origin only kaolinite is present. In the mines studied the presence of dickite, pyrophyllite with alunite and diaspore in Loma Blanca, indicate a hydrothermal origin. In order to estimate the isotopic composition of fluids responsible for the kaolinization process a temperature of 350 °C was assumed. Higher temperatures would not be reasonable as kaolinite coexists with pyropyillite and lower values would not be compatible with the presence of pyrophillite. Numerous authors have synthesized pyrophyllite at temperatures above 260°C (Grim, 1969; Roy & Osborn, 1954; Hemley, 1959; Reed & Hemley, 1967; Tzuzuki & Mizutani, 1971).

The coexistence of pyrophyllite and kaolinite allows us to estimate a maximum temperature of 350°C. Diaspore can be recognized frequently in kaolin deposits of hydrothermal origin. According to Roy & Osborn (1954), the bohemite-diaspore transformation temperature is between 270 and 300°C a t 175-1500 atm. water pressure. (Marfil et al. 2010)

A schematic summary of the paragenetic sequence from Loma Blanca mine is given in Table 5. The main Al minerals of higher temperature occur predominantly in the central area below the silicified zone, which is responsible for protecting the deposit from subsequent erosion processes. Dickite and nacrite commonly occur in deposits of hydrothermal origin (Murray, 1984). Numerous authors have synthesized pyrophyllite at temperatures above 260°C (Grim, 1969; Roy & Osborn, 1954; Hemley, 1959; Reed & Hemley, 1967; and Tzuzuki & Mizutani, 1971). The coexistence of pyrophyllite and kaolinite allows us to estimate a maximum temperature of 350°C. Diaspore can be recognized frequently in kaolin deposits of hydrothermal origin. According to Roy & Osborn (1954), the boehmite-diaspore transformation temperature is between 270 and 300°Ca at 175-1500 atm. water pressure.

Thus, using the O isotope fractionation equation of Shepard & Gilg (1996) indicate that O and H isotope composition of the fluids involved in the kaolinization process are of magmatic origin or of superficial origin but isotopically equilibrated with magmatic rocks at magmatic temperatures. (Marfil et al. 2005).

Table 5: Summary of the paragenetic sequence. (Marfil et al. 2010)

Mineral	Silicified Zone	Advanced argillic zone	Intermediate argillic zone	Sericitic zone	Propylitic zone
Quartz					
Diaspore					
Natroalunite					
Pyrophillite					
Dickite					
Kaolinite					
Illite					
Chlorite					

CONCLUSIONS

- The deposits studied consist of a set of irregular bodies and veins developed on tuffs of rhyolitic and andesitic composition.

- The alteration minerals present include kaolinite, dickite, pyrophyllite, alunite, diaspore, illite and chlorite.

- The trace elements contents of the kaolin samples in all the deposits studied are very similar. Most of the comparative diagrams (P vs. S, Zr vs. Ti, (Cr + Nb) vs. (Ti + Fe) and (Ce + Y + La) vs. (Ba + Sr) suggest that kaolinite formed from hydrothermal alteration of volcanic rocks. However the high (Nb + Cr) and (Ce + Y + La) contents do not exclude the presence of kaolinite of residual origin.

- The mineral assemblage (dickite-natroalunitepyrophyllite-diaspore), the alteration zonation pattern, the geochemistry of trace elements, the relation between LREE and HREE and the small δ18O values suggest that the deposits studied were formed by hydrothermal processes.

- Mineralization developed in two stages: the first stage, of higher temperature, led to the formation of dickite, pyrophyllite and diaspore. Then the temperature decreased due to contamination with meteoric water, and alunite and kaolinite developed.

- The mineral assemblage allows us to estimate a formation temperature between 270 and 350°C.
- The hypogene deposits are characterized by greater contents of Sr, Pb, V, S and P2O5, all of them, apart from V, increasing as alteration proceeds. REE, probably forming part of the alunite structure, are fractionated during alteration.
- The combination of chemical composition, mineralogical association and O and D isotopes data suggest that the deposits studied were formed because the circulation of hydrothermal fluids, discounting superficial processes at low temperatures.

ACKNOWLEDGMENTS

The authors wish to thank the Geology Department of the Universidad Nacional del Sur, the Comisión de Investigaciones Científicas de la Provincia de Buenos Aires and CONICET - INGEOSUR for their helpful support during the research.

REFERENCES

1. Condie K.C., Dengate J. & Cullers R.J. (1995) Behavior of rare earth elements in a paleoweathering profile on granodiorite in the Front Range, USA. Geochimica et Cosmochimica Acta, 59, 279-294.

2. Cravero M.F., Domínguez E. & Murray H.H. (1991) Valores δO18 y δDen caolinitas, indicadores de un clima templado moderado durante el Jurásico SuperiorCretácico Inferior de la Patagonia, Argentina. Revista Asociación Geológica Argentina, 46, 20-25.

3. Cravero F., Domínguez E. & Iglesias C. (2001) Genesis and applications of the Cerro Rubio kaolin deposit, Patagonia (Argentina). Applied Clay Science, 18, 157-172.

4. Cravero F., Marfil S. and Maiza P. (2010). Statistical analysis of geochemical data: A tool: to discriminate between kaolin deposits of hypogene and supergen origin. Patagonia, Argentina. Clay Minerals, 183-196.

5. De Groot P. & Baker J.H. (1992) High element mobility in 1.9-1.86 Ga hydrothermal alteration zones, Bergslagen, central Sweden: relationships with exhalative Fe-ore mineralizations. Precambrian Research, 54, 109-130.

6. Dill H., Bosse R., Henning H. & Fricke A. (1997) Mineralogical and chemical variations in hypogene and supergene kaolin deposits in a

mobile fold belt the Central Andes of northwestern Peru. Mineralium Deposita, 32, 149-163.

7.   Dill H.G., Bosse H.R. & Kassbohm J. (2000) Mineralogical and chemical studies of volcanicrelated argillaceous industrial minerals of the Central American Cordillera (western El Salvador).Economic Geology, 95, 517-538.

8.   Domínguez E. & Murray H.H. (1995) Genesis of the Chubut river valley kaolin deposits, and their industrial applications. Pp. 129-134 in: Proceedings of the 10th International Clay Conference, 1993 (G.J. Churchman, R.W. Fitzpatrick & R.A. Eggleton, editors) CSIRO Publishing, Melbourne, Australia.

9.   Domínguez E. & Murray, H.H. (1997) The Lote 8 Kaolin Deposit, Santa Cruz, Argentina. Genesis and paper industrial application. Pp. 57-64 in: Proceedings of the 11th International Clay Conference (H. Kodama, A.M. Mermut & J.K. Torrance, editors) Ottawa, Canada.

10. Domínguez E., Iglesias C. & Dondi M. (2008) The geology and mineralogy of a range of kaolins from the Santa Cruz and Chubut Provinces, Patagonia (Argentina). Applied Clay Science, 40, 124-142.

11. Ece O., Schroeder P.A., Smilley M.J. & Wampler J.M. (2008) Acid-sulphate hydrothermal alteration of andesitic tuffs and genesis of halloysite and alunite deposits in the Biga Peninsula, Turkey. Clay Minerals, 43, 281-315.

12. Galán E., Aparicio P., González I. & Miras A. (1998) Contribution of multivariate analysis to the correlation of some properties of kaolin with its mineralogical and chemical composition. Clay Minerals, 33, 66-75.

13. Galán E., Fernández-Caliani J.C., Miras A., Aparicio P. & Márquez M.G. (2007) Residence and fractionation of rare earth elements during kaolinization of alkaline peraluminous granites in NW Spain. Clay Minerals, 42, 341-352.

14. Gouveia M.A., Prudencio M.I., Figueiredo M.O., Pereira L.C.J., Waerenborgh J.C., Morgado I., Pena T. & Lopes A. (1993) Behavior of REE and other trace and major elements during weathering of granitic rocks, E´ vora, Portugal. Chemical Geology, 107, 293- 296.

15. Grim H. (1969) Clay Mineralogy. 2nd Edition. McGraw Hill Book Co, New York, USA. 131- 137.

16. Hayase, K. & Maiza, P.J. (1970) Génesis del yacimiento de caolín Mina Equivocada, Los Menucos, Prov. de Río Negro, República Argentina. Revista de la Asociación Argentina de Mineralogía Petrología y Secimentología, I, 1-2, 33-34.

17. Hayase, K. & Maiza, P.J. (1972a) Génesis del yacimiento de caolín Mina Loma Blanca, Los Menucos, Prov. de Río Negro, Argentina. 5° Congreso Geológico Argentino, Actas 2: 139-151.

18. Hayase, K. & Maiza, P.J. (1972b) Presencia de dickita en yacimientos de caolín de la Patagonia, Argentina. 5° Congreso Geológico Argentino, Actas 1: 153-170.

19. Hayase K. & Manera T. (1973) A statistical analysis of experimental data on filling temperature of fluid inclusions in fluorite from fluorite deposits of Patagonia Argentina. Mining Geology, Japan, 23, 1-2.

20. Hayase, K., Schincariol, O. & Maiza, P.J. (1971) Ocurrencia de alunita en cinco yacimientos de la Patagonia, Mina Equivocada, Mina Loma Blanca, Mina Estrella Gaucha, Mina Gato y Camarones, República Argentina. Revista de la Asociación Argentina de Mineralogía, Petrología y Sedimentología, 2, 49- 72.

21. Hemley J.J. (1959) Some mineralogical equilibria in the system K2O-Al2O3-SiO2-H2O. American Journal of Science, 257, 241-70.

22. Labudia C. & Hayase K. (1975) Relaciones entre las rocas y las mineralizaciones de Pb-CuZn, fluorita y caolín de los alrededores de Los Menucos, Prov. de Río Negro, Argentina. Sexto Congreso Geológico Argentino, Actas, Bahía Blanca, III, 69-80. Labudia C. & Bjerg E. (1994) Geología del sector oriental de la hoja Bajo Hondo (39e),

23. Provincia de Río Negro. Revista de la Asociación Geológica Argentina, 49, 284-296. Losada, O., Gelós, E., Maiza, P. y Bengochea, A. (1975) Geología de los afloramientos de caolín de la zona del arroyo Chilquirihuín, Prov. de Neuquén. Revista de la Asociación Geológica Argentina 30: 5-16.

24. Maiza, P. (1972) Los yacimientos de caolín originados por la actividad hidrotermal en los principales distritos caoliníferos de la Patagonia, República Argentina. Tesis Doctoral, Universidad Nacional del Sur, (inédita). Bahía Blanca. Argentina. 136 pp.

25. Maiza, P.J. (1981) Estudio de los yacimientos de caolín del oeste de la Provincia del Chubut, República Argentina, Minas Susana, Gato y Estrella Gaucha. 8° Congreso Geológico Argentino, Actas, 4, 471 484.

26. Maiza, P.J. & Hayase, K. (1975) Los yacimientos de caolín de la Patagonia. República Argentina. 2° Congreso Ibero Americano deGeología Económica, Actas, 2, 365 383, Buenos Aires.

27. Maiza, P; S. Marfil; E. Cardellach & J. Zunino. (2009) Geoquímica de la zona caolinizada de Mina Estrella Gaucha (Prov. de Chubut, Argentina).

Revista de la Asociación Geológica Argentina Vol 64. N° 3. ISSN 0004-4822, 426-432.

28. Maiza P. & Mas G. (1980) Estudio de los sulfatos alunita-natroalunita. Síntesis de la serie. Revista de la Asociación Argentina de Mineralogía, Sedimentología y Petrología, 11, 32-41

29. Maiza, P.J. & G. Mas (1981) Presencia de natroalunita en Muna Equivocada, río Negro. Su significado. VIII Congreso Geológico Argentino, San Luis. Actas, IV, 285-292.

30. Malvicini L & Llambías E. (1974) In: Malvicini L. & Vallés J. M. (1984) Metalogénesis. Capítulo III-5. Geología y recursos naturales de la Provincia de Río Negro.

31. Relatorio del IX Congreso Geológico Argentino, San Carlos de Bariloche. Río Negro, 649-662.

32. Manera T. (1972) La mineralización de los yacimientos de fluorita de la Provincia de Río Negro. Tesis Doctoral, Universidad Nacional del Sur, Bahía Blanca, Argentina.

33. Marfil S.A., Maiza P.J., Cardellach E. & Corbella M. (2005) Origin of kaolin deposits in the 'Los Menucos', Río Negro Province, Argentina. Clay Minerals, 40, 283-293.

34. Marfil, S. A., Maiza, P. J and Montecchiari, N. (2010) Alteration zonation in Loma Blanca Kaolin deposit, Los Menucos, province of Rio Negro, Argentina. Clay Minerals, 45, 157-169.

35. Murray H. & Janssen J. (1984) Oxigen isotopes – indicators of kaolin genesis?. Procedings of the 27th International Geological Congress, 15, 287-303.

36. Pandarinath K., Dulski P., Torres Alvarado I.S. & Verma S.P. (2008) Element mobility during the hydrothermal alteration of rhyolitic rocks of the Los Azufres geothermal field, Mexico. Geothermics, 37, 53-72.

37. Pankurst R.J.C., Rapela W., Caminos R., Llambías E. & Parica C. (1992) A revised age for the granites of the central Somoncura Batholith, North Patagonian Massif. Journal of South American Earth Sciences, 5, 321-325.

38. Pankhurst R.J. & Rapela W. (1995) Production of Jurassic rhyolite by anatexis in the lower crust of Patagonia. Earth and Planetary Science Letters, 134, 23-26.

39. Pankhurst R.J., Leat P.T., Sruoga P., Rapela C.W., Márquez M., Storey B.C. & Riley T.R. (1998) The Con Aike province of Patagonia and related rocks in West Antartica: A silicic large igneous province. Journal of Volcanology and Geothermal Research, 81, 113- 136.

40. Papoulis D. & Tsolis-Katagas P. (2008) Formation of alteration zones and kaolin genesis, Limnos Island, northeast Aegean Sea, Greece. Clay Minerals, 43, 631-646.

41. Papoulis D., Tsolis-Katagas P. & Katagas C. (2004). Monazite alteration mechanisms and depletion measurements in kaolins. Applied Clay Science, 24. 271-285.

42. Parsapoor A., Kahlili M. & Mackinzadeh H.A. (2009) The behaviour of trace and rare earth elements (REE) during hydrothermal alteration in the Rangan area (central Iran). Journal of Asian Earth Sciences, 34, 123-134.

43. Ploszkiewicz, J.V. y Ramos, V.A. (1977) Estratigrafía y tectónica de la Sierra de Payaniyeu (Provincia del Chubut). Revista Asociación Geológica Argentina 32: 209-226.

44. Reed, M. H. (1997) Hydrothermal Mineral Deposits: What we do adn don´t know. Chapter 1. In: Geochemistry of hydrothermal ore deposit. Third Edition. Ed. H. L. Barnes. 303- 358.

45. Reed B.L. & Hemley J.J. (1967) Ocurrence of pyrophyllite in Kekiktuk Conglomerate. In: Book Range, Northeastern Alaska. U.S. Geological Survey, 162-166.

46. Rollinson H. (1992) Using Geochemical Data: Evaluation, Presentation, Interpretation. University of Zimbabwe.

47. Roy R. & Osborn E. (1954) The system Al2O3-SiO2-H2O. American Mineralogist, 39, 853-85. Savin, S.M & Lee, S. (1988) Isotopic studies of phillosilicates. In: Hydrous Phyllosilicates (Exclusive of Micas). (Bailey S.W.) Editor). Reviews in Mineralogy. Mineralogical Society of America, 19. Washington DC, 189-223.

48. Shepard, S.M.F & Gilg h. A. (1996) Stable isotope geochemistry of clay minerals. Clay Minerals, 31, 1-24.

49. Skinner, B. (1997) Hydrothermal alteration and its relationship to ore fluid composición. Chapter 7. In: Geochemistry of hydrothermal ore deposit. Third Edition. Ed. H. L. Barnes. 303-358.

50. Sturchio N.C., Muehlenbchs K. & Meitz M. (1986) Element redistribution during hydrothermal alteration of rhyolite in an active geothermal system: Yellowstone drill cores Y-7 and Y-8. Geochimica et Cosmochimica Acta, 50, 1619-1631.

51. Terakado Y. & Fujitani T. (1998) Behavior of the rare earth elements and other trace elements during interactions between acidic hydrothermal solutions and silicic volcanic rocks, southwestern Japan. Geochimica et Cosmochimica Acta, 62, 1903-1998.

52. Tzuzuki Y. & Mizutani S. (1971) A study of rock alternation process based on kinetics of hydrothermal experiment. Contributions to Mineralogy and Petrology, 30, 15-33.

53. van der Weijden C.H. & van der Weijden R.D. (1995) Mobility of major, minor and some redox-sensitive trace elements and rare-earth elements during weathering of four granitoids in central Portugal. Chemical Geology, 125, 149-167.

CITATION

CHAPTER 1

Tsuyoshi Ishikawa, Masaharu Tanimizu, Kazuya Nagaishi, Jun Matsuoka, Osamu Tadai, Masumi Sakaguchi, Tetsuro Hirono, Toshiaki Mishima, Wataru Tanikawa, Weiren Lin, Hiroyuki Kikuta, Wonn Soh1 & Sheng-Rong Song; Coseismic fluid–rock interactions at high temperatures in the Chelungpu fault; doi:10.1038/ngeo308

CHAPTER 2

Charles Fairhurst (2013). Fractures and Fracturing - Hydraulic fracturing in Jointed Rock, Effective and Sustainable Hydraulic Fracturing, Dr. Rob Jeffrey (Ed.), ISBN: 978-953-51-1137-5, InTech, DOI: 10.5772/56366.

CHAPTER 3

Sebastian Brenne, Michael Molenda, Ferdinand Stöckhert and Michael Alber (2013). Hydraulic and Sleeve Fracturing Laboratory Experiments on 6 Rock Types, Effective and Sustainable Hydraulic Fracturing, Dr. Rob Jeffrey (Ed.), ISBN: 978-953-51-1137-5, InTech, DOI: 10.5772/56301.

CHAPTER 4

B. Damjanac, C. Detournay, P.A. Cundall and Varun (2013). Three-Dimensional Numerical Model of Hydraulic Fracturing in Fractured Rock Masses, Effective and Sustainable Hydraulic Fracturing, Dr. Rob Jeffrey (Ed.), ISBN: 978-953-51-1137-5, InTech, DOI: 10.5772/56313.

CHAPTER 5

Xu-Guang Chen, Qiang-Yong Zhang, Yuan Wang, De-Jun Liu, and Ning Zhang, "Model Test of Anchoring Effect on Zonal Disintegration in Deep Surrounding Rock Masses," The Scientific World Journal, vol. 2013, Article ID 935148, 16 pages, 2013. doi:10.1155/2013/935148.

CHAPTER 6

Rosper M. Nude, Kodjopa Attoh, John W. Shervais and Gordon Foli (2012). Petrological and Geochemical Characteristics of Mafic Granulites Associated with Alkaline Rocks in the Pan-African Dahomeyide Suture Zone, Southeastern Ghana, Petrology - New Perspectives and Applications, Prof. Ali Al-Juboury (Ed.), ISBN: 978-953-307-800-7, InTech, DOI: 10.5772/24563.

CHAPTER 7

X. Pan, Z. Feng, G. Dai and H. Liu, "Roughness Research of Center Profile Curve on Rock Fracture Surface Based on Statistical Method," Geomaterials, Vol. 3 No. 2, 2013, pp. 47-53. doi: 10.4236/gm.2013.32006.

CHAPTER 8

Yonglai Zheng and Shuxin Deng, "Failure Probability Model considering the Effect of Intermediate Principal Stress on Rock Strength," Mathematical Problems in Engineering, vol. 2015, Article ID 960973, 7 pages, 2015. doi:10.1155/2015/960973

CHAPTER 9

Shuangshuang Xiao, Kemin Li, Xiaohua Ding, and Tong Liu, "Rock Mass Blastability Classification Using Fuzzy Pattern Recognition and the Combination Weight Method," Mathematical Problems in Engineering, vol. 2015, Article ID 724619, 11 pages, 2015. doi:10.1155/2015/724619

CHAPTER 10

Sasaoka, T. , Takahashi, Y. , Sugeng, W. , Hamanaka, A. , Shimada, H. , Matsui, K. and Kubota, S. (2015) Effects of Rock Mass Conditions and Blasting Standard on Fragmentation Size at Limestone Quarries. Open Journal of Geology, 5, 331-339. doi: 10.4236/ojg.2015.55030.

CHAPTER 11

Zhibin Zhong, Ronggui Deng, Fang Lin, et al., "Study on Hysteretic Fracture of Naturally Cracked Surrounding Rock," Mathematical Problems in Engineering, vol. 2015, Article ID 569385, 11 pages, 2015. doi:10.1155/2015/569385

CHAPTER 12

Chong Zhang, Zhechao Wang, and Qi Wang, "Deformation and Failure Characteristics of the Rock Masses around Deep Underground Caverns," Mathematical Problems in Engineering, vol. 2015, Article ID 230126, 13 pages, 2015. doi:10.1155/2015/230126

CHAPTER 13

Yijiang Peng, Qing Guo, Zhaofeng Zhang, and Yanyan Shan, "Application of Base Force Element Method on Complementary Energy Principle to Rock Mechanics Problems," Mathematical Problems in Engineering, vol. 2015, Article ID 292809, 16 pages, 2015. doi:10.1155/2015/292809

CHAPTER 14

Silvina Marfil and Pedro Maiza (2012). Geochemistry of Hydrothermal Alteration in Volcanic Rocks, Geochemistry - Earth's System Processes, Dr. Dionisios Panagiotaras (Ed.), ISBN: 978-953-51-0586-2, InTech, DOI: 10.5772/31517.

INDEX

A

Acoustic Emission (AE) 56
Analysis hierarchy process (AHP) 172

B

Barite-iron-sand (BISA) 81
Barites-iron-sand cementation analogical (BISA) 80
Base force element method (BFEM) 249, 250, 277, 281
Bonded particle model (BPM) 64, 65
Breccia zone (BrZ) 2

C

Combination weight method (CWM) 172
Computerized numerical control (CNC) 85
Consistency index CI 172
Consistency ratio CR 172

D

Data acquisition system (DAS) 86
Diederichs 225, 247
Differential thermal analysis (DTA) 291
Discrete fracture network (DFN) 63
Discrete fracture networks (DFN's) 23

E

Elasto-hydrodynamic 2
Energy dispersive spectrometry (EDS) 115
Enhanced Geothermal Energy (EGS) 16
Enhanced Geothermal Systems (EGS 14
Entropy method (EM) 172

F

Fiber bragg grating (FBG) 89
Finite element method (FEM) 249
Fracture-damaged zone (FDZ) 2

G

Game theory (GT) 172
Geoenvironment Protection 199, 203
Geohazard Prevention 199, 203
Grey gouge zone (GGZ) 2

H

Hydraulic fracture (HF) 63, 64
Hydraulic Fracturing (HF) 20

I

Inductively Coupled Plasma Mass Spectroscopy (ICP-MS) 118
Island arc theoleiitic (IAT) 109

J

Joint Roughness Coefficients (JRC) 132

K

Kpong complex (KC) 109

L

Linearly Elastic Fracture Mechanics
    (LEFM) 18

M

Material heterogeneity 148
Microcracks 199
MTS (Mechanics Test Systems) 131, 132

N

Naturally fractured reservoirs (NFRs) 64

P

Particle flow code (PFC) 67
Particle Flow Code (PFC) code) 23
Physical phenomena 226
Pressurization 1, 2, 8, 10, 12

Q

Quadrilateral reduced integration ele-
    ment (Q4R 261, 264, 268, 277

R

Rare earth elements (REE) 118
Rare-earth elements (REE) 285
Rock Biaxial Compressive Test System
    (RBCTS) 203

S

Signal translating system (STS) 86
Smooth joint model (SJM) 64, 65
Synthetic rock mass (SRM) 64

T

Thermogravimetric (TG) 291
Traditional considerations 147
Trans-Saharan belt (TSB) 110
Triaxial test 152, 159

U

Unconstrained compressive strength
    (UCS) 83

W

West African craton (WAC) 110

X

X-ray diffraction (XRD) 291